Springer Proceedings in Mathematics & Statistics

Volume 254

Springer Proceedings in Mathematics & Statistics

This book series features volumes composed of selected contributions from workshops and conferences in all areas of current research in mathematics and statistics, including operation research and optimization. In addition to an overall evaluation of the interest, scientific quality, and timeliness of each proposal at the hands of the publisher, individual contributions are all refereed to the high quality standards of leading journals in the field. Thus, this series provides the research community with well-edited, authoritative reports on developments in the most exciting areas of mathematical and statistical research today.

More information about this series at http://www.springer.com/series/10533

Philippe Drobinski · Mathilde Mougeot
Dominique Picard · Riwal Plougonven
Peter Tankov
Editors

Renewable Energy: Forecasting and Risk Management

Paris, France, June 7–9, 2017

Springer

Editors
Philippe Drobinski
Laboratoire de Méteorologie Dynamique
CNRS
Palaiseau, France

Riwal Plougonven
Laboratoire de Météorologie Dynamique
Ecole Polytechnique
Palaiseau, France

Mathilde Mougeot
UFR de Mathématiques
Université Paris Diderot
Paris, France

Peter Tankov
CREST—ENSAE Paris Tech
Palaiseau, France

Dominique Picard
UFR de Mathématiques
Université Paris Diderot
Paris, France

ISSN 2194-1009 ISSN 2194-1017 (electronic)
Springer Proceedings in Mathematics & Statistics
ISBN 978-3-319-99051-4 ISBN 978-3-319-99052-1 (eBook)
https://doi.org/10.1007/978-3-319-99052-1

Library of Congress Control Number: 2018951904

Mathematics Subject Classification (2010): 62P12, 62P30

This Springer imprint is published by the registered company Springer Nature Switzerland AG
The registered company address is: Gewerbestrasse 11, 6330 Cham, Switzerland

Preface

This book contains a selection of 12 papers presented at the workshop "Forecasting and Risk Management for Renewable Energy", which took place in Paris in June 2017, and was organized in the framework of the project FOREWER (Forecasting and Risk Management of Wind Energy Production), funded by the French National Research Agency (ANR).

For reasons of environment protection and energy security, the share of renewable resources in the global energy supply is now rising at an overwhelming rate. The European Commission has set the target to reach a 20% share of energy from renewable sources by 2020 and further increases of this already ambitious objective will follow. The production of electricity from renewable resources such as wind and solar is both spatially distributed and highly dependent on atmospheric conditions and thus intermittent in nature, leading to challenging planning and risk management problems for the stakeholders of the wind energy industry.

These new challenges, in particular, those related to investment planning and grid integration under the conditions of large-scale renewable generation, call for better understanding of the spatial and temporal distribution of the renewable resource and power production with the help of precise statistical and probabilistic models. Besides, recent advances in climatology show that it may be possible to develop medium and long-term (seasonal to decadal) probabilistic forecasts of the renewable power output with a better performance than that of forecasts based on climatological averages, leading to improved risk management tools for power producers and grid operators.

The aim of the interdisciplinary workshop "Forecasting and Risk Management for Renewable Energy" was thus to bring together statisticians, probabilists, meteorologists, economists, and engineers working on various aspects of renewable energy, from production forecasting to optimal storage management, to discuss together quantitative methods for renewable energy risk management and forecasting. The specific focus was on wind energy with some talks addressing photovoltaic energy as well.

This book contains a representative selection of papers discussed at the workshop. It is loosely divided into two parts. The first part focuses on modeling and forecasting for renewable energy, with mostly statistical tools. In addition to wind forecasting, two papers focus on the forecasting of electric demand, which is the second major source of randomness in electric systems. The second part, on the other hand, adopts a "risk management" point of view and analyzes the issues such as network security with high renewable penetration; management of energy storage in the presence of renewable assets and the effect of renewable penetration on electricity prices.

We hope that this selection of papers will give the readers a taste of the truly multidisciplinary collaborations which are required to make progress on the key scientific challenges of energy transition, and stimulate new projects and partnerships in this field.

Paris, France
June 2018

Philippe Drobinski
Mathilde Mougeot
Dominique Picard
Riwal Plougonven
Peter Tankov

Acknowledgements

We gratefully acknowledge the support of Agence Nationale de Recherche (grant ANR-14-CE05-0028, project FOREWER) for the conference "Forecasting and Risk Management for Renewable Energy".

Contents

Contributors

Bastien Alonzo LMD/IPSL, École Polytechnique, Université Paris Saclay, ENS, PSL Research University, Sorbonne Universités, UPMC Univ Paris 06, CNRS, Palaiseau, France

Jordi Badosa LMD/IPSL, Ecole Polytechnique, Palaiseau Cedex, France

Alain Bensoussan International Center for Risk and Decision Analysis, Jindal School of Management, The University of Texas, Dallas, TX, USA; Department SEEM, City University of Hong Kong, Hong Kong, People's Republic of China

Mireille Bossy Université Côte d'Azur, Inria, Sophia Antipolis, France

Christian Briard Zephyr ENR, Saint-Avertin, France

Alexandre Brouste Institut du Risque et de l'Assurance du Mans, Laboratoire Manceau de Mathématiques, Le Mans, France

Alain Burtin EDF R&D, Palaiseau, France

Andrés Castrillejo IESTA, Fac. CCEE, Universidad de la Republica, Montevideo, Uruguay

Jérôme Collet EdF R&D, Palaiseau, France

James Cruise School of Mathematical and Computer Sciences, Heriot-Watt University, Edinburgh, UK

Jairo Cugliari Université de Lyon, Lyon, France

Thomas Deschatre Université Paris-Dauphine, PSL Research University, CNRS, Ceremade, Paris, France; EDF Lab, OSIRIS, Palaiseau, France

Philippe Drobinski LMD/IPSL, École polytechnique, Université Paris Saclay, ENS, PSL Research University, Sorbonne Universités, UPMC Univ Paris 06, CNRS, Palaiseau, France

Aurore Dupré LMD/IPSL, École Polytechnique, Université Paris Saclay, ENS, PSL Research University, Sorbonne Universités, UPMC Univ Paris 06, CNRS, Palaiseau, France

Jun Fei Department of Applied Mathematics & Statistics and Advanced Energy Center, Stony Brook University, Stony Brook, NY, USA

Eugene A. Feinberg Department of Applied Mathematics & Statistics and Advanced Energy Center, Stony Brook University, Stony Brook, NY, USA

Olivier Féron EDF Lab, Palaiseau, France

Aurélie Fischer Laboratoire de Probabilités, Statistique et Modélisation, Université Paris Diderot—Paris 7, Paris, France

Paul Fourment EDF R&D, Palaiseau, France

Caroline Gatti-Bono EDF R&D, Palaiseau, France

Emmanuel Gobet CMAP, Ecole Polytechnique, Palaiseau Cedex, France

Maxime Grangereau CMAP, Ecole Polytechnique, Palaiseau Cedex, France

Timothee Hinchliffe EDF R&D, Palaiseau, France

Daeyoung Kim Ecole Polytechnique, Palaiseau Cedex, France

Vincent Lefieux RTE-EPT & UPMC-ISUP, La Défense Cedex, France

Miguel López-Botet Zulueta EDF R&D, Palaiseau, France

Miranda Marchand RTE-R&DI, La Défense Cedex, France

Fernando Massa IESTA, Fac. CCEE, Universidad de la Republica, Montevideo, Uruguay

Mathilde Mougeot Laboratoire de Probabilités, Statistique et Modélisation, Université Paris Diderot—Paris 7, Paris, France; Université Paris Diderot, LPSM UMR 8001, Sorbonne Paris Cité, Paris, France

Dominique Picard Université Paris Diderot, LPSM UMR 8001, Sorbonne Paris Cité, Paris, France

Riwal Plougonven LMD/IPSL, École Polytechnique, Université Paris Saclay, ENS, PSL Research University, Sorbonne Universités, UPMC Univ Paris 06, CNRS, Palaiseau, France

Ignacio Ramirez IEE, Fac. de Ingeniería, Universidad de la Republica, Montevideo, Uruguay

Michael Richard EdF R&D, Palaiseau, France; University of Orléans, Orléans, France

Vera Silva EDF R&D, Palaiseau, France

Peter Tankov CREST-ENSAE Paris Tech, Palaiseau, France

Almut E. D. Veraart Department of Mathematics, Imperial College London, London, UK

Laurent Violeau Université Côte d'Azur, Inria, Sophia Antipolis, France

Ye Wang EDF R&D, Palaiseau, France

Stan Zachary School of Mathematical and Computer Sciences, Heriot-Watt University, Edinburgh, UK

Part I
Renewable Energy: Modeling and Forecasting

Chapter 1
Marginal Weibull Diffusion Model for Wind Speed Modeling and Short-Term Forecasting

Alain Bensoussan and Alexandre Brouste

Abstract We propose a dynamical model for the wind speed which is a Markov diffusion process with Weibull marginal distribution. It presents several advantages, namely nice modeling features both in terms of marginal probability density function and temporal correlation. The characteristics can be interpreted in terms of shape and scale parameters of a Weibull law which is convenient for practitioners to analyze the results. We calibrate the parameters with the maximum quasi-likelihood method and use the model to generate and forecast the wind speed. We have tested the model on wind-speed datasets provided by the National Renewable Energy Laboratory. The model fits well the data and we obtain a very good performance in point and probabilistic forecasting in the short-term in comparison to the benchmark.

Keywords Statistical modeling · Ergodic diffusions · Wind speed forecasts

1.1 Introduction

The two-parameter Weibull probability density function has become widely used to fit wind speed datasets in the literature of wind energy (see [5, 9] and the reference therein). It has been included in regulations concerning wind energy and in most popular software on wind modeling like HOMER and WAsP.[1]

[1]One can find more informations on software respectively at https://analysis.nrel.gov/homer/ and http://www.wasp.dk/.

A. Bensoussan
International Center for Risk and Decision Analysis, Jindal School of Management, The University of Texas, Dallas, TX, USA
e-mail: alain.bensoussan@utdallas.edu

A. Bensoussan
Department SEEM, City University of Hong Kong, Hong Kong, People's Republic of China

A. Brouste (✉)
Institut du Risque et de l'Assurance du Mans, Laboratoire Manceau de Mathématiques, Le Mans Université, Le Mans, France
e-mail: alexandre.brouste@univ-lemans.fr

P. Drobinski et al. (eds.), *Renewable Energy: Forecasting and Risk Management*, Springer Proceedings in Mathematics & Statistics 254,
https://doi.org/10.1007/978-3-319-99052-1_1

The energy production of a wind farm is related in particular to the wind speed on the site through the power transfer function of wind turbines. Forecasting production on timescales from minutes to days improves the operational management of a wind farm. From seconds to minutes, this forecast allows an accurate indication to achieve to control the turbine and to smooth the production on the electricity grid by storage. From hours to half a day, it can be used to decide whether or not to store for efficient trading on electricity markets. Finally, forecasting production on timescales of a day to several days helps in scheduling maintenance operations. These problems belong to the class of stochastic optimization problems due to the intermittent nature of the production (or the wind) and the solutions depend mainly on the underlying model used for the production (or the wind speed).

Several dynamical models such as classical time series (ARMA, FARIMA, ...), Markov chains [6, 13, 18, 20], semi-Markov chains [7, 8] and neural networks [17] have been considered in the literature for both modeling and short-term forecasting. Short-term forecasting corresponds in this paper to timescales of half an hour to half a day and can be distinguished for long-term forecasting where numerical weather prediction models are involved in addition to statistical methods.

We have considered the Cox–Ingersoll–Ross (CIR) diffusion in [2] to model the square of the wind speed. Diffusion processes in general provide efficient point and probabilistic forecasts obtained from transition probability density functions. For instance, the CIR model outperforms the persistence benchmark (i.e. last measured value as forecast value) in terms of one-step forecast mean square error (MSE). Moreover, stochastic optimization problems with such continuous Markov processes are handled conveniently.

In this paper, we propose a diffusion process for the wind speed whose marginal law is Weibull. Temporal correlation structure of the stochastic process is also parametrized. We show that the model is interesting as a wind generator model. Indeed, the fact that characteristics are understandable in terms of shape and scale parameters of a Weibull law helps practitioners analyzing the results. This model also shows good performances in short-term forecasting for a large class of datasets provided by the National Renewable Energy Laboratory (NREL). The model is compared to the persistence benchmark and the Ornstein–Uhlenbeck diffusion process both in terms of MSE (for point forecasts) and continuous ranked probability score (CRPS, for probabilistic forecasts) for short-term lead times (see also [12] for definitions of other score and other benchmark for one-step ahead forecasting). Contrary to the CIR process, the transition probability density function is no longer available in closed form. Consequently, new calibration and forecast methods are presented in this paper which differ from our previous work [2].

In Sect. 1.2, the dynamic marginal Weibull diffusion model is presented. Point forecast and probabilistic forecast methods are developed in Sect. 1.3. In Sect. 1.4, the estimation method of the parameters of the aforementioned model is presented and the dynamic model is calibrated on NREL dataset. Forecasting performances of the different models are also summarized.

1.2 Dynamic Model for the Wind Speed

Let $\vartheta = (\vartheta_1, \vartheta_2, \vartheta_3)$ be a 3 dimensional parameter in $(\mathbb{R}_*^+)^3$. Practitioners consider the wind speed as a Weibull law whose probability density function is

$$f(z, \vartheta) = \frac{\vartheta_3}{\vartheta_2}\left(\frac{z}{\vartheta_2}\right)^{\vartheta_3 - 1} \exp\left(-\left(\frac{z}{\vartheta_2}\right)^{\vartheta_3}\right), \quad z \geq 0. \tag{1.1}$$

For this distribution, the mean is directly given by

$$\mu_W = \vartheta_2 \Gamma\left(1 + \frac{1}{\vartheta_3}\right)$$

and the standard deviation by

$$\sigma_W = \sqrt{\vartheta_2^2 \Gamma\left(1 + \frac{2}{\vartheta_3}\right) - \mu_W^2}.$$

In this section, we describe the model mentioned in the introduction. The three-parameter marginal Weibull diffusion process $(Z_t, t \geq 0)$ is the solution of the stochastic differential equation (sde)

$$Z_t = z_0 + \int_0^t v_0(Z_s, \vartheta)ds + \int_0^t v_1(Z_s, \vartheta)d\beta_s, \quad t \geq 0, \tag{1.2}$$

where $(\beta_t, t \geq 0)$ is a standard Wiener process. Moreover the drift coefficient $v_0(\cdot, \cdot)$ and the diffusion coefficient $v_1(\cdot, \cdot)$ are known functions that are described below. The stochastic process $(Z_t, t \geq 0)$ is an homogeneous Markov diffusion process which can be characterized by its transition probability density function $p(t, \cdot; x, \vartheta)$ representing the conditional probability density function of Z_{s+t} given $Z_s = x$ (for more information on Markov diffusion processes, see [1] or [11]). The long term law of Z_t (as $t \to \infty$) is called the stationary distribution or invariant distribution. It is denoted $p(\cdot; \vartheta)$. If the initial condition z_0 is random and follows the invariant distribution, then the law of Z_t is still the invariant distribution for all $t \geq 0$. In this setting, the stationary distribution of our model is Weibull with scale parameter ϑ_2 and shape parameter ϑ_3 (see Eq. (1.1)).

In this model, we fix the drift term equal to

$$v_0(z, \vartheta) = \vartheta_1\left(\vartheta_2 \Gamma\left(1 + \frac{1}{\vartheta_3}\right) - z\right) \tag{1.3}$$

where Γ is the gamma function. Here, the parameter ϑ_1 is the temporal correlation parameter of the process. It can be shown, if the initial condition is the invariant distribution, that the correlation structure is given by

$$\mathrm{corr}(Z_s, Z_t) = e^{-\vartheta_1(t-s)}. \tag{1.4}$$

For more information, see Appendix 1. Also, the computation of the proper diffusion coefficient $v_1(z, \vartheta)$ to obtain Weibull probability density function as marginal distribution given the previously fixed correlation structure is postponed to Appendix 2. Namely, the diffusion coefficient is given by

$$v_1^2(z, \vartheta) = \frac{2\vartheta_1\vartheta_2\Gamma\left(1+\frac{1}{\vartheta_3}\right)}{f(z, \vartheta)}\left(\left(1-e^{-\left(\frac{z}{\vartheta_2}\right)^{\vartheta_3}}\right) - \frac{1}{\Gamma\left(1+\frac{1}{\vartheta_3}\right)}\int_0^{\left(\frac{z}{\vartheta_2}\right)^{\vartheta_3}} u^{\left(1+\frac{1}{\vartheta_3}\right)-1} e^{-u} du\right). \tag{1.5}$$

For the marginal Weibull diffusion model the transition probability density function $p(t, y; x, \vartheta)$ cannot be obtained in closed form. Fortunately, it is the solution of the Fokker–Planck (FP) equation

$$\frac{\partial}{\partial t}p(t, y; x, \vartheta) = -\frac{\partial}{\partial y}\left(v_0(y, \vartheta)p(t, y; x, \vartheta)\right) + \frac{1}{2}\frac{\partial^2}{\partial y^2}\left(v_1(y, \vartheta)^2 p(t, y; x, \vartheta)\right) \tag{1.6}$$

with initial condition $p(0, y; x, \vartheta) = \delta_x(y)$ where $\delta_x(y)$ represents a Dirac distribution at point x.

1.3 One-Step Forecasting

The operator of a wind farm is interested in forecasting the production in the next hours to provide the information to the entity in charge of the electric grid and trade on electricity markets. From the transfer function of the wind turbine, the production is directly related to the wind speed. So the problem boils down to the short-term forecasting of the wind speed.

1.3.1 Definitions

Suppose that we fix the present time at $t = 0$ and that the initial observed wind speed is $\tilde{Z}_0$. Let us denote $\tilde{Z}_t$ the true (random and unknown) value of the wind speed at time $t > 0$ and $\tilde{\psi}(t, \cdot)$ its (unknown) probability density function (see Fig. 1.1).

In this paper, a point forecast is an estimator of the wind speed $\tilde{Z}_t$ given the knowledge of $\tilde{Z}_0$. It is denoted by $\pi(Z_t^{\tilde{Z}_0})$. We call persistence benchmark the current knowledge

$$\pi_{per}(Z_t^{\tilde{Z}_0}) = \tilde{Z}_0. \tag{1.7}$$

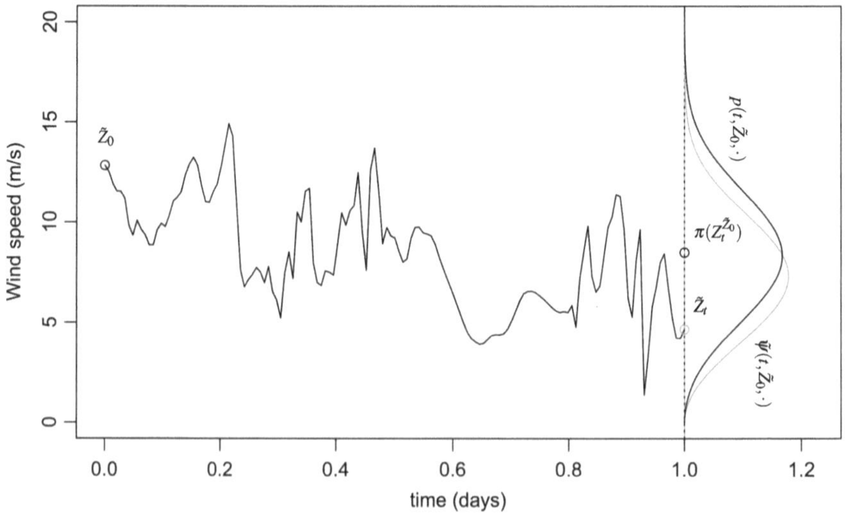

Fig. 1.1 One-step ahead (short-term) forecasting error

Following the diffusion model, the forecast value is defined by

$$\pi_d(Z_t^{\tilde{Z}_0}) = \mathbf{E}_\vartheta\,(Z_t)$$

where Z_t is the solution of (1.2) with initial condition $z_0 = \tilde{Z}_0$.

But, a probabilistic forecast can also be proposed. It consists in defining an estimator of the probability density function $\tilde{\psi}(t,\cdot)$ of $\tilde{Z}_t$ given the knowledge of $\tilde{Z}_0$. It is denoted $\widehat{\psi}(t,\cdot;\tilde{Z}_0)$. It is worth mentioning that no probabilistic forecasts can be specified in the basic persistence benchmark. In the diffusion model, it is natural to define the probabilistic forecast as the transition probability density function, namely

$$\widehat{\psi}(t,\cdot;\tilde{Z}_0) = p(t,.;\tilde{Z}_0,\vartheta)$$

where p is the solution of (1.6).

1.3.2 *Point Forecasts and Mean-Square Error*

1.3.2.1 Point Forecasts

For diffusion models, the point forecast is

$$\pi_d(Z_t^{\tilde{Z}_0}) = \mathbf{E}_\vartheta\,(Z_t) = \int_{\mathbb{R}} y\,p(t,y;\tilde{Z}_0,\vartheta)dy \tag{1.8}$$

where Z_t is the solution of (1.2) with initial condition $z_0 = \tilde{Z}_0$. In the particular case of diffusion processes with a linear drift $v_0(z, \vartheta) = \vartheta_1(\alpha - z)$, it can be shown (see Appendix 1) that the point forecast has the following closed-form

$$\pi_d(Z_t^{\tilde{Z}_0}) = \alpha + \left(\tilde{Z}_0 - \alpha\right) e^{-\vartheta_1 t}. \tag{1.9}$$

The parameter α is the mean value of the stationary distribution, namely

$$\alpha = \int_{\mathbb{R}} y p(y; \vartheta) dy.$$

We recall that the parameter ϑ_1 parametrizes the correlation structure and stands for the mean-reverting speed. For the marginal Weibull diffusion model

$$\alpha = \vartheta_2 \Gamma \left(1 + \frac{1}{\vartheta_3}\right).$$

It is worth mentioning that the Ornstein–Uhlenbeck process with

$$v_0(z, \vartheta) = \vartheta_1(\alpha - z) \quad \text{and} \quad v_1(z, \vartheta) = \sigma \tag{1.10}$$

and the Cox–Ingersoll–Ross process with

$$v_0(z, \vartheta) = \vartheta_1(\alpha - z) \quad \text{and} \quad v_1(z, \vartheta) = \sigma\sqrt{z} \tag{1.11}$$

have also linear drift (see also [2]) and consequently generate similar point forecasts.

1.3.2.2 Mean-Square Error

Given a point forecast $\pi(Z_t^{\tilde{Z}_0})$, the mean-square forecasting error is defined by

$$MSE(t) = \mathbf{E}\left(\left(\pi(Z_t^{\tilde{Z}_0}) - \tilde{Z}_t\right)^2\right). \tag{1.12}$$

This indicator makes it possible to compare persistence model and diffusion model point forecasts. For the persistence model defined in (1.7),

$$MSE_{per}(t) = \mathbf{E}\left(\left(\tilde{Z}_0 - \tilde{Z}_t\right)^2\right).$$

Diffusion models (1.2) have the property that the well-specified MSE (namely MSE (1.12) where $\widetilde{Z}_t = Z_t$) can be written as

$$MSE(t) = \mathbf{E}_\vartheta\left(Z_t^2\right) - \left(\mathbf{E}_\vartheta(Z_t)\right)^2 = u(t, Z_0) - \left(\mathbf{E}_\vartheta(Z_t)\right)^2$$

in which $u(t, x)$ solves the Feynman–Kac pde, i.e.

$$\frac{\partial u}{\partial t} = v_0(x, \vartheta)\frac{\partial u}{\partial x} + \frac{v_1^2(x, \vartheta)}{2}\frac{\partial^2 u}{\partial x^2}$$

with

$$u(0, x) = x^2.$$

For instance, the MSE can be obtained in closed form for the Ornstein–Uhlenbeck process and the Cox–Ingersoll–Ross process (see [2]). Direct computations lead to

$$MSE(t) = \frac{\sigma^2(1 - e^{-2\vartheta_1 t})}{2\vartheta_1}$$

for the Ornstein–Uhlenbeck process (1.10) with

$$\lim_{t\to\infty} MSE(t) = \frac{\sigma^2}{2\vartheta_1} \quad \text{and} \quad MSE(t) \sim_0 \sigma^2 t,$$

and, comparatively,

$$MSE(t) = \frac{Z_0\sigma^2(e^{-\vartheta_1 t} - e^{-2\vartheta_1 t})}{\vartheta_1} + \frac{\alpha\sigma^2(1 - e^{-2\vartheta_1 t})^2}{2\vartheta_1}$$

for the Cox–Ingersoll–Ross process (1.11) with

$$\lim_{t\to\infty} MSE(t) = \frac{\alpha\sigma^2}{2\vartheta_1} \quad \text{and} \quad MSE(t) \sim \sigma^2 Z_0 t \quad \text{as } t \to 0.$$

In the marginal Weibull setting, no closed form are available but numerical computations can be performed for the MSE with $MSE(t) \sim v_1^2(Z_0, \vartheta)t$ as $t \to 0$.

1.3.3 *Probabilistic Forecasts and Continuous Ranked Probability Score*

1.3.3.1 Probabilistic Forecasts

For a diffusion model in general, the probabilistic forecast is given by its transition probability density function

$$\widehat{\psi}(t, \cdot; \tilde{Z}_0) = p(t, .; \tilde{Z}_0, \vartheta). \tag{1.13}$$

The corresponding forecasting empirical cumulative distribution function is defined by

$$\widehat{F}(u) = \int_{-\infty}^{u} p(t, y; \tilde{Z}_0, \vartheta)dy.$$

1.3.3.2 Continuous Ranked Probability Score

Different probabilistic forecasts can be compared in term of Continuous Ranked Probability Score (CRPS). Let $\hat{\psi}(t, \cdot; \tilde{Z}_0, \vartheta)$ be a probabilistic forecast and $\widehat{F}$ its corresponding cumulative distribution function. For any $x \in \mathbb{R}$, continuous ranked probability function can be defined by

$$\mathrm{CRP}(\widehat{F}, x) = \int_{\mathbb{R}} \left(\widehat{F}(u) - \mathbb{1}_{\{u \geq x\}}\right)^2 du. \tag{1.14}$$

The continuous ranked probability score (CRPS, see [10]) is defined by

$$\mathrm{CRPS}(t) = \mathbf{E}\left(\mathrm{CRP}(\widehat{F}, \tilde{Z}_t)\right). \tag{1.15}$$

For the Gaussian distribution, the previous function is explicitly computable, namely

$$\mathrm{CRP}(\mathcal{N}(\mu, \sigma^2), x) = -\sigma\left(\frac{1}{\sqrt{\pi}} - 2\varphi\left(\frac{x-\mu}{\sigma}\right) - \frac{x-\mu}{\sigma}\left(2\Phi\left(\frac{x-\mu}{\sigma}\right) - 1\right)\right) \tag{1.16}$$

where φ and Φ are respectively the probability distribution function and the cumulative probability distribution function of a standard Gaussian random variable. For a deterministic forecast,

$$\mathrm{CRP}(\delta_y, x) = \int_{\mathbb{R}} \left(\mathbb{1}_{\{u \geq y\}} - \mathbb{1}_{\{u \geq x\}}\right)^2 du = |y - x|. \tag{1.17}$$

It is worth mentioning that for deterministic forecast, thanks to (1.17) the CRPS is reduced to the mean absolute error (MAE) defined by

$$MAE(t) = \mathbf{E}\left(\left|\pi(Z_t^{\tilde{Z}_0}) - \tilde{Z}_t\right|\right). \tag{1.18}$$

Conversely to the CIR model [2], transition (forecast) probability density functions are not in closed form. Consequently, computations of the CRPS requires using the Fokker–Planck finite element numerical solving scheme which is sophisticated (see Sect. 1.3.3.3). In the following, we propose probabilistic forecasts methods based on the approximation of the transition probability density function $p(t, \cdot; x, \vartheta)$.

1.3.3.3 Finite-Element-Method Approximation of the Transition Probability Densities

Finite-Element-Method (FEM) scheme (see Appendix 3) computes the numerical approximation of the transition probability density $p(t, \cdot; \tilde{Z}_0, \vartheta)$ on a regular grid on a compact $\Omega \subset \mathbb{R}$ (the distribution is supposed to be zero outside Ω).

1.3.3.4 Gaussian Approximations of the Transition Probability Densities

For a very short lead time $t > 0$, the Gaussian approximation for transition probability density function is valid. Consequently, we use the Gaussian distribution as the probabilistic forecast with the first order or second order Itò-Taylor expansion characteristics. For the first order expansion, we recall that the mean is given by

$$m_x = x + v_0(x, \vartheta)t$$

and variance by

$$\sigma_x^2 = v_1^2(x, \vartheta)t.$$

Here v_0 and v_1 are given respectively by (1.3) and (1.5). For the second order expansion, the mean is given by

$$m_x = x + v_0(x, \vartheta)t + \left(v_0(x, \vartheta)\frac{\partial}{\partial x}v_0(x, \vartheta) + \frac{1}{2}v_1^2(x, \vartheta)\frac{\partial^2}{\partial x^2}v_0(x, \vartheta) \right)\frac{t^2}{2}$$

and the variance (see [14]) by

$$\begin{aligned}\sigma_x^2 = x^2 + \left(2v_0(x, \vartheta) + v_1^2(x, \vartheta)\right)t + \Bigg(2v_0(x, \vartheta)\left(x\frac{\partial}{\partial x}v_0(x, \vartheta) + v_0(x, \vartheta)\right.\\ \left. v_1(x, \vartheta)\frac{\partial}{\partial x}v_1(x, \vartheta)\right) + v_1^2(x, \vartheta_2)\left(x\frac{\partial^2}{\partial x^2}v_0(x, \vartheta) + 2\frac{\partial}{\partial x}v_0(x, \vartheta)\right.\\ \left. + \frac{\partial}{\partial x}v_1^2(x, \vartheta) + v_1(x, \vartheta)\frac{\partial^2}{\partial x^2}v_1(x, \vartheta)\right)\Bigg)\frac{t^2}{2} - m_x^2.\end{aligned}$$

For the marginal Weibull diffusion model and linear drift diffusion models,

$$\frac{\partial^2}{\partial x^2}v_0(x, \vartheta) = 0$$

and corresponding terms in the previous equation disappear.

1.4 Application to Wind Speed Modeling and Forecast

A large class of datasets is provided by the National Renewable Energy Laboratory (NREL). It offers from http://wind.nrel.gov wind-speed datasets for several years for more than 25000 locations.

The considered dataset is a time series that includes wind speeds at height (100 m) every hour from 2010 to 2012 in three different locations in the US (ID 24310 (Wyoming), ID 69759 (Nevada) and ID112269 (Oregon)). The dataset contains no null wind. We consider NEG 2MW turbine manufacturer's transfer function to compute the production.

Firstly, we present the calibration method of the three-parameter marginal Weibull diffusion model with a quasi-likelihood estimation procedure in Sect. 1.4.1. Then, we compare the three-parameter marginal Weibull diffusion with the (Gaussian) Ornstein–Uhlenbeck process in terms of modeling using Cramer–von Mises statistics on the three turbines for each year.

We also compare our model to the benchmark in terms of wind-speed forecasting for the ID 69759 wind turbine on the whole period 2010–2012 in Sect. 1.4.3. Finally, we compare our model in terms of production forecasts (with the transfer function) against an Ornstein–Uhlenbeck process calibrated directly on production in Sect. 1.4.4.

1.4.1 Calibration Method

Let Θ be a bounded open subset of $(\mathbb{R}_*^+)^3$. We present in this section the calibration method for the parameter $\vartheta = (\vartheta_1, \vartheta_2, \vartheta_3) \in \Theta$ in the marginal Weibull diffusion model. In order to obtain the best estimate we use the maximum quasi-likelihood estimator, that we describe now.

Let us consider an observation of the process $(Z_t, t \geq 0)$ on a (regular) discrete temporal grid

$$0 = t_0 < t_1 < \cdots < t_n.$$

The mesh is denoted $\Delta_n = \frac{t_n}{n}$. In the following, we denote $Z^{(n)} = \left(Z_{t_1}, \ldots, Z_{t_n}\right)$ the observation sample. The sequence $Z^{(n)}$ is a Markov chain and the corresponding loglikelihood is given by

$$\mathscr{L}(\vartheta, Z^{(n)}) = \sum_{i=1}^{n} \log p(\Delta_n, Z_{t_i}; Z_{t_{i-1}}, \vartheta) \tag{1.19}$$

where the transition probability density $p(t, y; x, \vartheta)$ is given by the solution of the Fokker–Planck equation (1.6).

In the Ornstein–Uhlenbeck process or CIR model presented in [2], a closed form of the transition probability density is known and maximum loglikelihood estimator

$$\widehat{\vartheta}_n = \max_{\vartheta \in \Theta} \mathscr{L}(\vartheta, Z^{(n)})$$

can be computed numerically. But, for the three-parameter marginal Weibull diffusion process, this is not the case anymore.

Several methods are available in this context depending on the discretization time scheme. For a large observation horizon t_n and large mesh size Δ_n, one can use an approximation of the likelihood function (numerical approximation of the Fokker–Planck equation in [15], Monte-Carlo simulation approximation [16], ...). A recent review of possible methods in this setting is proposed in [19].

When the mesh is relatively small (thereafter Δ_n will be equal to 1 hour), which is the case we are considering, it is possible to use the quasi-likelihood approach. For small Δ_n, under proper assumptions, it is possible to approximate the transition probability density function (and consequently the likelihood) by a Gaussian probability density function with the same mean and variance.

The simplest approximation is the Euler method,[2] in which the transition probability density $p(\Delta_n, y; x, \vartheta)$ is approximated by a Gaussian with mean

$$m_x = x + v_0(x, \vartheta)\Delta_n \tag{1.20}$$

and variance

$$\sigma_x^2 = v_1^2(x, \vartheta)\Delta_n. \tag{1.21}$$

Consequently, the quasi-loglikelihood is given by

$$\mathscr{L}^*(\vartheta, Z^{(n)}) = -\frac{n}{2}\log\left(2\pi v_1^2(Z_{t_{i-1}}, \vartheta)\Delta_n\right) - \frac{1}{2}\sum_{i=1}^{n} \frac{\left(Z_{t_i} - (Z_{t_{i-1}} + v_0(Z_{t_{i-1}}, \vartheta)\Delta_n)\right)^2}{v_1^2(Z_{t_{i-1}}, \vartheta)\Delta_n}. \tag{1.22}$$

The maximum quasi-likelihood (or quasi-loglikelihood) estimator is obtained by

$$\widehat{\vartheta}_n^* = \max_{\vartheta \in \Theta} \mathscr{L}^*(\vartheta, Z^{(n)}).$$

Although the estimator is not in a closed form, it can be computed numerically. This approximation is valid for a "rapidly increasing experimental design" which means $\Delta_n \to 0$, $n\Delta_n \to \infty$ and $n\Delta_n^2 \to 0$ (see [14] for details).

[2] The Gaussian approximation of the conditional density function $p(\Delta, y; x, \vartheta)$ proposed in [14] uses the higher order Itò-Taylor expansion to approximate the mean and the variance. It is worth emphasizing that the Euler method is the one order Itò-Taylor expansion.

1.4.2 Performance in Terms of Modeling

Contrary to the classical i.i.d. setting, classical goodness-of-fit tests (Kolmogorov–Smirnov, Cramer–von Mises, etc.) do not apply directly to the diffusion (Markov) setting. In order to evaluate the performance of the modeling, we propose the following methodology inspired by [9]: (1) calibration of the dynamical model (marginal distribution and correlation structure simultaneously as in Sect. 1.4.1 for the three-parameter marginal Weibull diffusion model and the (Gaussian) Ornstein–Uhlenbeck process on the one-year dataset, (2) computation of the Cramer–von Mises statistics

$$\begin{aligned}\omega_n^2 &= n\int_{\mathbb{R}}\left(F_n(x) - F(x,\widehat{\vartheta}_n)\right)^2 dF(x,\widehat{\vartheta}_n)\\ &= \frac{1}{2n} + \sum_{i=1}^{n}\left(F(x_i,\widehat{\vartheta}_n) - \frac{i-1/2}{n}\right)^2,\end{aligned}$$

with respect to the calibrated marginal distribution F (Weibull and Gaussian respectively) and (3) selection of the distribution corresponding to the smallest value of the computed statistic.

Stationary distribution for the marginal Weibull diffusion model and the Ornstein–Uhlenbeck process (respectively Weibull and Gaussian) are illustrated in Fig. 1.2 for the ID 24310 wind turbine in 2011. Correlation structure for the marginal Weibull diffusion model is also illustrated in Fig. 1.2.

The result of the selection for all wind turbines and all the years are given in the following Table 1.1.

The previous table shows for instance that the marginal Weibull diffusion model outperforms the (Gaussian) Ornstein–Uhlenbeck process in terms of wind-speed generation in 2011 in Fig. 1.2. This is merely confirming common practice of using the Weibull distribution rather than using a Gaussian to describe the distribution of the wind speed.

1.4.3 Wind-Speed Forecasting

Time is considered in days. We denote Δ_n the time mesh (thereafter $\Delta_n = \frac{1}{24}$). We consider rolling training datasets of 14 days (14×24 hourly measures) where the model is calibrated.

Let us fix the first measure of the testing dataset at time 0 and the horizon time $\tau = k\Delta_n$. In our dataset; for instance $k = 1$ ($\tau = 1$ h), $k = 2$ ($\tau = 2$ h), $k = 3$ ($\tau = 2$ h) and $k = 6$ ($\tau = 6$ h) are considered for short-term forecasting.

For successive times $t_j = j\Delta_n, j = 0, \ldots, N-k$, we want to compute the forecast of $\tilde{Z}_{t_j+\tau}$ at time $t_j + \tau$. We denote

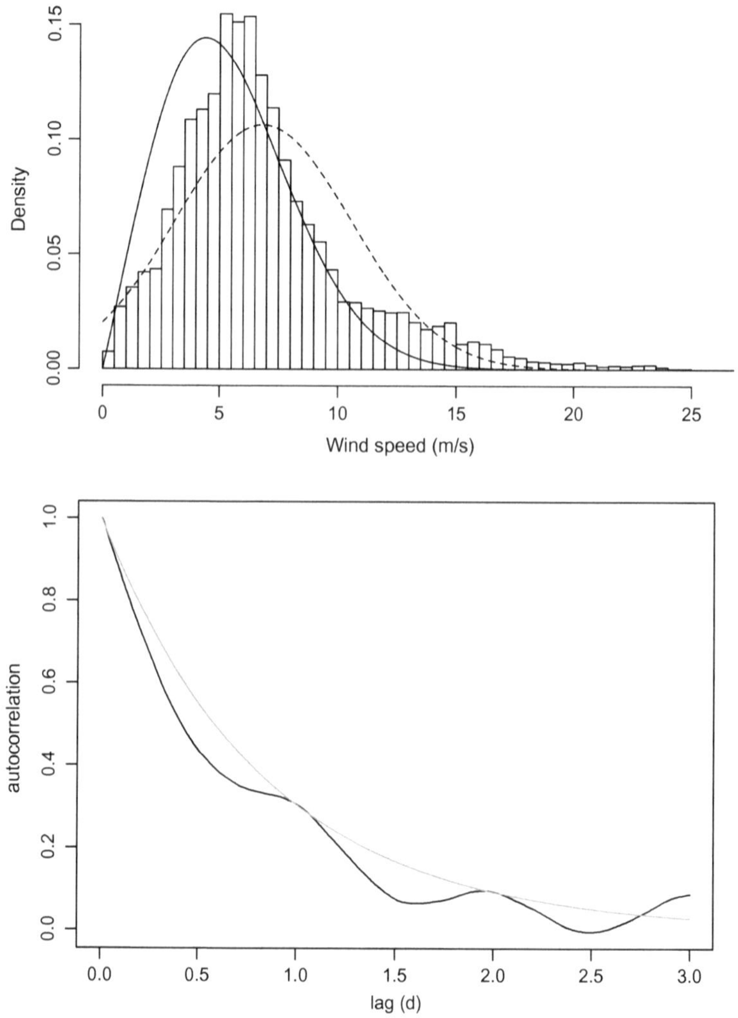

Fig. 1.2 On the top, histogram of the training NREL wind speed dataset (see Sect. 1.4). Stationary distribution for the marginal Weibull diffusion model (plain line) and Gaussian Ornstein–Uhlenbeck process (dashed line) fitted on the training dataset are superposed. On the bottom, empirical autocorrelation of the training NREL wind speed dataset (plain black line) and calibrated correlation structure of the marginal Weibull diffusion model (plain gray line)

$$\pi\left(Z_{t_j+\tau}^{t_j,z_j}\right)$$

the forecast value knowing that $\tilde{Z}_{t_j} = z_j$. The empirical mean square error (eMSE) is given by

Table 1.1 Selected model (mW or OU) for different wind turbines and different years

ID/year	2010	2011	2012
24310	OU	mW	OU
69759	mW	mW	mW
112269	mW	OU	mW

$$eMSE(\tau) = \frac{\sum_{j=0}^{N-1-k} \left(\pi\left(Z_{t_j+\tau}^{t_j,z_j}\right) - z_{j+k}\right)^2}{N-k}.$$

We compute the empirical MSE for all models described above. Namely,

1. For the persistence benchmark, $\pi\left(Z_{t_j+\tau}^{t_j,z_j}\right) = z_j$;
2. For the dynamic 3 parameters marginal Weibull diffusion model with linear drift, the point forecast is given by (1.9) or

$$\pi\left(Z_{t_j+\tau}^{t_j,z_j}\right) = \vartheta_2\Gamma\left(1+\frac{1}{\vartheta_3}\right) + \left(z_j - \vartheta_2\Gamma\left(1+\frac{1}{\vartheta_3}\right)\right)e^{-\vartheta_1\tau}$$

where, in practice, $\vartheta = \widehat{\vartheta}^*$ is the maximum quasi-likelihood estimator of ϑ obtained as indicated in Sect. 1.4.1.

We summarize the results on bias and RSME (root square of the MSE) in the Table 1.2 containing lead times of relevance (1, 2, 3 and 6 h).

In term of RMSE, the results are comparable for forecast below one hour lead time. Marginal Weibull diffusion model outperforms the persistence benchmark for forecasting over 1h. Finally, on this data set, this bias is negligeable under 6 h lead time.

In the next Table 1.3, we summarized the result of CRPS and MAE computed on the Weibull diffusion model with 1st-order Itò probabilistic forecast and FEM probabilistic forecast both presented in Sects. 1.3.3.3 and 1.3.3.4. The computation has been done on a subset of the dataset (every 25 hourly measurements).

Here again we can notice similar results for the three-parameter Weibull diffusion model in term of CRPS and MAE at very-short term. The marginal Weibull diffusion model shows a better result at medium range lead times (from 3 h to 6 h). It is worth emphasizing that the first order approximation is valid at lead times less than 3 h.

Table 1.2 Bias and RMSE in parenthesis (all measures in m/s) for wind-speed forecasts for lead times of relevance

	$\tau = 1$ h	$\tau = 2$ h	$\tau = 3$ h	$\tau = 6$ h
Weibull diffusion	0.00 (1.97)	0.00 (2.59)	0.00 (2.97)	0.00 (3.58)
Persistence benchmark	0.00 (2.04)	0.00 (2.73)	0.00 (3.19)	0.00 (4.04)

Table 1.3 CRPS and MAE in brackets for wind-speed forecasts for lead times of relevance

	$\tau = 1\,\text{h}$	$\tau = 2\,\text{h}$	$\tau = 3\,\text{h}$	$\tau = 6\,\text{h}$
FEM	0.25 [0.34]	0.36 [0.49]	0.41 [0.56]	0.51 [0.71]
1st order TI Gaussian	0.25 [0.34]	0.36 [0.49]	0.44 [0.57]	0.61 [0.71]
Persistence benchmark	[0.33]	[0.50]	[0.58]	[0.78]

Table 1.4 Bias, RMSE in parenthesis, RMSE in % of the nominal power for production forecasts (all measures in kWh) for lead times of relevance

	$\tau = 1\,\text{h}$	$\tau = 2\,\text{h}$	$\tau = 3\,\text{h}$	$\tau = 6\,\text{h}$
Weibull diffusion	−10 (171) [9%]	−13 (229) [11%]	−20 (261) [13%]	−29 (310) [16%]
Ornstein–Uhlenbeck	−23 (192) [10%]	−31 (253) [12%]	−51 (289) [14%]	−75 (339) [17%]

1.4.4 Production Forecasting

In this section, we compare the production forecast given, on the one hand, by the three-parameter marginal Weibull diffusion model calibrated on the wind speed and the computation of the corresponding production through the manufacturer power function and, on the other hand, by an Ornstein–Uhlenbeck directly calibrated on the production dataset. Production point forecasting (in kWh) performance is given in terms of RSME in the following Table 1.4. The computation has been done on a subset of the dataset (every 25 hourly measurements).

It is worth emphasizing that, considering no errors on the manufacturer transfer function and a transfer function only depending on the wind speed, forecasting using the wind speed outperforms the Ornstein–Uhlenbeck model on production.

1.5 Conclusion

Cox–ingersoll–Ross process and marginal Weibull diffusion model, presented in [2] and in this paper respectively, fit the wind speed data and are able to forecast at short term. Considering some improvements, they could provide a wind generator. For instance, it is possible to consider a model

$$Y_t = f(t)Z_t + g(t)$$

where $f(\cdot)$ and $g(\cdot)$ are periodic functions (see for instance [3] for a related work where the diurnal cycle is taken into account). The characteristics of the marginal Weibull diffusion model are understandable in terms of shape and scale parameter of a Weibull law that is appreciated by practitioners to analyze the results.

Properties of diffusion process are important to study the problem of optimal storage, optimal trading and optimal maintenance where wind speed and energy production are the stochastic entries.

It is possible to extend results on the diffusion models to any marginal distribution (given a fixed drift term). Measurements errors of the anemometers behind the blades can also be handled with non-linear filtering methods. Let us also mention the fact that errors on the transfer function are not taken into account and should be addressed for a better estimation of the production in the operational management problems.

Acknowledgements We would like to thank the anonymous referees for their comments that improves our original work. This research has been supported by a grant from EREN Groupe and partially by NSF grant DMS-1612880, Research Grants Council of HKSAR 11303316 and ANR project CAESARS (ANR-15-CE05-0024).

Appendix 1: Autocorrelation Structure of Homogeneous Diffusion Processes with Linear Drift

Let $(Z_s, s \geq 0)$ be the solution of Eq. (1.2). Let us denote $m(t, z_0) = \mathbf{E}\,(Z_s)$. Since $v_0(z, \vartheta) = \vartheta_1\,(\alpha - z)$ is linear in z with

$$\alpha = \vartheta_2 \Gamma\left(1 + \frac{1}{\vartheta_3}\right),$$

we have

$$dm(t, z_0) = \vartheta_1\,(\alpha - m(t, z_0))\,dt$$

with $m(z_0, 0) = z_0$. Consequently,

$$m(z_0, t) = \alpha + (z_0 - \alpha)\,e^{-\vartheta_1 t}. \tag{1.23}$$

The joint distribution $f(y, z)$ of the pair (Z_s, Z_t), $s \leq t$, is given by

$$f(y, z) = p(t - s, z; y, \vartheta) p(s, y; z_0, \vartheta)$$

due to the Markov property. Consequently,

$$\begin{aligned} \operatorname{cov}(Z_s, Z_t) &= \int_{\mathbb{R}^2} (y - \mathbf{E}Z_s)(z - \mathbf{E}Z_t) f(y, z)\,dydz \\ &= \int_{\mathbb{R}} (y - \mathbf{E}Z_s) p(s, z_0, y) \underbrace{\left(\int_{\mathbb{R}} (z - \mathbf{E}Z_t) p(t - s, y, z)dz\right)}_{e^{-\vartheta_1(t-s)}(y - \mathbf{E}Z_s)} dy \\ &= e^{-\vartheta_1(t-s)} \operatorname{var}(Z_s) \end{aligned}$$

which proves the result. In the case of the stationary distribution,

$$\mathrm{corr}(Z_s, Z_t) = e^{-\vartheta_1(t-s)}$$

is obtained.

Appendix 2: Marginal Weibull Diffusion Coefficient

Following [4], it is possible to construct a marginal Weibull diffusion process, solution of the stochastic differential equation

$$Z_t = z_0 + \int_0^t \nu_0(Z_s, \vartheta)ds + \int_0^t \nu_1(Z_s, \vartheta)d\beta_s, \quad t \geq 0, \tag{1.24}$$

where $(\beta_t, t \geq 0)$ is a Wiener process. Fixing the drift coefficient

$$\nu_0(z, \vartheta) = \vartheta_1 \left(\vartheta_2 \Gamma \left(1 + \frac{1}{\vartheta_3} \right) - z \right),$$

the solution of Eq. (1.24) has an invariant density equals to the Weibull probability density function

$$f(z, \vartheta) = \frac{\vartheta_3}{\vartheta_2} \left(\frac{z}{\vartheta_2} \right)^{\vartheta_3 - 1} \exp\left(- \left(\frac{z}{\vartheta_2} \right)^{\vartheta_3} \right), \quad z \geq 0.$$

If the diffusion coefficient satisfies

$$\nu_1^2(z, \vartheta) = \frac{2\vartheta_1}{f(z)} \left(\vartheta_2 \Gamma \left(1 + \frac{1}{\vartheta_3} \right) F(z) - \int_0^z yf(y)dy \right) \tag{1.25}$$

where F is the distribution function associated to the density f. Direct computations lead to

$$F(z) = 1 - \exp\left(- \left(\frac{z}{\vartheta_2} \right)^{\vartheta_3} \right)$$

and

$$\int_0^z yf(y)dy = \vartheta_2 \int_0^{\left(\frac{z}{\vartheta_2}\right)^{\vartheta_3}} u^{\frac{1}{\vartheta_3}} e^{-u} du \quad \left(\text{change of variable} \quad u = \left(\frac{y}{\vartheta_2} \right)^{\vartheta_3} \right).$$

Appendix 3: Finite Element Method for Fokker–Planck Equation

Let us approximate the solution $(p(t,y), t \geq 0, y \in \mathbb{R})$ satisfying

$$\frac{\partial p}{\partial t} = -\frac{\partial}{\partial y}(a(y)p) + \frac{1}{2}\frac{\partial^2}{\partial y^2}(b(y)p) \tag{1.26}$$

with the initial condition $p(0,y) = \delta_x(y), x \in \mathbb{R}$. For our problem $a = v_0$ and $b = v_1^2$. By integrating (1.26) with respect to y, we get

$$\begin{aligned}\frac{\partial}{\partial t}\int_{\mathbb{R}} p(t,y)dy &= \int_{\mathbb{R}}\left(-\frac{\partial}{\partial y}(a(y)p) + \frac{1}{2}\frac{\partial^2}{\partial y^2}(b(y)p)\right)dy \\ &= -a(y)p(t,y) + \frac{1}{2}\frac{\partial}{\partial y}(b(y)p)\Big|_{-\infty}^{\infty} \\ &= \left(\frac{1}{2}b'(y) - a(y)\right)p(t,y)\Big|_{-\infty}^{\infty} + \frac{1}{2}b(y)\frac{\partial}{\partial y}p(t,y)\Big|_{-\infty}^{\infty}.\end{aligned}$$

Consequently, using boundary conditions

$$\lim_{y\to\pm\infty} p(t,y) = 0 \quad \text{and} \quad \lim_{y\to\pm\infty}\frac{\partial}{\partial y}p(t,y) = 0$$

dominating possible growth of $a(\cdot)$ and $b(\cdot)$ we get that the solution is norm preserving, namely

$$\int_{\mathbb{R}} p(t,y)dy = 1, \quad t \geq 0.$$

Let us consider the approximation

$$\widehat{u}(t,y) = \sum_{j=1}^{N-1}\lambda_j(t)\varphi_j(y)$$

where φ_i are the sequence of finite elements. This approximation satisfies (1.26) on the compact $\Omega = \mathbf{supp}(\widehat{u}) \subset \mathbb{R}$. Computing the variational formulation with respect to the test function $\varphi_i(y)$, $i = 1, \ldots, N-1$, we obtain

$$\sum_{j=1}^{N-1}\lambda_j'(t)C_{i,j} - \lambda_j(t)K_{i,j} = 0 \tag{1.27}$$

with

$$C_{i,j} = \int_{\Omega} \varphi_i(y)\varphi_j(y)dy$$

and

$$\begin{aligned}
K_{i,j} &= \int_{\Omega} \left(-\frac{\partial}{\partial y}\left(a(y)\varphi_j(y)\right) + \frac{1}{2}\frac{\partial^2}{\partial y^2}\left(b(y)\varphi_j(y)\right) \right) \varphi_i(y)dy \\
&= -a(y)\varphi_j(y)\varphi_i(y)\Big|_{\partial\Omega} + \int_{\Omega} a(y)\varphi_i'(y)\varphi_j(y)dy \\
&\quad + \frac{1}{2}\frac{\partial}{\partial y}\left(b(y)\varphi_j(y)\right)\varphi_i(y)\Big|_{\partial\Omega} - \frac{1}{2}\int_{\mathbb{R}} \varphi_i'(y)\frac{\partial}{\partial y}\left(b(y)\varphi_j(y)\right)dy \\
&= \int_{\mathbb{R}} a(y)\varphi_i'(y)\varphi_j(y)dy - \frac{1}{2}\int_{\mathbb{R}} \varphi_i'(y)\frac{\partial}{\partial y}\left(b(y)\varphi_j(y)\right)dy \\
&= \int_{\mathbb{R}} \left(a(y) - \frac{1}{2}b'(y) \right) \varphi_i'(y)\varphi_j(y)dy - \frac{1}{2}\int_{\mathbb{R}} b(y)\varphi_i'(y)\varphi_j'(y)dy \qquad (1.28)
\end{aligned}$$

considering that $\varphi_i(y) = 0$ for $y \in \partial\Omega$. It is worth mentioning that the first derivative of v_1 is needed to compute (1.28).

Consequently, Eq. (1.27) can be rewritten as a multidimensional first order linear EDO

$$C\lambda'(t) - K\lambda(t) = 0 \qquad (1.29)$$

with the proper initial condition $\lambda(0)$. Namely, in order to mimic the true initial condition δ_x, $x \in \mathbb{R}$, we consider for $i = 1, \ldots, N-1$

$$\int_{\Omega} \widehat{u}(0, y)\varphi_i(y)dy = \varphi_i(x).$$

This can be written as

$$\sum_{j=1}^{N-1} \lambda_j(0) \int_{\Omega} \varphi_j(y)\varphi_i(y)dy = \varphi_i(x)$$

or

$$C\lambda(0) = \Phi_x \quad \text{where} \quad \Phi_x = \begin{pmatrix} \varphi_1(x) \\ \vdots \\ \varphi_{N-1}(x) \end{pmatrix}.$$

Finally, fixing $R \in \mathbb{N}$, we can apply an implicit Euler scheme to (1.29) on $[0, \tau]$, $\tau > 0$ with the mesh size $\frac{\tau}{R}$.

References

1. F. Baudoin, *Diffusion Processes and Stochastic Calculus*. EMS Textbooks in Mathematics (2014)
2. A. Bensoussan, A. Brouste, Cox–Ingersoll–Ross model for wind speed modeling and forecasting. Wind Energy **19**(7), 1355–1365 (2016)
3. A. Bensoussan, A. Brouste, P. Bertrand, A generalized linear model approach to seasonal aspects of wind speed modeling. J. Appl. Stat. **41**(8), 1694–1707 (2014)
4. B. Bibby, I. Skovgaard, M. Sorensen, Diffusion-type models with given marginal distribution and autocorrelation function. Bernoulli **1**, 191–220 (2003)
5. J. Carta, P. Ramirez, S. Velasquez, A review of wind speed probability distributions used in wind energy analysis, case studies in the Canary Islands. Renew. Sustain. Energy Rev. **13**, 933–955 (2009)
6. F. Castino, R. Festa, C. Ratto, Stochastic modeling of wind velocities time series. J. Wind Eng. Ind. Aerodyn. **74**, 141–151 (1998)
7. G. D'Amico, F. Petroni, F. Prattico, First and second order semi-Markov chains for wind speed modeling. Phys. Stat. Mech. Appl. **392**(5), 1194–1201 (2013)
8. G. D'Amico, F. Petroni, F. Prattico, Wind speed modeled as an indexed semi-Markov process. Environmetrics **24**(6), 367–376 (2013)
9. P. Drobinski, C. Coulais, B. Jourdier, Surface wind-speed statistics modelling: alternatives to the weibull distribution and performance evaluation. Bound. Layer Meteorol. **157**, 97–123 (2015)
10. T. Gneiting, A. Raftery, Strictly proper scoring rules, prediction, and estimation. J. Am. Stat. Assoc. **102**(477), 359–378 (2007)
11. N. Ikeda, S. Watanabe, *Stochastic Differential Equations and Diffusion Processes* (North Holland, New York, 1981)
12. T. Jonsonn, P. Pinson, H. Madsen, H. Nielsen, Predictive densities for day-ahead electricity prices using time adaptative quantile regression. Energies **7**, 5523–5547 (2014)
13. H. Kantz, D. Holstein, M. Ragwitz, N.K. Vitanov, Markov chain model for turbulent wind speed data. Phys. Stat. Mech. Appl. **342**(1–2), 315–321 (2004)
14. M. Kessler, Estimation of an ergodic diffusion from discrete observations. Scand. J. Stat. **24**, 211–229 (1997)
15. A. Lo, Maximum likelihood estimation of generalized Itô processes with discretely sampled data. Econom. Theory **4**, 231–247 (1988)
16. A. Pedersen, A new approach to maximum-likelihood estimation for stochastic differential equations based on discrete observations. Scand. J. Stat. **22**, 55–71 (1995)
17. K. Pourmousavi, M. Ardehali, Very short-term wind speed prediction: a new artificial neural network-Markov chain model. Energy Convers. Manag. **52**(1), 738–745 (2011)
18. A.D. Sahin, Z. Sen, First-order Markov chain approach to wind speed modeling. J. Wind Eng. Ind. Aerodyn. **89**(3–4), 263–269 (2001)
19. H. Sorensen, Parametric inference for diffusion processes observed at discrete points of time: a survey. Int. Stat. Rev. **72**(3), 337–354 (2004)
20. J. Tang, A. Brouste, K. Tsui, Some improvements of wind speed Markov chain modeling. Renew. Energy **81**, 52–56 (2015)

Chapter 2
From Numerical Weather Prediction Outputs to Accurate Local Surface Wind Speed: Statistical Modeling and Forecasts

Bastien Alonzo, Riwal Plougonven, Mathilde Mougeot, Aurélie Fischer, Aurore Dupré and Philippe Drobinski

Abstract Downscaling a meteorological quantity at a specific location from outputs of Numerical Weather Prediction models is a vast field of research with continuous improvement. The need to provide accurate forecasts of the surface wind speed at specific locations of wind farms has become critical for wind energy application. While classical statistical methods like multiple linear regression have been often used in order to reconstruct wind speed from Numerical Weather Prediction model outputs, machine learning methods, like Random Forests, are not as widespread in this field of research. In this paper, we compare the performances of two downscaling statistical methods for reconstructing and forecasting wind speed at a specific location from the European Center of Medium-range Weather Forecasts (ECMWF) model outputs. The assessment of ECMWF shows for 10 m wind speed displays a systematic bias, while at 100 m, the wind speed is better represented. Our study shows that both classical and machine learning methods lead to comparable results. However, the time needed to pre-process and to calibrate the models is very different in both cases. The multiple linear model associated with a wise pre-processing and variable selection

B. Alonzo (✉) · R. Plougonven · A. Dupré · P. Drobinski
LMD/IPSL, École Polytechnique, Université Paris Saclay, ENS,
PSL Research University, Sorbonne Universités,
UPMC Univ Paris 06, CNRS, Palaiseau, France
e-mail: bastien.alonzo@lmd.polytechnique.fr

R. Plougonven
e-mail: riwal.plougonven@lmd.polytechnique.fr

A. Dupré
e-mail: aurore.dupre@lmd.polytechnique.fr

P. Drobinski
e-mail: philippe.drobinski@lmd.polytechnique.fr

M. Mougeot · A. Fischer
Laboratoire de Probabilités, Statistique et Modélisation,
Université Paris Diderot - Paris 7, Paris, France
e-mail: mathilde.mougeot@univ-paris-diderot.fr

A. Fischer
e-mail: aurelie.fischer@univ-paris-diderot.fr

P. Drobinski et al. (eds.), *Renewable Energy: Forecasting and Risk Management*,
Springer Proceedings in Mathematics & Statistics 254,
https://doi.org/10.1007/978-3-319-99052-1_2

shows performances that are slightly better, compared to Random Forest models. Finally, we highlight the added value of using past observed local information for forecasting the wind speed on the short term.

Keywords Local wind speed · Downscaling · Statistical modeling · Numerical weather prediction model · Wind speed forecasts

2.1 Introduction

The wind energy sector has seen a very sharp growth in recent years. According to the Global Wind Energy Council (GWEC), 54 GW has been installed in 2016, corresponding to an increase of 12.6% of the total installed capacity [11]. Worldwide, the number of wind farms increases each year and feeds the electrical network with a larger amount of energy. For instance, in 2016, France has seen its highest capacity growth rate ever recorded. This sharp increase of connected wind power has for instance allowed the network to receive 8.6 GW from wind power plants, on November 20th, corresponding to 17.9% of the energy produced this day [19]. The need to have access to accurate wind forecasts on several timescales is thus becoming crucial for the wind energy producer and grid operator, in order to anticipate the energy production, to plan maintenance operations and to handle balance between energy production and consumption. Changing regulations of the energy market with the end of feeding-in tariffs make this anticipation vital for wind energy producers. Finally, a related but different topic consists in the estimation of the wind resource of its long-term (multi-year) variability and trends mainly for prospecting purposes.

The increasing need for accurate forecasts of the surface wind speed fortunately comes with the improvement of the Numerical Weather Prediction models (NWP) describing and forecasting atmospheric motions. Indeed, they constitute a key source of information for surface wind speed forecasts all the more so as their realism, accuracy and resolution have increased steadily over the years [2].

Nevertheless, these models are not necessarily performing uniformly well for all atmospheric variables. Their astonishing performances are evaluated on variables such as mid-tropospheric pressure which reflect the large-scale mass distribution, which is effectively well understood physically (see, e.g., [23]) and efficiently modeled numerically. Variables tied to phenomena occurring on smaller scales (such as cloud-cover or near-surface winds) depend much more directly on processes that are *parameterized* (e.g, not resolved). In contrast to large-scale motions (governed by the Navier–Stokes equations), parameterizations are generally partly rooted in physical arguments, but also in large part empirical. When comparing output from a numerical model to a local measurement, there will therefore always be several sources of error: representativity error (contrast between the value over a grid-box and the value at a specific point), numerical error (even if we were describing only processes governed by well-established physical laws, discretization is unavoidable), and error tied to the physics described (because processes, especially parameterized ones, are not well modeled). To reduce representativity error and to better represent small-scale processes, in particular those tied to topography and surface roughness, one strategy consists in downscaling with models that describe the atmospheric flow on finer scales (see, e.g., [24]). One disadvantage of this approach is the numerical cost, and one limitation is the need for finer observations to initialize the state of the atmosphere, if details of the flow other than those directly implied by the topography and surface condition are sought for.

Strategies to estimate surface winds, or other meteorological variables, from the output of Numerical Weather Prediction models (NWP) or climate models have been developed in several contexts, with different motivations, and leading to different methodologies and applications.

Model Output Statistics (MOS) has been developed in weather forecasting for several decades to estimate the *weather related* variability of a physical quantity, based on NWP model output [10]. NWP models perform now very well in predicting large-scale systems. Relations thus can be derived to link the latters to local variables at an observation site. Linear models are generally used, with the outcome now expanded over a wider area than only the location of stations where observations are available [27].

In the context of climate change, downscaling a meteorological quantity at a given location in order to produce time series which have plausible statistical characteristics under climate change has for long been investigated [26]. The challenge is here to capture appropriately the relation between large-scale flow (as it can be described by a model with a moderate or low resolution) and a variable at a specific location (e.g. wind, temperature, precipitation) and then use climate models to provide a description of the large-scale atmospheric state under climate change. Local time series with appropriate variability and consistent with this large-scale state of the atmosphere are then generated, e.g. [17, 20, 25].

Wind energy domain is nowadays a very active branch in downscaling techniques because of the need for accurate forecasts at specific location of a wind farm. For describing winds close to the surface, 10 m wind speed is often a convenient variable as it has been for decades a reference observed variable and also now a reference NWP model output. In the case of wind energy, the wind speed then needs to be extrapolated at the hub height to have access to wind power, leading to an increase of the error on the predicted power [13, 16, 18]. Wind speed at the hub height (typically 100 m) is a variable of interest as it allows to avoid vertical extrapolation errors [4], but it is rarely available in observations. Different outputs of NWP models can be used as explanatory variables of the near surface wind speed. It seems that there is no strong consensus on the predictors to use, mainly because relations between predictors and predictand should differ from one location to the other. However, different studies have shown the importance of a certain set of variables to predict surface wind speed. Amongst them, markers of large-scale systems (geopotential height, pressure fields) and boundary layer stability drivers (surface temperature, boundary layer height, wind and temperature gradient) can be cited [5, 6, 20]. In terms of methodology, several models have already been studied, including Linear regression, Support Vector Models (SVM) or Artificial Neural Network (ANN) [14, 21].

The model of the European Center for Medium-range Weather Forecasts (ECMWF) has reached a resolution of about 9 km in the horizontal. In addition, ECMWF analyses and forecasts now give access to 100 m wind speed output, developed mainly for wind energy applications. If we can be very confident in the ability of NWP models to represent several variables, some others may not be so reliable. This is especially the case for surface variables such as 10 and 100 m wind speed. Consequently, using the robust information given by some variables to correct surface wind speed is straightforward. We have access to surface wind speed observed at 10, 100 m over a long period of 5 years at SIRTA observation platform [12]. The aim of this project is, in particular, to explore how different statistical models perform in forecasting the 10 and 100 m wind speed using informations of ECMWF analyses and forecasts outputs at different horizons. We choose multiple linear regression because it is a widely used technique, and Random Forests which have not been, to our knowledge, deeply studied in the framework of downscaling surface wind speed. For multiple linear regression, variable selection is a very important step for calibrating the statistical models, whereas Random Forests handle variables automatically. Moreover, Random Forests can handle nonlinear relations very well. Therefore, the comparison of those very different statistical models, as well as the information used by each of them, should be very instructive.

The paper is organized in 5 parts. The next section describes together the data and the statistical models to be used. In Sect. 2.3, the training dataset is explored, and used to calibrate the statistical models. In Sect. 2.4, forecasts of 10 and 100 m wind speed are run to downscale wind speed at the observation site. In the last section, we discuss the results, conclude and give perspectives to this work.

2.2 Data and Methodology

2.2.1 Data

Observed Wind Speed

In this paper, we use observations of the wind speed at the SIRTA observation platform [12]. Surface wind speed at 10 m height from anemometer recording is available at the 5-min frequency. The wind speed at 100 m height from Lidar recording is available at 10-min frequency. Both data span for 5 years from 2011 to 2015. We filter observations by a sinusoidal function over a 6-h window centered at 00, 06, 12 and 18 h to obtain a 6-h observed wind speed to be compared to the NWP model outputs available at this time frequency. We found that the resulting time series are not sensitive to the filter function. We also try different filtering windows, concluding that 6-h is the best to compare to the NWP model outputs. Due to some missing data,

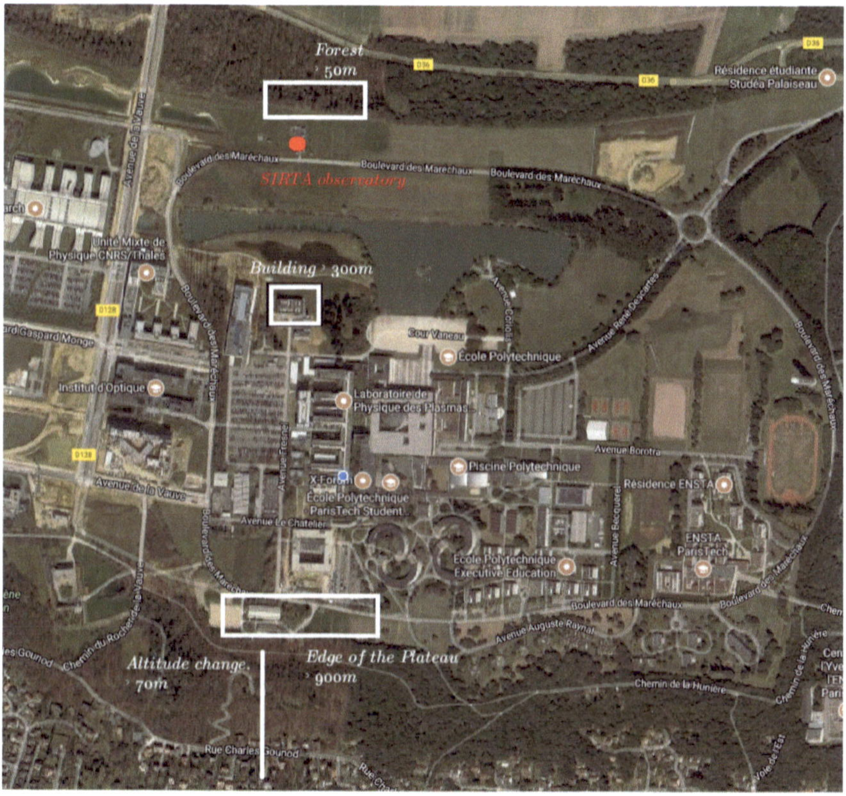

Fig. 2.1 Map of the SIRTA observation platform and its surroundings

two final time series of 5049 filtered observations are computed (over 7304 if all data were available).

SIRTA observatory is based 20 Km in the South of Paris on the Saclay plateau (48.7°N and 2.2°E). Figure 2.1 shows the SIRTA observation platform location, marked by the red point on the map, and its close environment. Regarding the relief near SIRTA, observe that a forest is located at about 50 m north to the measurement devices. South, buildings can be found at about 300 m from the SIRTA observatory. In the East-West axis, no close obstacle are encountered. Further south, the edge of the Saclay plateau shows a vertical drop of about 70 m, from 160 m on top to 90 m at the bottom.

NWP Model Outputs - ECMWF Analyses

Variables are retrieved from ECMWF analyses at 4 points around the SIRTA platform. The spatial resolution of ECMWF analyses is of about 16 km (0.125° in latitude and longitude). Topography is thus smoothed compared to the real one. As the surface wind speed is very influenced by the terrain, the modeled surface wind speed is not necessarily close to the observed wind speed. The data spans from the 01/01/2011 to

Table 2.1 Surface variables

Altitude (m)	Variable	Unit	Name
10/100 m	Norm of the wind speed	$m{\cdot}s^{-1}$	F
10/100 m	Zonal wind speed	$m{\cdot}s^{-1}$	U
10/100 m	Meridional wind speed	$m{\cdot}s^{-1}$	V
2 m	Temperature	K	T
2 m	Dew point temperature	K	Dp
Surface	Skin temperature	K	skt
Surface	Mean sea level pressure	Pa	msl
Surface	Surface pressure	Pa	sp
–	Boundary layer height	m	blh
–	Boundary layer dissipation	$J{\cdot}m^{-2}$	bld
Surface	Surface latent heat flux	$J{\cdot}m^{-2}$	slhf
Surface	Surface sensible heat flux	$J{\cdot}m^{-2}$	sshf

31/12/2015 at the 6-h frequency. It is sampled at each date where a filtered sampled observation is available.

The near surface wind speed at a given location can be linked to different phenomena. The large-scale circulation brings the flow to the given location explaining the slowly varying wind speed. The wind speed in altitude, the geopotential height, the vorticity, the flow divergence, sometimes the temperature can be markers of large systems like depressions, fronts, storms, or high pressure systems explaining a large part of the low frequency variations of the surface wind speed (Table 2.2). At a finer scale, what is happening in the boundary layer is very important to explain the intraday variations of the wind speed. The state and stability of the boundary layer can be derived from surface variables describing the exchanges inside the layer. Exchanges are driven mostly by temperature gradient and wind shear that develop turbulent flow (Table 2.3). Thermodynamical variables like surface, skin, and dew point temperatures and surface heat fluxes can also inform on the stability of the boundary layer, as well as its height and dissipation on its state (Table 2.1). In the end, 20 output variables are retrieved from ECMWF analyses at the 4 points around the SIRTA observatory and at different pressure levels. Note that we restrict the study to local variables (at the location of measurements or in the column above). It might also be possible to take advantage from larger scale information [5, 27]. The choice of taking 4 points around the SIRTA platform has the advantage of being simple and straightforward. Providing instead the explanatory variables by their interpolated value at SIRTA and the two components of their gradient does not lead to significantly different results.

ECMWF Deterministic Forecasts

The year 2015 of deterministic forecasts is retrieved from ECMWF model. A forecast is launched every day at 00:00:00 UTC. The time resolution retained is of 3 h and the maximum lead-time is 5 days. The same variables as for the analyses are retrieved at the same points around the SIRTA platform.

Table 2.2 Altitude variables

Pressure level (hPa)	Variable	Unit	Name
1000 hPa/925 hPa/850 hPa/700 hPa/500 hPa	Zonal wind speed	$m{\cdot}s^{-1}$	U
1000 hPa/925 hPa/850 hPa/700 hPa/500 hPa	Meridional wind speed	$m{\cdot}s^{-1}$	V
1000 hPa/925 hPa/850 hPa/700 hPa/500 hPa	Geopotential height	$m^2{\cdot}s^{-2}$	Z
1000 hPa/925 hPa/850 hPa/700 hPa/500 hPa	Divergence	s^{-1}	Di
1000 hPa/925 hPa/850 hPa/700 hPa/500 hPa	Vorticity	s^{-1}	Vo
1000 hPa/925 hPa/850 hPa/700 hPa/500 hPa	Temperature	K	T

Table 2.3 Computed variables

Pressure level (hPa)	Variable	Unit	Name
10 m to 925 hPa	Wind shear	$m{\cdot}s^{-1}$	ΔF
10 m to 925 hPa	Temperature gradient	K	ΔT

2.2.2 *Methodology*

Our aim is to model the real observed wind speed from outputs of NWP model described above. More specifically, we use ECMWF analyses i.e the best estimate of the atmospheric state at a given time using a model and observations [15]. In what follows, the observed wind speed is the target and the analysed variables are potential explanatory features. Because of the complexity of meteorological phenomena, statistical modeling provides an appropriate framework for corrections of representativity errors and the modeling of site dependent variability. In this context, two main directions may be as usual investigated, *parametric* and *nonparametric* models.

Parametric models assume that the underlying relation between the target variable and the explanatory variables has, relatively to a certain noise, a particular analytical shape depending on some parameters, which need to be estimated through the data. Among this family of models, the linear model with a Gaussian noise is widely used, mostly thanks to its simplicity [8]. Associated to an adequate variable selection, it may be very effective.

Nonparametric models do not suppose in advance a specific relation between the variables: instead, they try to learn this complex link directly from the data itself. As such, they are very flexible, but their performance usually strongly depends on regularization parameters. The family of nonparametric models is quite large: among others, one may cite the nearest neighbors rule, the kernel rule, neural networks, support vector machines, regression trees, random forests... Regression trees, which have the advantage of being easily interpretable, show to be particularly effective when associated to a procedure allowing to reduce their variance as for the Random Forest Algorithm.

Let us describe the linear model and random forests in our context with more details. The linear model supposes a relation between the target Y_t, observed wind speed at time t, and explanatory variables $X_t^1, \dots, X_t^d$, available from the ECMWF, at this time t. For lightening the notation, we omit the index t in the next equation. The linear model is given by

$$Y = \beta_0 + \sum_{j=1}^{d} \beta_j X_j + \varepsilon,$$

where the β_j's are coefficients to be estimated using least-square criterion minimization method, and $\varepsilon \sim \mathcal{N}(0, \sigma^2)$ represents the noise. Among the meteorological variables $X_1, \dots, X_d$, some of them provide more important information linked to the target than others, and some of them may be correlated. In this case, the stepwise variable selection method is useful to keep only the most important uncorrelated variables [8]. Denoting by $\beta_0, \dots, \beta_d$ the final coefficients obtained this way (some of them are zero), the estimated wind $\hat{Y}$ is then given by

$$\hat{Y} = \beta_0 + \sum_{j=1}^{d} \beta_j X_j. \tag{2.1}$$

An alternative approach to perform variable selection and regularization is to use the Lasso method (see for instance [22]), relying on minimization of the least square criterion penalized by the ℓ^1 norm of the coefficients $\beta_1, \dots, \beta_d$. More specifically, for this model, the predicted wind speed at time t is a linear combination of all the previous variables as in Eq. (2.1), the coefficients $\beta_1, \dots, \hat{\beta}_d$ being estimated using the least square procedure, under the constraint $\sum_{j=1}^{d} |\beta_j| \leq \kappa$ for some constant $\kappa > 0$.

Regression trees are binary trees built by choosing at each step the cut minimizing the intra-node variance, over all explanatory variables $X_1, \dots, X_d$ and all possible thresholds (denoted by S_j hereafter). More specifically, the intra-node variance, usually called deviance, is defined by

$$D(X_j, S_j) = \sum_{X_j < S_j} (Y_s - \overline{Y}^-)^2 + \sum_{X_j \geq S_j} (Y_s - \overline{Y}^+)^2,$$

where $\overline{Y}^-$ (respectively $\overline{Y}^+$) denotes the average of the observed wind speed in the area $\{X_j < S_j\}$ (respectively $\{X_j \geq S_j\}$). Then, the selected j_0 variable and associated threshold is given by $(X_{j_0}, S_{j_0}) = \arg\min_{j,S_j} D(X_j, S_j)$. The prediction is provided by the value associated to the leaf in which the observation falls.

To reduce variance and avoid over-fitting, it may be interesting to generate several bootstrap samples, fitting then a tree on every sample and averaging the predictions, which leads to the so-called Bagging procedure [3]. More precisely, for B bootstrap samples, the predicted power is given by

$$\hat{Y} = \frac{1}{B}\sum_{b=1}^{B}\hat{Y}^b, \tag{2.2}$$

where $\hat{Y}^b$ denotes the wind speed predicted by the regression tree associated with the bth bootstrap sample. To produce more diversity in the trees to be averaged, an additional random step may be introduced in the previous procedure, leading to Random Forests, where the best cut is chosen among a smaller subset of randomly chosen variables. The predicted value is the mean of the predictions of the trees, as in Eq. (2.2).

2.3 The Relationship Between Analysed and Observed Winds

2.3.1 10/100 m Wind Speed Variability Comparison

In this section we compare the observed wind speed at 10 and 100 m with the 10 and 100 m wind speed output of the ECMWF analyses at the closest grid point, respectively. No significant difference can be found when using other grid points, or the mean of the four surrounding locations.

Figure 2.2 shows the Probability Density Function (PDF) of the wind speed coming from ECMWF analyses and observations, and also for illustration an example of a time series of corresponding wind speeds. It appears that the 10 m wind speed from ECMWF analyses displays a systematic bias by overestimating the 10 m observed wind speed (Fig. 2.2a, b). The wind at 100 m is comparatively well modeled in terms of variations in the time series, but also in terms of distribution (Fig. 2.2c, d). It seems that the errors mainly come from the overestimation of peaked wind speeds and the underestimation of low wind speeds (Fig. 2.2c, d). As 10 m wind speed is very influenced by even low topography and surrounding obstacles, which are smoothed or not represented in ECMWF analyses, some of its variations are not well described, and even a quite systematic bias is displayed. The effect of the topography and terrain specificity have less impact on the 100 m wind speed, so that it is much better represented in ECMWF analyses.

The ability of the model to represent the observed wind speed is quantified in Table 2.4 by the deviation, the Root Mean Square Error (RMSE), and the Pearson correlation which formula are given by Eqs. (2.3), (2.5), and (6.8) respectively.

$$\text{Deviation for the } i\text{th observation} = (y_i - x_i) \tag{2.3}$$

$$\text{RMSE} = \sqrt{\frac{\sum_{i=1}^{n}(x_i - y_i)^2}{n}} \tag{2.4}$$

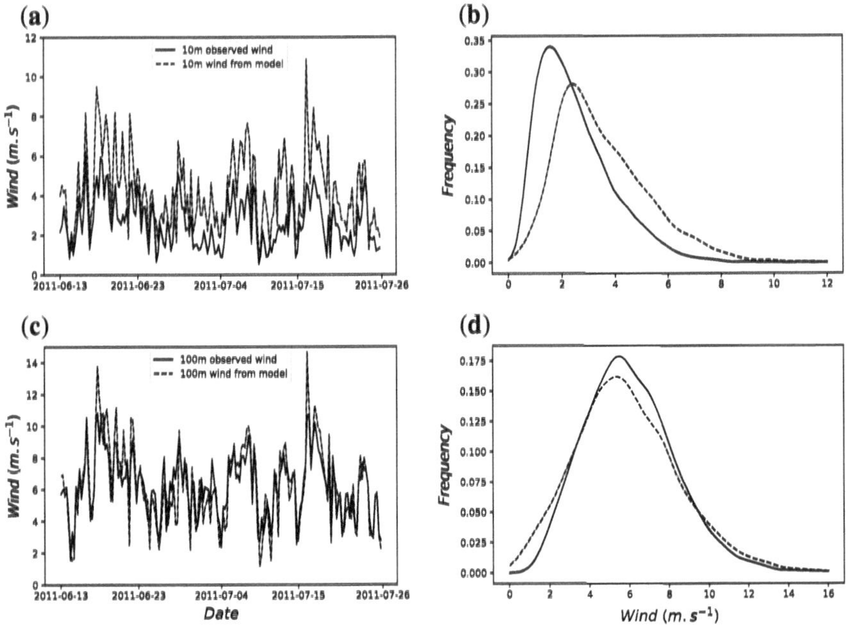

Fig. 2.2 10 m (top) and 100 m (bottom) wind speed time series in summer 2011 (panels **a** and **c**, respectively) and the respective probability density function estimated over the 5 years sample wind speed (panels **b** and **d**)

Table 2.4 MeanDeviation, RMSE, and correlation performed by ECMWF analyses for modeling the 10 and 100 m wind speed

Periods	Deviation		RMSE		Correlation	
	F10	F100	F10	F100	F10	F100
2011–2015	−1.00	0.14	1.41	1.01	0.82	0.93
2011	−1.19	0.04	1.59	1.06	0.80	0.91
2012	−0.94	0.23	1.31	1.03	0.85	0.92
2013	−1.13	0.06	1.52	0.93	0.82	0.94
2014	−0.88	0.26	1.30	1.00	0.80	0.93
2015	−0.87	0.14	1.30	0.97	0.82	0.94
Winter	−0.97	0.04	1.41	0.97	0.83	0.94
Spring	−1.11	0.27	1.56	1.02	0.71	0.90
Summer	−0.92	0.33	1.31	1.04	0.80	0.91
Fall	−1.04	−0.10	1.36	1.00	0.87	0.93

$$\text{Correlation} = \frac{\sum_{i=1}^{n}(x_i - \bar{x})(y_i - \bar{y})}{\sqrt{\sum_{i=1}^{n}(x_i - \bar{x})^2}\sqrt{\sum_{i=1}^{n}(y_i - \bar{y})^2}}, \qquad (2.5)$$

where x_i is the wind speed from the NWP model and y_i the observed wind speed; n is the number of samples (x_i, y_i) and $\bar{x} = \frac{1}{n}\sum_{i=1}^{n} x_i$ (the sample mean) and analogously for $\bar{y}$.

No clear improvement of the ECMWF analyses over the years from 2011 to 2015 can be detected in Table 2.4. The correlation stays quite constant over the years for both 10 and 100 m wind speeds. The Deviation and RMSE seem to decrease for the 10 m wind speed but it cannot be confirmed because of the good score performed in 2012. The variations of performance may only come from changes in the predictability of the weather over Europe [7]. Seasonal variations of the performance of ECMWF analyses can be seen, especially on the correlation between the observed and modeled wind speed. At both 10 and 100 m, the analysed wind speed is better correlated with the observations in winter and fall than in spring and summer. In all cases, the scores shown are better for the 100 m wind speed than for the 10 m wind speed.

Variations of the performance of the ECMWF analyses in representing the observed wind speed are evidenced by Fig. 2.3. The figure shows the 10 m wind speed from ECMWF analyses as a function of the 10 m observed wind speed for

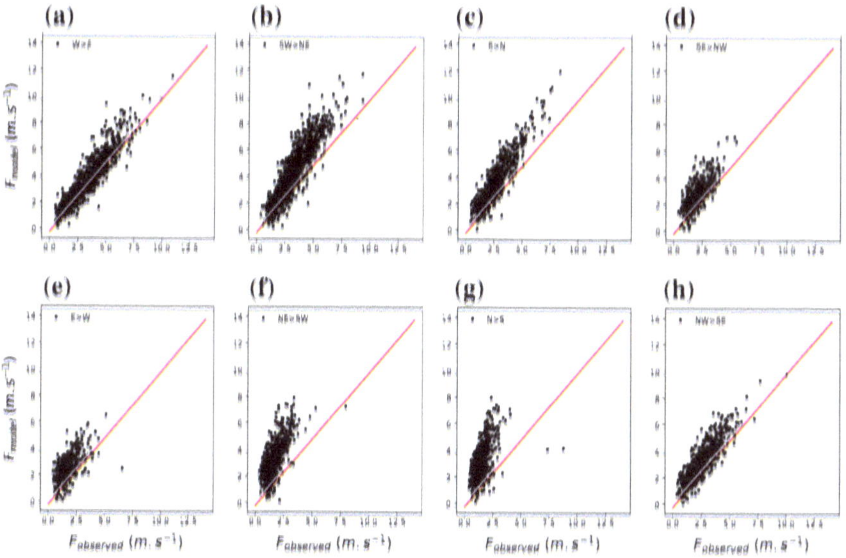

Fig. 2.3 10 m wind speed from ECMWF analyses as function of the 10 m observed wind speed given cardinal directions. Panels correspond to a direction modeled by ECMWF analyses; the wind blows from **a** West, **b** Southwest, **c** South, **d** Southeast, **e** East, **f** Northeast, **g** North, **h** Northwest

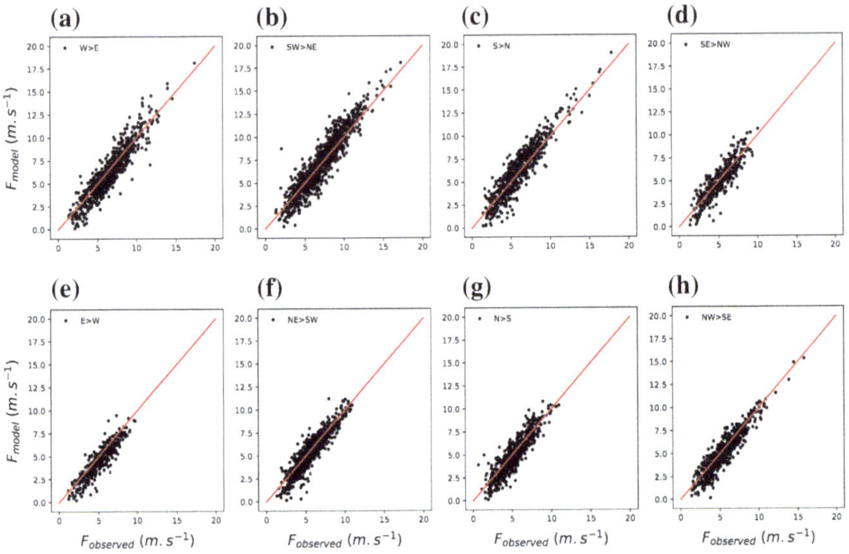

Fig. 2.4 Same as Fig. 2.3 but for 100 m wind speed

different directions of the analysed wind. It is obvious that the errors made by the numerical model differ regarding the direction of the wind. For instance, when the wind comes from the West (Fig. 2.3a), the correlation is very close to one, but for a wind coming from the North/Northeast (Fig. 2.3f, g), it is very low, and the model highly overestimates the 10 m wind speed. It can be easily linked to the specificity of the terrain. Indeed, when a northerly wind is recorded, it has been blocked by the forest north of the anemometer. The same happened for southerlies with the building situated further and which influence is thus not as substantial as the forest. Figure 2.4 displays the same as Fig. 2.3 but for the 100 m wind speed. It seems that there is no more dependence of the performance of the ECMWF analyses regarding the direction of the 100 m wind speed; it appears to be not significantly impacted by the surrounding forests and buildings.

2.3.2 Reconstruction of the 10/100 m Observed Wind Speed Using NWP Outputs

In the sequel, a k-fold cross validation is performed over 10 different periods taken within the 5-years of analyses and observation. Each time, statistical downscaling models are trained on a given period of about 4500 data points and applied over the remaining period of about 500 data points to reconstruct the 10 and 100 m wind speed, so that it results in 10 reconstructions that span the 5 years of data. Table 2.5 enumerates the statistical downscaling models assessed in this study. Models differ

Table 2.5 Statistical models used to downscale 10 and 100 m wind speed

Model type	Explanatory variables	Direction dependance	Name
Linear	F10	No	LR_F
Linear	All	No	LR_A
Linear	Stepwise	No	LR_{SW}
Linear	Lasso	No	LR_{La}
Linear	F10	Yes	LR_F^{dir}
Linear	All	Yes	LR_A^{dir}
Linear	Stepwise	Yes	LR_{SW}^{dir}
Random forest	All	No	RF_A
Random forest	All	Yes	RF_A^{dir}

by their types (Linear Regression and Random Forests), the explanatory variable selection, and whether a model is conditionally fitted regarding the direction of the wind speed or not. We evaluate the different statistical models in terms of RMSE and Correlation with the observed wind speed on the reconstruction period.

10 m Wind Speed Reconstruction

Figure 2.5 shows results for the reconstruction of the 10 m wind speed. Each box contains the 10th reconstructed k-fold periods. First, by using only wind speed with a linear model LR_F, RMSE is reduced by about 40%, but the correlation stays constant. Adding other variables to linear model (i.e. LR_A, LR_{SW} and LR_{La}) allows to reduce the RMSE by 60%, and to significantly improve correlation from 0.80 to 0.91 between reconstructed wind speed and observed one. Using stepwise selection of variables, the Lasso penalization or all variables does not change results in this case, showing that only a part of the information is useful. Using variable selection as stepwise or ℓ_1 penalization (Lasso) avoids over-fitting. Random Forests models perform slightly better than linear models without defining one given model per cardinal wind directions. Variables selected stepwise are very diverse (wind speed, large scale variables, boundary layer state drivers), while Lasso technique mainly selects wind speed and wind component, thus using redundant information. Analyzing the main variables used by Random Forests shows that this methods seems to put much weight on wind component first, highlighting the dependence of the error on the 10 m wind speed regarding its direction.

By fitting a linear model in each direction (noted with 'dir') we manually introduce a relevant information, especially for 10 m wind speed (Fig. 2.3). The model is however more exposed to under-fitting as the sample size of the training data in one direction can be low. Nevertheless, LR_{SW}^{dir} performs better than all other models. Indeed, stepwise choice is made for each direction so that the model is deeply adapted to each direction. This method results in a significant improvement of the RMSE and correlation scores. As expected regarding Fig. 2.3g, the best improvement is retrieved for northerly wind speed and is of more than $0.1\,\mathrm{m{\cdot}s^{-1}}$ compared

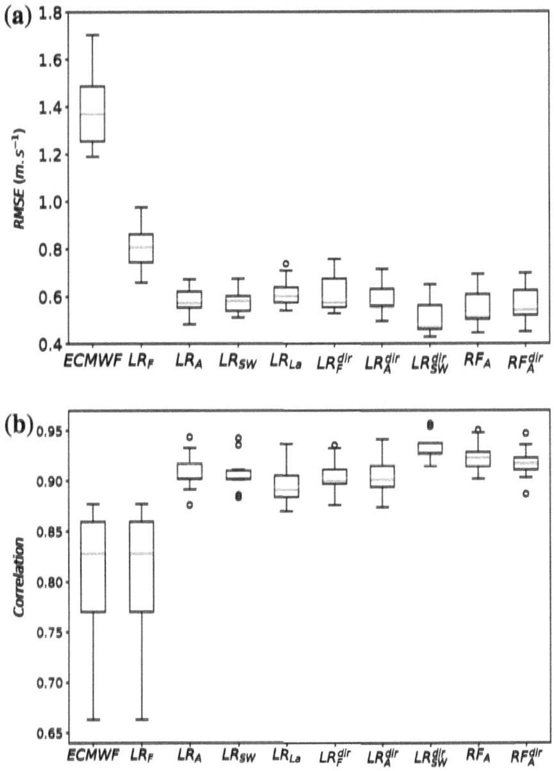

Fig. 2.5 RMSE and correlation results when reconstructing 10 m wind speed with models described in Table 2.5. The first boxes stand for the ECMWF analyses 10 m wind speed

to LR_{SW}. No improvement is found for easterlies, surely because the number of data is too small. Fitting a Random Forest in each direction does not improve results, maybe because the direction is already well handled by this model by using the zonal and meridional component of the wind. So one big advantage of Random Forests over linear regression is that it does not require to explore previously deeply the data for extracting appropriate and relevant features as inputs to the model. Figure 2.6 shows time series of 10 m observed wind speed, NWP model wind speed output over summer period of 2011 (panel a) and the probability density function corresponding to the entire period, 2011–2015 (panel b). Panels c and e show respectively time series of the reconstructed 10 m wind speed by LR_{SW}^{dir} (red line) and LR_{SW} (blue line), and by RF_A^{dir} (magenta line) and RF_A (cyan line). Panels d and f show the corresponding probability density functions. All statistical models allow for a good bias correction. All models underestimate the small quantiles of the distribution and give a distribution very peaked around the mean observed wind speed. High percentiles are however well reconstructed. This is encouraging because this part of the distribution is important in terms of energy production. We can nevertheless expect an overestimation of the wind energy production with those models because of the underestimation of small percentiles.

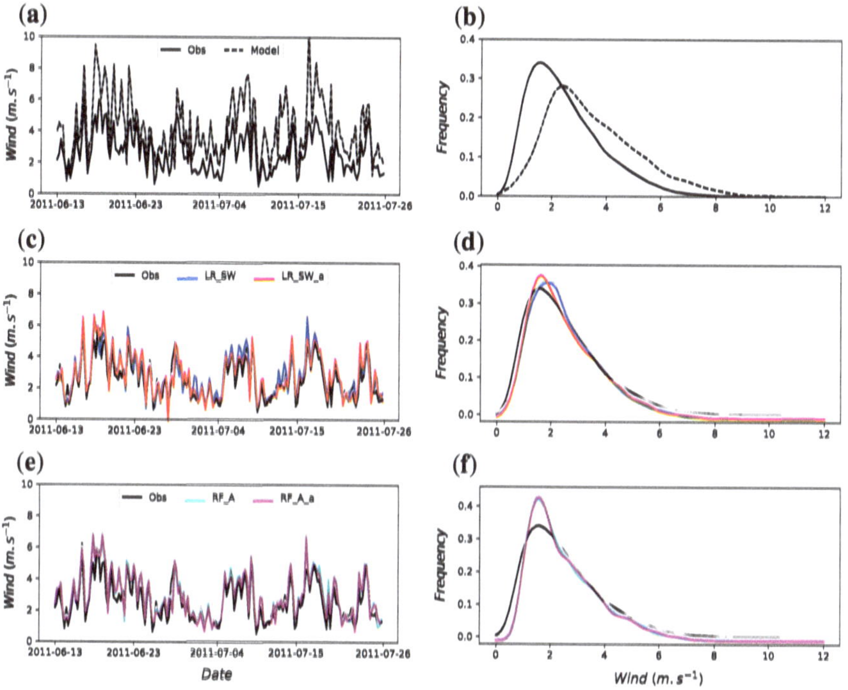

Fig. 2.6 Timeseries (left) and PDF (right) of the observed 10 m wind speed (straight black line), and 10 m wind speed from ECMWF (dotted black line) (**a** and **b**), Linear models (LR_{SW} (blue) and LR^{dir}_{SW} (red)) (**c** and **d**), Random Forest models (RF_A (cyan) and RF^{dir}_A (magenta)) (**e** and **f**)

100 m Wind Speed Reconstruction

Figure 2.7 shows results of the reconstruction of 100 m wind speed with statistical models described in Table 2.5. LR_F allows a reduction of the RMSE of about 15% corresponding to 0.14 m·s^{-1} and the best model LR^{dir}_{SW} reduces the RMSE by 23% corresponding to 0.23 m·s^{-1}. The correlation is improved from 0.92 to 0.94. Adding the direction dependence to linear model with only 100 m wind speed (i.e. LR^{dir}_F) does not improve results regarding LR_F. Indeed, the error on the 100 m wind speed does not depend on the direction. Using all explanatory variables (i.e. LR^{dir}_A) leads to a strong over-fitting. Surprisingly, the linear model using stepwise selection of explanatory variables in each direction (i.e. LR^{dir}_{SW}) recovers an important information as it performs significantly better than the other. Again, its adaptability may be the cause of its good performance. In the case of 100 m wind speed, the best improvement is found for easterly wind speeds. The information on the direction in Random Forests does not improve the results like for 10 m wind speed reconstruction. The more important variables for Random forests and stepwise choice are mainly the 100 m wind speed, but also the wind shear in the boundary layer. Lasso technique selects mainly 100 m wind speed.

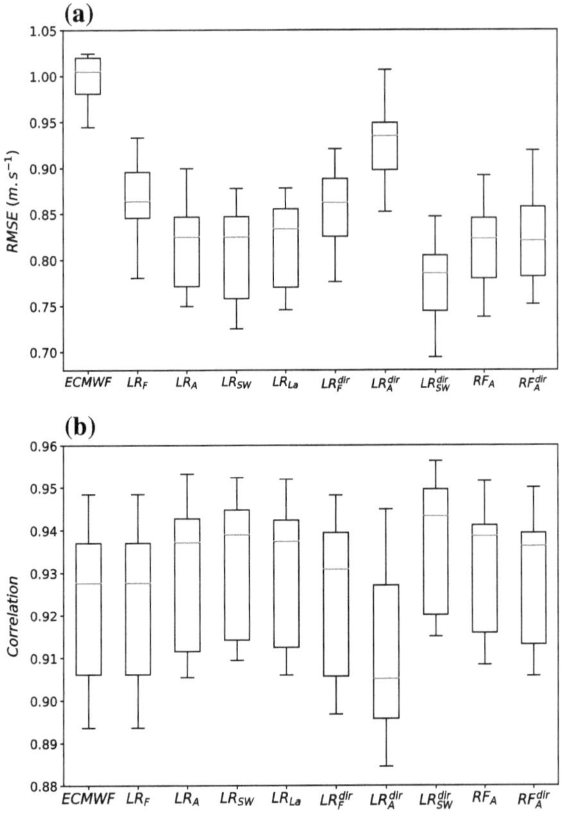

Fig. 2.7 Same as Fig. 2.7, for 100 m wind speed

Figure 2.8 shows time series of 100 m observed wind speed, NWP model wind speed output over summer period of 2011 (panel a) and the probability density function corresponding to the entire period from 2011 to 2015 (panel b). panel c and e show respectively time series of the reconstructed 100 m wind speed by LR_{SW}^{dir} (red line) and LR_{SW} (blue line), and by RF_A^{dir} (magenta line) and RF_A (cyan line). Panels d and f show the corresponding probability density functions. Some peaked wind speeds are less overestimated after statistical downscaling. As for the 10 m wind speed, statistical models underestimate the small quantiles of the distribution and give a distribution peaked around the mean observed wind speed.

To conclude, we built different statistical models to improve the representation of the 10 and 100 m wind speed of the ECMWF analyses. It has been shown that the 100 m wind speed in ECMWF analyses is already well represented as it displays no systematic bias and a good correlation. Nevertheless the RMSE computed for the period 2011–2015 is still of 1.0 m$\cdot$s^{-1}. Statistical models reduces the RMSE on the 10 m wind speed between 40 and 65%, and between 15 and 23% for the 100 m wind speed. They improve at the same time the correlation between the observed wind speed and the reconstructed one. For linear models, the variables selection is

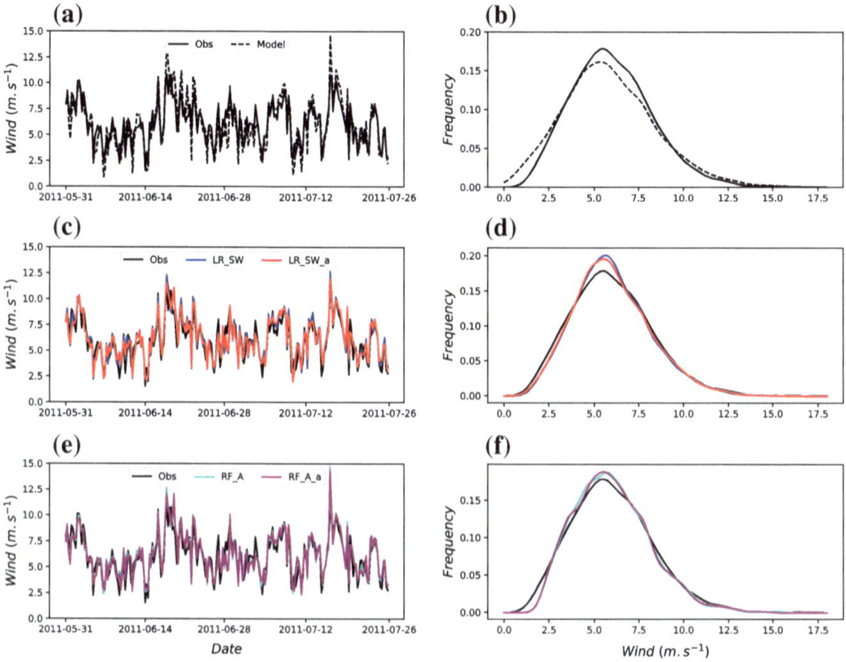

Fig. 2.8 Same as Fig. 2.6, for 100 m wind speed

of great importance to avoid over-fitting, and an exploration step allows to improves results significantly. Random Forests give quite comparable results as the best linear models, without needing variable selection and a preliminary exploration of the data.

2.4 Forecasts of Surface Winds

In this section we use the previous statistical models based on the knowledge of the observed wind speed and the outputs of ECMWF analyses to forecast wind speed at five days horizon with a frequency of 3 h. We have access to 1 year of forecasts in 2015. We train these statistical models on ECMWF analyses from 2011 to 2014, and apply the resulting model to the forecasts. Figures 2.9 and 2.10 show respectively the RMSE averaged over the 365 forecasts for the 10 and 100 m wind speed. A strong diurnal cycle of the performances of both ECMWF forecasts and downscaled statistical predictions of the 10 m wind speed is evidenced. This diurnal cycle seems to be observed also for 100 m wind speed forecasts, but with a less important amplitude. As the dataset is trained on the ECMWF analyses, we can affirm that diurnal cycle is better represented in the ECMWF analyses than in ECMWF forecasts. This could be indeed explained by the data assimilation system that may help to correct errors coming from NWP model parameterizations.

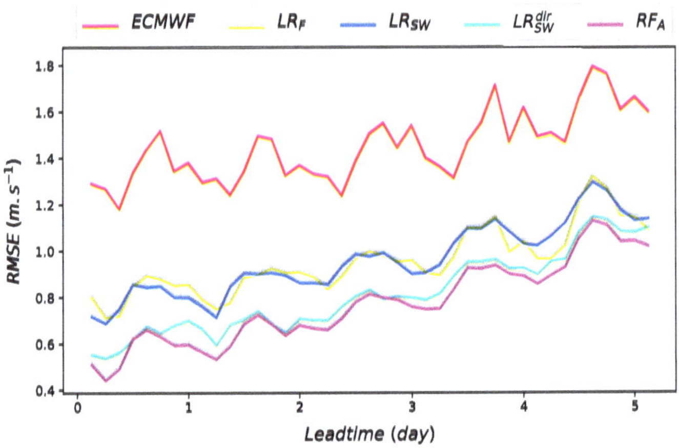

Fig. 2.9 RMSE, computed between the 10 m observed wind speed, and the 10 m forecast wind speed, averaged over the entire set of forecasts

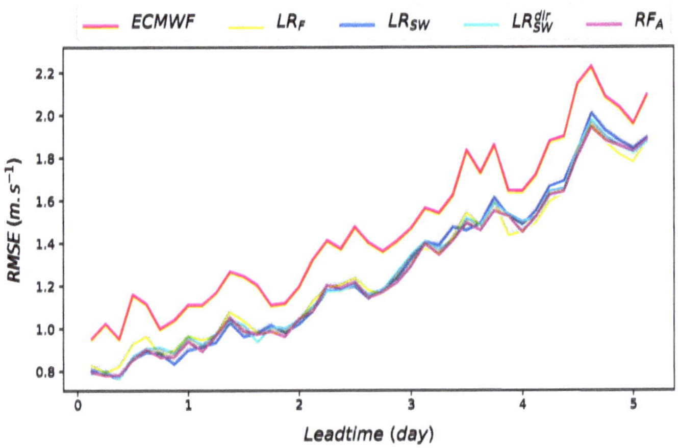

Fig. 2.10 RMSE, computed between the 100 m observed wind speed, and the 100 m forecast wind speed, averaged over the entire set of forecasts

An interesting result shown in Fig. 2.9 is that performance of the LR_F statistical model which is comparable to linear model LR_{SW}, showing that the added value of other explanatory variables is valuable mainly for small lead-times in this case. Adding the dependence with the direction (i.e. LR_{SW}^{dir}) allows for a significant reduction of the RMSE. Random Forests RF_A shows the best performance. In addition to the simplicity to fit this model, its robustness seems to overcome linear regression models. For 100 m wind speed forecasts (Fig. 2.10), Linear models LR_F, LR_{SW}, and LR_{SW}^{dir} and Random Forest RF_A are comparable.

For both 10 and 100 m wind speed forecasts, statistical downscaling models allow for significant improvements regarding ECMWF predicted wind speed, at any lead-time from 3 h to 5 days. Training dataset on the analyses of ECMWF may not be optimal. Indeed, training a statistical model for each lead-time separately should deeply improve results. First, this could help to remove the displayed diurnal cycle, but may also let the increase in RMSE with the lead-time be less sharp.

2.5 Summary and Concluding Remarks

We have used statistical models to evaluate 10 and 100 m wind speed at a given location from output of a NWP model. Comparison of the observed wind speed and ECMWF wind speed output at 10 and 100 m within the 5 years of data show that ECMWF analyses well represent 100 m wind speed. The computed RMSE is of $1.0\,\mathrm{m{\cdot}s^{-1}}$ (the mean wind speed being of $5.8\,\mathrm{m{\cdot}s^{-1}}$) and no systematic bias is displayed. On the contrary, 10 m wind speed output from ECMWF analyses displays a systematic overestimation the observed wind speed. The computed RMSE is of $1.4\,\mathrm{m{\cdot}s^{-1}}$ (the mean wind speed being of $2.4\,\mathrm{m{\cdot}s^{-1}}$).

By applying linear regression between a certain amount of selected variables and observed wind speed, we reduce the RMSE for the 10 and 100 m reconstructed wind speed up to 65 and 23%, respectively. Those good results have been achieved by fitting a linear model in 8 directions and by automatic selection of valuable variables in those directions. Building such a model thus requires a special treatment and a good knowledge of the specific site so that it cannot be systematically applied to another site. Very interestingly, using Random Forests to reconstruct 10 and 100 m wind speed gives comparable results as the best linear models (about 57 and 20%, respectively), while their performance is not sensitive to any preparation of the data. Computing time is a bit longer than simple linear models, but it is quite similar when a linear model is fitted in each direction.

In a second step, we applied the statistical models to forecast up to 5 days. Note that statistical models are trained on past analyses. Applying it on forecasts will work 'as well' only if the relationship between NWP outputs and observed wind speed does not change with lead-time. This is not a-priori guaranteed as the analyses incorporate information from observation via data assimilation. Results are encouraging, because the RMSE between forecast wind speed and observed wind speed is significantly reduced compared to ECMWF forecasts whatever the lead-time, and for both 10 and 100 m wind speeds. Interestingly, Random Forests perform the same or better than linear models when applied to forecast 10 m or 100 m wind speed.

The results obtained for the forecasts are very encouraging: even though the training only involved analyses, the reduction in RMSE persisted for lead-times up to 5 days. Promisingly, there are evident changes to be tried which should lead to improvements of the performances. As a first, training statistical downscaling models directly on ECMWF forecasts makes sense as a transfer function adapted to each lead-time should take into account the reduced performance of ECMWF forecasts around 15 pm and thus correct systematic errors in the representation of the diurnal cycle. Moreover, training dataset for each lead-time separately should also help to reduce the increase of RMSE with lead-time by adapting the explanatory variables to forecasts performance. For instance, for short lead-time, statistical models may find out that surface wind speed in ECMWF forecasts gather valuable information so that this information would be used. It may nevertheless not be the case at longer timescales, so that statistical models would prefer using information from large-scale circulation (e.g. pressure) which is well modeled by ECMWF forecasts, even at lead-time up to 5 days. Secondly, the good performance of Random Forests together with linear regression models denotes that it is possible to reconstruct the wind speed with very different relations. Model aggregation may thus be a path to retrieve more information than when using a single statistical model. It also seems that using statistical downscaling techniques results in a more peaked distribution around the mean, whereas the ECMWF forecasted 100 m wind speed overestimates the extremes. As a consequence, a properly weighted mean of the two different forecasts could improve results as well.

In this study, we choose to use only local informations coming from NWP outputs. Additive valuable informations may be retrieved from larger-scale NWP outputs such as large-scale horizontal gradients of the pressure. However, the discussion on the added value of any other NWP outputs is site dependent, and is already part of research matters. For instance, it has been proved that large scale circulation patterns give valuable information at timescales up to months in some regions of France [1].

A wind farm is often equipped with many anemometers situated at 10 m and at the hub height, so that local intime observations are easily available as well as wind power output. Forecasting wind speed using only NWP outputs is a good way to improve forecasts, but local past observations may also be used as explanatory variable. Indeed, at very short lead-time (3-h), we can assume that the error the NWP model make at t_{0h} (corresponding to the analyses) may be correlated to the future error at time t_{3h}. We could also imagine to create a direct link between NWP outputs and wind energy production as in [9], using in addition the information on the close past wind energy production at the considered wind farm.

As a preliminary illustration of the benefit of such an approach, we train Random Forests and Linear Regression with stepwise selection of variables to forecast 10 and 100 m wind speed at time t_{3h} only, and add the error on the wind speed at time t_{0h} as an explanatory variable of the future wind speed at time t_{3h}. We perform a k-fold of 10 forecasts over the year 2015. Results are represented in Fig. 2.11. When forecasting 10 m wind speed at t_{3h}, using the error at time t_{0h} allows for a reduction of the RMSE of 5% with Random Forests and of 10% with linear model compared to Random Forest without the observation at time t_{0h}. When forecasting 100 m wind speed at t_{3h}, using the information on the 10 m wind speed observed at t_{0h} allows for an improvement of 2 – 6%. Adding the information on the 100 m wind speed at time t_{0h} spectacularly improves results by 18% with linear regression model.

In addition of the good results obtained when reconstructing 10 and 100 m wind speed, we also showed encouraging results when forecasting wind speed up to 5 days. By using very different statistical models, we highlight advantages of Random Forests over multiple linear regression, e.g. simplicity and robustness. Finally, very

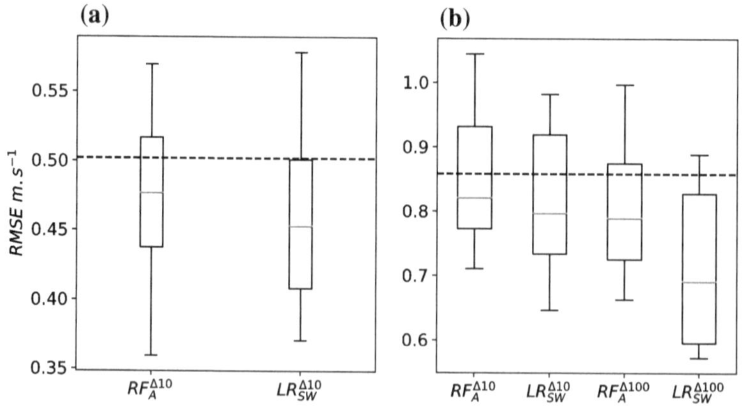

Fig. 2.11 RMSE computed over 10 k-fold forecasts of 10 m (**a**) and 100 m (**b**) wind speed at 3 h lead-time, using the error on the 10 and 100 m wind speed at time t_{0h} (denoted by $\Delta 10$ and $\Delta 100$, respectively) as an explanatory variable. The dashed line represent the averaged RMSE of Random Forest without using observations at t_{0h}, and boxes represents the RMSE over 10 k-fold forecasts

promising perspective for improving downscaling at short-term horizon is exposed; it involves a pseudo-assimilation of a local past observed information into the statistical downscaling model.

Acknowledgements This research was supported by the ANR project FOREWER (ANR-14-538 CE05-0028). This work also contributes to TREND-X program on energy transition at Ecole Polytechnique. The authors thank Côme De Lassus Saint Geniès and Medhi Kechiar who produced preliminary investigations for this study.

References

1. B. Alonzo, H. Ringkjob, B. Jourdier, P. Drobinski, R. Plougonven, P. Tankov, Modelling the variability of the wind energy resource on monthly and seasonal timescales. Renew. Energy **113**, 1434–1446 (2017)
2. P. Bauer, A. Thorpe, G. Brunet, The quiet revolution of numerical weather prediction. Nature **525**, 47–55 (2015)
3. L. Breiman, Bagging predictors. Mach. Learn. 123–140 (1996)
4. F. Cassola, M. Burlando, Wind speed and wind energy forecast through Kalman filtering of numerical weather prediction model output. Appl. Energy **99**, 154–166 (2012)
5. R.J. Davy, M.J. Woods, C.J. Russell, P.A. Coppin, Statistical downscaling of wind variability from meteorological fields. Bound. Layer Meteorol. **175**, 161–175 (2010)
6. A. Devis, N. van Lipzig, M. Demuzere, A new statistical approach to downscale wind speed distribution at a site in Northern Europe. J. Geophys. Res. Atmos. **118**, 2272–2283 (2013)
7. C.K. Folland, A.A. Scaife, J. Lindesay, D.B. Stephenson, How potentially predictable is Northern European winter climate a season ahead? Int. J. Climatol. **32**, 801–818 (2012)
8. J. Friedman, T. Hastie, R. Tibshirani, *The Elements of Statistical Learning*. Springer Series in Statistics, vol. 1 (Springer, New York, 2001)

9. M.G.D. Giorgi, A. Ficarella, M. Tarantino, Assessment of the benets of numerical weather predictions in wind power forecasting based on statistical methods. Energy **36**, 3968–3978 (2011)
10. H.R. Glahn, D.A. Lowry, The use of model output statistics (MOS) in objective weather forecasting. J. Appl. Meteorol. **11**, 1203–1211 (1972)
11. GWEC, Global wind statistics (2016)
12. M. Haeffelin, L. Barthes, O. Bock, C. Boitel, S. Bony, D. Bouniol, H. Chepfer, M. Chiriaco, J. Cuesta, J. Delanoe, P. Drobinski, J.-L. Dufresne, C. Flamant, M. Grall, A. Hodzic, F. Hourdin, F. Lapouge, Y. Lemaitre, A. Mathieu, Y. Morille, C. Naud, V. Noel, W. OHirok, J. Pelon, C. Pietras, A. Protat, B. Romand, G. Scialom, R. Vautard, SIRTA, a ground-based atmospheric observatory for cloud and aerosol research. Ann. Geophys. **23**, 1–23 (2005)
13. T. Howard, P. Clark, Correction and downscaling of NWP wind speed forecasts. Meteorol. Appl. **14**, 105–116 (2007)
14. J. Jung, R.P. Broadwater, Current status and future advances for wind speed and powerforecasting. Renew. Sustain. Energy Rev. **31**, 762–777 (2014)
15. E. Kalnay, *Atmospheric Modeling, Data Assimilation and Predictability* (Cambridge University Press, Cambridge, 2003)
16. M.L. Kubik, P.J. Coker, C. Hunt, Using meteorological wind data to estimate turbine generation output: a sensitivity analysis, in *Proceedings of Renewable Energy Congress* (2011), pp. 4074–4081
17. D. Maraun, F. Wetterhall, M. Ireson, E. Chandler, J. Kendon, M. Widmann, S. Brienen, H.W. Rust, T. Sauter, M. Themel, V.K.C. Venema, K.P. Chun, C.M. Goodess, R.G. Jones, C. Onof, M. Vrac, I. ThieleEich, Precipitation downscaling under climate change: recent developments to bridge the gap between dynamical models and the end user. Rev. Geophys. **48**, RG3003 (2010)
18. M. Mohandes, S. Rehmanb, S. Rahmand, Estimation of wind speed prole using adaptive neuro-fuzzy inference system (ANFIS). Appl. Energy **88**, 4024–4032 (2011)
19. RTE, Annual electricity report (2016)
20. T. Salameh, P. Drobinski, M. Vrac, P. Naveau, Statistical downscaling of near-surface wind over complex terrain in Southern France. Meteorol. Atmos. Phys. **103**, 253–265 (2009)
21. S.S. Soman, H. Zareipour, O. Malik, O., and P. Mandal, A review of wind power and wind speed forecasting methods with different time horizons. In *North American power symposium (NAPS)* (2010), pp. 1–8
22. R. Tibshirani, Regression shrinkage and selection via the lasso. J. R. Stat. Soc. Ser. B **58**, 267–288 (1994)
23. G.K. Vallis, *Atmospheric and Oceanic Fluid Dynamics* (Cambridge University Press, Cambridge, 2006)
24. N.S. Wagenbrenner, J.M. Forthofer, B.K. Lamb, K.S. Shannon, B.W. Butler, Downscaling surface wind predictions from numerical weather prediction models in complex terrain with WindNinja. Atmos. Chem. Phys. **16**, 5229–5241 (2016)
25. R.L. Wilby, C.W. Dawson, The statistical downscaling model: insights from one decade of application. Int. J. Climatol. **33**, 1707–1719 (2013)
26. R.L. Wilby, T.M.L. Wigley, D. Conway, P.D. Jones, B. Hewitson, J. Main, D.S. Wilks, Statistical downscaling of general circulation model output: a comparison of methods. Water Resour. Res. **34**, 2995–3008 (1998)
27. M. Zamo, L. Bel, O. Mestre, J. Stein, Improved gridded wind speed forecasts by statistical postprocessing of numerical models with block regression. Weather Forecast. **31**, 1929–1945 (2016)

Chapter 3
Stochastic Lagrangian Approach for Wind Farm Simulation

Mireille Bossy, Aurore Dupré, Philippe Drobinski, Laurent Violeau and Christian Briard

Abstract We present a stochastic Lagrangian approach for atmospheric boundary layer simulation. Based on a turbulent-fluid-particle model, a stochastic Lagrangian particle approach could be an advantageous alternative for some applications, in particular in the context of down-scaling simulation and wind farm simulation. This paper presents two recent advances in this direction, first the analysis of an optimal rate of convergence result for the particle approximation method that grounds the space discretisation of the Lagrangian model, and second a preliminary illustration of our methodology based on the simulation of a Zephyr ENR wind farm of six turbines.

Keywords Stochastic Lagrangian models · Numerical analysis · Wind farm simulation

3.1 Introduction

The stakes of the simulation of wind farm production are growing with the development of renewable energies. The various time scales involved (from wind potential evaluation, to short-term production forecast), the mix of various constraints on existing sites or on new projects are all issues where numerical simulations can bring quantified answers.

M. Bossy (✉) · L. Violeau
Université Côte d'Azur, Inria, Sophia Antipolis, France
e-mail: mireille.bossy@inria.fr

A. Dupré · P. Drobinski
LMD/IPSL, École polytechnique, Université Paris Saclay, ENS,
PSL Research University, Sorbonne Universités, UPMC Univ Paris 06, CNRS,
Palaiseau, France

C. Briard
Zephyr ENR, Saint-Avertin, France

P. Drobinski et al. (eds.), *Renewable Energy: Forecasting and Risk Management*,
Springer Proceedings in Mathematics & Statistics 254,
https://doi.org/10.1007/978-3-319-99052-1_3

Although some computational fluid dynamics models, together with wind turbine models, and software are already established in this sector of activity (see e.g. Sørensen [17], Niayifar and Porté-Agel [11], and the references cited therein), the question of how to enrich and refine a wind simulation (from a meteorological forecast, or from a larger scale information, eventually combined with measurements) remains largely open. This is particularly true at the scale of a wind farm, regarding the production estimation of a given site, wind turbine by wind turbine. Among various existing approaches for wind farm simulation we can distinguish

- wind extrapolation methods, and parametrization of wake effect for real-time simulation response,
- fluid and structure interaction models for wake computations, with often laminar flow hypothesis and rather simple terrain description,
- Large eddy simulation (LES) models for turbulent flows, including turbine contribution forces related to actuator disc modeling.

The turbulent nature of the atmospheric boundary layer (ABL) contributes to the uncertainty of the wind energy estimation. This has to be taken into account in the modeling approach when assessing the wind power production. This paper is devoted to a downscaling approach that typically aims to compute the wind at a refined scale in the ABL, from a coarse wind computation obtained with a mesoscale meteorological solver. This is the purpose of the Stochastic Downscaling Model (SDM) presented here.

The main features of SDM reside in the choice of a fully Lagrangian viewpoint for the turbulent flow modeling. This is allowed by stochastic Lagrangian modeling (SLM) approaches that adopt the viewpoint of a fluid-particle dynamics in a flow. Such methods are computationally inexpensive when one need to refine the spatial scale. This is a main advantage of the SDM approach, as particles methods are free of numerical constraints (such as the Courant Friedrichs Lewy condition that imposes a limit to the size of the time step for the convergence of many explicit time-marching numerical methods).

The development of SDM is a collaborative long term task (see [1, 2, 5] for detailed presentation), that addresses jointly mathematical and modeling issues with the elaboration of a numerical solver. It is an interdisciplinary work involving disciplines such as stochastic analysis and numerical analysis for the design and the optimal use of the Lagrangian particle solver, physics of the ABL for the calibration and validation of SDM equations and boundary conditions, and engineering for the Lagrangian adaptation of actuator disk model for the turbine wake effect.

This paper presents two recent advances in these directions:

- Section 3.2 is dedicated to the convergence rate analysis of the stochastic particle algorithm used in SDM. We analyse the convergence rate with numericals experiments and check its adequacy with the theoretical optimal rate of convergence result obtained in [4] for the particle approximation method that grounds the SDM numerical algorithm.

- Section 3.3 presents some first SDM simulation, by computing the wind energy production of an existing wind farm: the *Parc de Bonneval* operated by Zephyr ENR. With the initial and boundary conditions generated from the MERRA reanalysis, we evaluate SDM result against measurements collected at the wind farm. This numerical experiment is representative of the SDM capabilities to refine the spatial scale of the wind computation up to the scale of the wind farm: starting from the MERRA wind profile computed on a horizontal grid of 60 by 60 km, SDM is refining the wind computation on a spatial grid of 40 by 60 m, during a computational time interval of 24 h.

3.2 Stochastic Lagrangian Models

Lagrangian approaches for turbulent flow are already well established for turbulent subgrid-scale modeling. This refers to the representation of the small-scales of the flow that cannot be adequately resolved solely on a computational mesh. In the context of atmospheric flow, the so-called Lagrangian Particle Dispersion Models (LPDM) are widely used for the analysis of air pollutants dispersion (see e.g. Stohl [18] and the references therein). Such method adopts perspective of a 'air parcel' by tracking a number of fictitious particles (with position X_t) released into a flow field:

$$dX_t = \overline{U}(t, X_t)dt + u(t)dt \tag{3.1}$$

where $u(t)$ is a random fluctuation of the mean velocity $\overline{U}$, given for example by a LES computation. The velocity fluctuation is modeled with stochastic differential equation (SDE) of various degrees of complexity according to the involved representations, but generally starting from the simplest Langevin model

$$du(t) = -\frac{u(t)}{T}dt + \sqrt{C_0 \varepsilon(t, X_t)}dW_t \tag{3.2}$$

where the stochastic (or fast) part of the motion is described by the 3-dimensional Brownian motion W, amplified with the turbulent pseudo dissipation of the flow ε. Stochastic description of particles in turbulent flows are also well established in the case of disperse two-phase flows and may concern many other applications (see e.g. Minier [9]).

The SDM methodology also makes use of the air parcel viewpoint. But now the mean velocity (in the particle velocity dynamics (3.1)) is not given any more but has to be computed as a statistical mean velocity $\langle U \rangle$ by solving locally a Lagrangian probability density function (PDF) model. This approach relies on the so-called fluid particle approach developed in the seminal work of S. Pope ([12], see also [10] and the references therein). In this approach, a fluid-particle, or virtual fluid parcel with a position, an instantaneous velocity and a temperature state (X_t, U_t, θ_t) is described as the solution of a stochastic differential equation (SDE), generically of the form

$$
\begin{aligned}
dX_t &= U_t dt, \\
dU_t &= -\frac{1}{\rho}\nabla_x\langle\mathscr{P}\rangle(t, X_t)dt - G(t, X_t)\left(U_t - \langle U\rangle(t, X_t)\right)dt \\
&\quad + F_t dt + \sqrt{C(t, X_t)\varepsilon(t, X_t)}dW_t, \\
d\theta_t &= D_1(t, X_t, \theta_t)dt + D_2(t, X_t, \theta_t)d\widetilde{W}_t.
\end{aligned}
\tag{3.3}
$$

$(W, \widetilde{W})$ is a $(3d \times 1d)$-Brownian motion. From a SDE like (3.3), it is always possible to write (at least formally) the partial differential equation (PDE) of its density function, and from that to recover the dynamics of the associated velocity field. Equation (3.3) is in the just enough detailed form that allows to recognize/intensify the corresponding coefficients in a given targeted Navier Stokes equation combined with a chosen turbulence modeling (we refer the reader to [2] for details). Except for the mean gradient pressure term $-\frac{1}{\rho}\nabla_x\langle\mathscr{P}\rangle$, the choice of the coefficients in the right-hand side of (3.3) corresponds to the choice of the turbulence closure. In particular, the chosen coefficients and forces in (3.3) for SDM in the ABL are described in Sect. 3.3.1.

All computational approaches in turbulence modeling are focused on the computation of the Eulerian statistical average of the velocity and of other associated quantities. This averaging operator is classically represented by the $\langle U\rangle$ in Reynolds-averaged Navier–Stokes (RANS) approaches, by $\widetilde{U}$ or $\overline{U}$ in LES approaches. In SDM, the Eulerian average is recovered as the probabilistic conditional expectation[1] of the particle velocity U_t, knowing that its position X_t is at point x. Denoting $\mathbb{P}$ the probability of the model (3.3), provided with expectation symbol $\mathbb{E}$, the mathematical definition of Eulerian average in SDM is

$$\langle U\rangle(t, x) := \mathbb{E}\left[U_t | X_t = x\right], \tag{3.4}$$

More generally, for any integrable function f, we set

$$\langle f(U, \theta)\rangle(t, x) := \mathbb{E}\left[f(U_t, \theta_t) | X_t = x\right]. \tag{3.5}$$

Equivalently, in term of PDF approach (see [15] for further details), denoting $\gamma(t, \cdot, \cdot, \cdot)$ the probability density law of the random variable (X_t, U_t, θ_t), and $\rho(t, x) = \int_{\mathbb{R}^3\times\mathbb{R}} \gamma(t, x, u, \theta)d\theta dx$ the renormalized mass, the statistical average also writes

$$\langle f(U, \theta)\rangle(x, t) = \frac{\int_{\mathbb{R}^3\times\mathbb{R}} f(u, \theta)\gamma(t, x, u, \theta)du d\theta}{\rho(t, x)}.$$

Thus, the coefficients of the stochastic equation (3.3) are (function of, or derivatives of) statistical averages $\langle u^{(i)}\rangle$, $\langle u^{(i)}u^{(j)}\rangle$, defined as in (3.5). Here and in the sequel, we make use of the notation $U_t = (u_t^{(1)}, u_t^{(2)}, u_t^{(3)})$.

---

[1] We consider here only the case of constant mass density flow, for the sake of clarity.

3.2.1 Numerical Analysis of SLM: Particle Approximation

Solution of nonlinear SDE, with coefficients depending on expectations of the unknowns, can be constructed (under some appropriated regularity hypotheses) as the mean field limit of a linear system of N-interacting particles, as N tends to infinity. Such particle approximation principle is at the basis of the SDM numerical method. (see e.g. [3] for an introductory review). We detail this principle in the simplified prototype equation

$$\begin{aligned} X_t &= X_0 + \int_0^t U_s ds \\ U_t &= U_0 + \int_0^t \mathbb{E}[b(U_s)|X_s]ds + \sigma W_t, \end{aligned} \tag{3.6}$$

preferably to the complex model (3.3). In this section, we adopt a formal mathematical viewpoint to analyze numerical algorithms, and $u \mapsto b(u)$ in (3.6) is any generic function that can play role of the mean velocity field ($x \mapsto \mathbb{E}[b(U_t)|X_t = x] = \langle U \rangle(t, x)$), or turbulent kinetic energy, or more complex quantities appearing in the SDM model in (3.20), but the resulting algorithm remains similar. Particle approximation for the solution of (3.6) relies on a statistical estimator for the conditional expectation function $x \mapsto \mathbb{E}[b(U_t)|X_t = x]$. Typically, a conditional estimator uses local averaging estimates on the N-particle set $(X_t^i, U_t^i, i = 1, \dots, N, t \in [0, T])$:

$$\mathbb{E}[b(U_t)|X_t = x] \quad \text{is approximated by} \quad \sum_{i=1}^{N} \mathscr{W}_{N,i}(x) b(U_t^i). \tag{3.7}$$

Propositions for the weights $\mathscr{W}_{N,i}(x)$ are mainly of two kinds: the Nadaraya–Watson kernel estimator relies on a choice of a kernel function $K_\varepsilon(x) = K(\frac{x}{\varepsilon})$:

$$\mathscr{W}_{N,i}(x) = \frac{K_\varepsilon(x - X^i)}{\sum_{j=1}^N K_\varepsilon(x - X^j)}, \tag{3.8}$$

while partitioning (or mesh) estimator relies on a given M-partition $\mathscr{P}_M = \{\mathscr{B}_{M,1}, \mathscr{B}_{M,2}, \dots, \mathscr{B}_{M,M}\}$ (or a mesh) of the space domain:

$$\mathscr{W}_{N,i}(x) = \frac{\mathbb{1}_{\{X^i \in \mathscr{B}_{M,j}\}}}{\sum_{k=1}^N \mathbb{1}_{\{X^k \in \mathscr{B}_{M,j}\}}}, \quad \text{for } x \in \mathscr{B}_{M,j}. \tag{3.9}$$

It is worth noting that the algorithm complexity of a particle system based on kernel estimator is up to $\mathscr{O}(N^2)$ whereas the partitioning estimator version is up to $\mathscr{O}(N)$ (see also Sect. 3.2.3). We retained this last solution for SDM together with some refinement of Particle-in-cell (PIC) technics (see further details in [2]).

The convergence and precision of a particle-based numerical algorithm for solving (3.6) is driven by N the number of particles to simulate and ε the characteristic size of the partition or the characteristic size of the support of the kernel K when it is applied on particles. In [4], Bossy and Violeau prove the theoretical rate of convergence for the particle approximation of the solution of (3.6). This result gives a relationship between the two parameters N and ε in order to achieve the optimal reduction of the error (or bias). This is the first mathematical result of this kind and to make the difficulty of the mathematical analysis more affordable, the boundary conditions are assumed periodic for simplicity. In a periodic box or torus domain equal to $\mathscr{D} = [0, 1]^d$, the Lagrangian model in (3.6) becomes:

$$\begin{aligned} X_t &= \left[X_0 + \int_0^t U_s\, ds\right] \mod 1 \\ U_t &= U_0 + \int_0^t B[X_s; \rho_s]\, ds + W_t, \quad \text{and } \rho_t \text{ is the density law of } (X_t, U_t), \end{aligned} \tag{3.10}$$

where, we have written $\mathbb{E}[b(U_t)|X_t]$ with its equivalent mathematical form $B[X_t; \rho_t]$, for $(x, \gamma) \mapsto B[x; \gamma]$ defined for all probability density function γ by

$$B[x; \gamma] = \frac{\int_{\mathbb{R}^d} b(v)\gamma(x, v)\, dv}{\int_{\mathbb{R}^d} \gamma(x, y) dv} \mathbb{1}_{\{\int_{\mathbb{R}^d} \gamma(x,y)dv>0\}}.$$

The associated particle approximation system $((X^{i,N}, U^{i,N}), N \geq 1)$ is defined as the solution to

$$\begin{aligned} X_t^{i,N} &= \left[X_0^i + \int_0^t U_s^{i,N} ds\right] \mod 1, \\ U_t^{i,N} &= U_0^i + \int_0^t B_\varepsilon[X_s^{i,N}; \bar{\mu}_s^{N,\varepsilon}]\, ds + W_t^i, \\ \bar{\mu}_t^{N,\varepsilon} &= \frac{1}{N} \sum_{j=1}^N \delta_{\{(X_t^{j,N}, U_t^{j,N})\}} \text{ is the particles empirical measure} \end{aligned} \tag{3.11}$$

where the kernel regression version B_ε of B, given by the approximation (3.7), (3.8), is defined for all density γ by

$$B_\varepsilon[x; \gamma] := \frac{\int_{[0,1]^d \times \mathbb{R}^d} b(v) K_\varepsilon(x - y)\, \gamma(y, v) dy\, dv}{\int_{[0,1]^d \times \mathbb{R}^d} K_\varepsilon(x - y)\, \gamma(y, v) dy\, dv} \mathbb{1}_{\{\int_{\mathbb{R}^d} \gamma(x,y)dv>0\}}.$$

The $(W_t^i, t \leq T, 1 \leq i \leq N)$ are independent Brownian motions valued in $\mathbb{R}^d$, and independent from the initial variables $(X_0^i, U_0^i, 1 \leq i \leq N)$, independent, identically distributed with initial law ρ_0. The nonlinear model (3.10) is thus approximated with the linear system (3.11) (of dimension $2dN$), easy to discretize in time with the help of a time-discretisation Euler scheme (see below (3.16)). This algorithm is at the basis of the so-called Stochastic Lagrangian numerical algorithm (see e.g. Pope [14] and for the SDM method [2]).

The Theoretical Convergence Analysis

In the algorithm (3.11), conditional expectation $\mathbb{E}[f(U_t)|X_t = x]$, for $f = b$, and more generally for any f measurable bounded on $\mathscr{D}$, is approximated by

$$x \mapsto F_\varepsilon[x; \bar{\mu}_t^{\varepsilon,N}] := \frac{\sum_{j=1}^N f(U_t^{j,N}) K_\varepsilon(x - X_t^{j,N})}{\sum_{j=1}^N K_\varepsilon(x - X_t^{j,N})},$$

the corresponding kernel approximation function, where $\bar{\mu}_t^{\varepsilon,N}$ is the empirical measure of particles as in (3.11). A pertinent criterion for the evaluation of the algorithm (3.11) is then the measure of the mean error on the conditional expectation used all along the time loop:

$$\mathbb{E}\Big|\mathbb{E}[f(U_t)|X_t = x] - F_\varepsilon[x; \bar{\mu}_t^{\varepsilon,N}]\Big|. \tag{3.12}$$

We reduce this error function by its L^1-norm on $\mathscr{D}$ weighted by the particles position distribution ρ_t), by considering:

$$\text{Error}_{L^1_{\rho_t}(\mathscr{D})} := \int_{\mathscr{D}} \mathbb{E}\Big|\mathbb{E}[f(U_t)|X_t = x] - F_\varepsilon[x; \bar{\mu}_t^{\varepsilon,N}]\Big|\rho_t(x)dx. \tag{3.13}$$

Theorem 3.1 (see Bossy Violeau [4]) *Assume the following:*

(i) f and b are smooth and bounded functions with bounded derivatives
(ii) the kernel K is positive and bounded, with compact support in $\{x;\ \|x\| \leq 1\}$
(iii) the initial density law ρ_0 is smooth and bounded below by a constant $\zeta > 0$.

Then for any $T > 0$, $1 < p < 1 + \frac{1}{1+3d}$ and $c > 0$, there exists a constant C such that for all $\varepsilon > 0$ and $N > 1$ satisfying $(\varepsilon^{(d+2)} N^{\frac{1}{p}})^{-1} \leq c$, we have for all $1 \leq i \leq N$,

$$\textit{Error}_{L^1_{\rho_T}(\mathscr{D})} \leq C\Big(\varepsilon + \frac{1}{\varepsilon^{(d+1)}N} + \frac{1}{\varepsilon^{(d+1)p}N} + \frac{1}{(\varepsilon^d N)^{\frac{1}{p}}} + \frac{1}{\varepsilon^{\frac{dp}{2}}\sqrt{N}}\Big). \tag{3.14}$$

The optimal rate of convergence is achieved for the choice $N = \varepsilon^{-(d+2)p}$ and

$$\textit{Error}_{L^1_{\rho_T}(\mathscr{D})} \leq CN^{-\frac{1}{(d+2)p}}. \tag{3.15}$$

Notice that p can be chosen almost equal to one. The global error given in (3.14) is a combination of several sources of approximations. First, the $\mathscr{O}(\varepsilon)$ term corresponds to the smoothing error for F. The $\mathscr{O}(\varepsilon^{-\frac{dp}{2}}\sqrt{N}^{-1})$ term is the Monte Carlo variance contribution to the error, next $\mathcal{O}((\varepsilon^d N)^{-\frac{1}{p}})$ is the error due to the replacement of the law ρ_t by the empirical measure $\bar{\mu}_t^{N,\varepsilon}$. There is also the approximation due to the replacement of the position of the exact process as the location where the conditioned expectation is computed by the position of a numerical particle. This is a part of the statistical error, (the use of the Nadaraya Watson estimator to compute the expectation) in $\mathcal{O}(\varepsilon + \frac{1}{\varepsilon^{d+1} N} + \frac{1}{\varepsilon^{(d+1)p} N})$.

3.2.2 Empirical Numerical Analysis

In this section, we measure and analyse the effective convergence of the algorithm with numerical experiments in order to verify and illustrate that the claimed convergence rate in Theorem 3.1 is optimal. For both computational time reason and clarity of the presented graphs, we limit our experiments to $d = 2$, (the wind farm simulation presented in Sect. 3.3.1 is a fully 3 dimensional case).

Numerical experiments proceed using an Euler scheme. We decompose the time interval $[0, T]$ into M time steps of length $\Delta t := \frac{T}{M}$ and we introduce the time discretization of the interacting particle process:

$$\begin{cases} X_t^{i,N,\Delta t} = \left[X_0^i + \int_0^t U_{\eta(s)}^{i,N,\Delta t} ds\right] \mod 1, \\ U_t^{i,N,\Delta t} = U_0^i + \int_0^t B_\varepsilon[X_{\eta(s)}^{i,N,\Delta t}; \bar{\mu}_{\eta(s)}^{N,\varepsilon,\Delta t}] ds + W_t^i, \quad \bar{\mu}_t^{N,\varepsilon,\Delta t} = \frac{1}{N} \sum_{j=1}^N \delta_{(X_t^{j,N,\Delta t}, U_t^{j,N,\Delta t})}, \end{cases} \tag{3.16}$$

for all $1 \leq i \leq N$ and $t \in [0, T]$ where $\eta(t) := \Delta t \left\lfloor \frac{t}{\Delta t} \right\rfloor$ is the Δt-step time function. For all time step $k\Delta t$, $0 \leq k \leq M$, each random variable $(X_{(k+1)\Delta t}^{i,N,\Delta t}, U_{(k+1)\Delta t}^{i,N,\Delta t})$ is computed from the values of all the variables $(X_{k\Delta t}^{j,N,\Delta t}, U_{k\Delta t}^{j,N,\Delta t})$, $1 \leq j \leq N$.

This algorithm has a total complexity of order $\mathcal{O}(M)\,\mathcal{O}(N^2)$. The major drawback of the kernel estimator method used here lies on the computation of the drift at any point x that requires a loop over all the N particles, even if they do not contribute to the final result. As we already mention, for this reason, we preferably use the alternative particle-mesh algorithm for SDM.

The Test Case Description

We introduce some nontrival behavior in the model (3.10) by adding a potential function $P(x, y)$ that models an external, but static in time, pressure force as

$$P(x, y) = \frac{1}{2\pi} \cos(2\pi x) \sin(2\pi y) - \frac{1}{2} x, \quad \text{for all } (x, y) \text{ in } \mathcal{D} = [0, 1]^2.$$

The drift $(x, u, \gamma) \mapsto B[x, u; \gamma]$ is a mean reverting term such that

$$B[x, u; \gamma] = \frac{\int_{\mathbb{R}^d} (v - 2u)\gamma(x, v)\, dv}{\int_{\mathbb{R}^d} \gamma(x, v)\, dv} \quad \text{for all } (x, u) \text{ in } \mathcal{D} \times \mathbb{R}^2 \text{ and all } \gamma \text{ in } \mathcal{P}(\mathcal{D} \times \mathbb{R}^2)$$

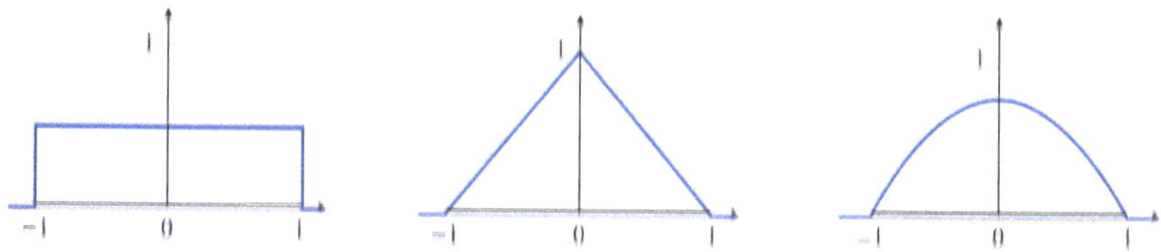

(a) Naive kernel: $K(x) = \frac{1}{2}\mathbb{1}_{\{\|x\|\leq 1\}}$

(b) Triangular kernel: $K(x) = (1 - \|x\|)\mathbb{1}_{\{\|x\|\leq 1\}}$

(c) Epanechnikov kernel: $K(x) = (1 - \|x\|^2)\mathbb{1}_{\{\|x\|\leq 1\}}$

Fig. 3.1 Some examples of normalized kernel functions K

with, for all (x, u) in $\mathscr{D} \times \mathbb{R}^d$,

$$B[x, u; \rho_t] = \mathbb{E}[U_t | X_t = x] - 2u, \text{ when } \rho_t \text{ is the density of } (X_t, U_t).$$

We solve for $t \leq T = 2$,

$$\begin{cases} X_t = \left[X_0 + \int_0^t U_s ds\right] \mod 1 \\ U_t = U_0 - \int_0^t \nabla P(X_s)\, ds + \int_0^t B[X_s, U_s; \rho_s]\, ds + W_t, \quad \rho_t \text{ is the density of } (X_t, U_t). \end{cases}$$

The initial distribution ρ_0 of (X_0, U_0) is such that X_0 has a Gaussian distribution on $\mathbb{T}^d$ with variance σ^2 (i.e. $X_0 = \sigma Z \mod 1$, $\sigma^2 = 0.3$) and U_0 is a centered Gaussian random variable independent from X_0, with variance $\nu^2 = 1$. On Fig. 3.2, we represent the time evolution of the *particles mass* density $\rho_t(x) = \int_{\mathbb{R}^2} \rho_t(x, u) du$ of the process X_t distributed in the torus (plot (a)), as well as the turbulent kinetic $tke(t, x) = \frac{1}{2}\mathbb{E}[(U_t - \mathbb{E}[U_t | X_t = x])^2 | X_t = x]$ (plot (b)). We can observe that the density is clearly non uniform in space, and we expect this should put some stress on the estimation of the mean fields in low density areas. Moreover, although starting from a Gaussian distribution, the density quickly converges in time to a stationary state and this allows to fix the final time to $T = 2$ for all the error analysis simulations, with $M = 128$ time steps. The kernel regression is performed with the Epanechnikov kernel (see Fig. 3.1c) and $\varepsilon = \frac{1}{16}$.

Expected L^1 Error of the Kernel Method

We focus our attention on the *expected L^1 error* defined in (3.13). In order to estimate this quantity, we need to proceed with some approximations on the integral. In the following, we write $\pi^{\Delta x}(g)$ for the spline-interpolated function g on a grid with mesh size Δx. The reference numerical solution for $\mathbb{E}[f(U_T)|X_T = x]$ is approximated by the splined mean fields defined by

$$\overline{F_{\underline{\varepsilon}}[x; \bar{\mu}_T^{\underline{\varepsilon}, \bar{N}}]}^{\Delta x} := \pi^{\Delta x}(\overline{F_{\underline{\varepsilon}}[\cdot; \bar{\mu}_T^{\underline{\varepsilon}, \bar{N}}]})(x) \tag{3.17}$$

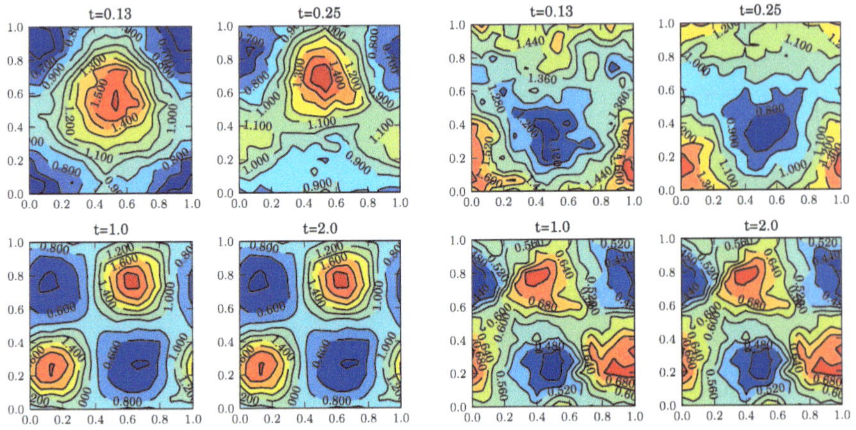

(a) Density (b) Turbulent kinetic energy

Fig. 3.2 Evolution of the density and TKE for (X_t, U_t), $[N = 10^5, \varepsilon = 16^{-1}]$

for a large number of particles $\bar{N}$ and a sufficiently small window parameter $\underline{\varepsilon}$. The numerical approximation is also splined to ease the integration step:

$$F_{\varepsilon}^{\Delta x}[x; \bar{\mu}_T^{\varepsilon,N}] := \pi^{\Delta x}\big(F_{\varepsilon}[\cdot; \bar{\mu}_T^{\varepsilon,N}](x)\big). \tag{3.18}$$

The reference mass density $\rho_T(x)$ is also estimated by using the Monte Carlo mean of kernel density estimation:

$$\bar{\rho}_T(x) := \frac{1}{N_{\mathrm{mc}}} \sum_{k=1}^{N_{\mathrm{mc}}} \frac{1}{\bar{N}} \sum_{j=1}^{\bar{N}} K_{\underline{\varepsilon}}(x - X_T^{j,\underline{\varepsilon},\bar{N}}(\omega_k)), \quad \text{and} \quad \bar{\rho}_T^{\Delta x}(x) := \pi^{\Delta x}(\bar{\rho}_T) \tag{3.19}$$

where the ω_k represent N_{mc} independent realizations of the simulation. The computation of the integral of splined functions can be carried out very precisely over regular grids with the help of numerical libraries. All that remains is to evaluate the expected splined L^1 error by means of a Monte Carlo simulation:

$$\mathrm{Error}_{L^1_{\rho_T}(\mathscr{D})} \sim \frac{1}{N_{mc}} \sum_{k=1}^{N_{mc}} \int_{\mathscr{D}} \overline{\big|F_{\underline{\varepsilon}}[x; \bar{\mu}_T^{\underline{\varepsilon},\bar{N}}]}^{\Delta x} - F_{\varepsilon}^{\Delta x}[x; \bar{\mu}_T^{\varepsilon,N}(\omega_k)]\big|\bar{\rho}_T^{\Delta x}(x)\,dx.$$

In Fig. 3.3, we plot the expected L^1 error calculated as above as a function of the window parameter ε for different total number of particles N: for each choice of N, we observe that the error is first decreasing with the value of ε (from right to the left) toward a minimum value, but next start to increase with two small values of ε: this is the effect of the competition between the terms ε and $\frac{1}{\varepsilon^{\alpha}}$ in the bias formula (3.14). This is effect is delayed by choosing larger values of N who reduces

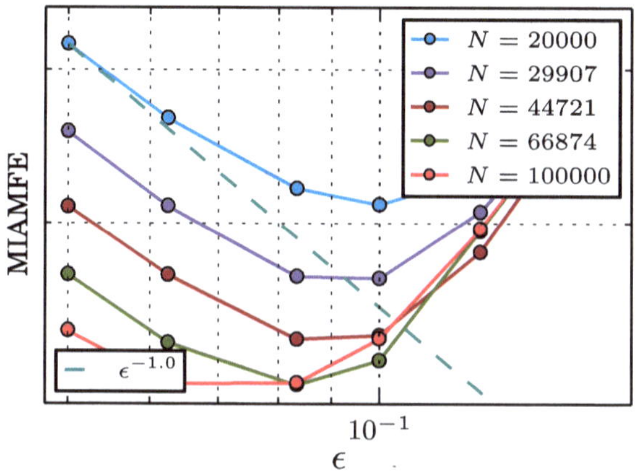

Fig. 3.3 L^1 error as a function of ε for different number of particles N

the variance in the computation. We can also notice that the asymptotic slope of the error when ε tends to zero is very close to -1 for a log-log scale (represented with a blue dashed line). We expect the error to behave like $\mathcal{O}(\varepsilon + \frac{C}{\varepsilon^3 N} + \frac{C}{\varepsilon\sqrt{N}})$. Then, it seems reasonable to infer that the term of order $\mathcal{O}(\frac{1}{\varepsilon\sqrt{N}})$ related to the variance of the stochastic integral in the model dominates the L^1 error. Recall, however, that our theoretical analysis of the error is valid under the constraint $\frac{1}{\varepsilon^{d+2}N^{1/p}} \leq c$, for some positive constant c, so we cannot rigorously extend the bound to an asymptotic analysis when ε decreases to zero. Finally, we can observe that the slope of L^1 is bounded by one when N is sufficiently large and ε becomes large. This is in complete agreement with the bounds in Theorem 3.1 although this figure does not explain the relative contribution of the smoothing error and the kernel estimation error in the total L^1 error.

We can also consider the expected L^1 error as a function of $\frac{N}{\varepsilon^d}$, as in Fig. 3.4. Note that $\frac{N}{\varepsilon^d}$ loosely represents the number of particles in interaction with a given particle (for compact support kernel functions), and is often referred to as the number of particle "per cell" (denoted N_{pc}), especially in the case of partitioning estimates. N_p here denotes the total number of particles. This Fig. 3.4 illustrates the concept of bias-variance trade-off and its relation with the number of particle per cell: for a given small number of particle per cell (compared to the optimal number of particle per cell), we can observe that the L^1 error is almost independent of the absolute value of ε. This clearly shows that the variance is directly related to the number of particles used in the computation of the estimator. On the contrary, when the number of particle per cell becomes large and the bias dominates, the L^1 error becomes smaller with ε. The convergence of the error with respect to the number of particles N $(= N_p)$ can be observed in Fig. 3.5. When ε is sufficiently small, we notice as expected a convergence of order $\mathcal{O}(\frac{1}{\sqrt{N}})$, related to the reduction of the variance component of the error. On the other hand, when ε is large, increasing the number of particle does not reduce the error as the bias dominates.

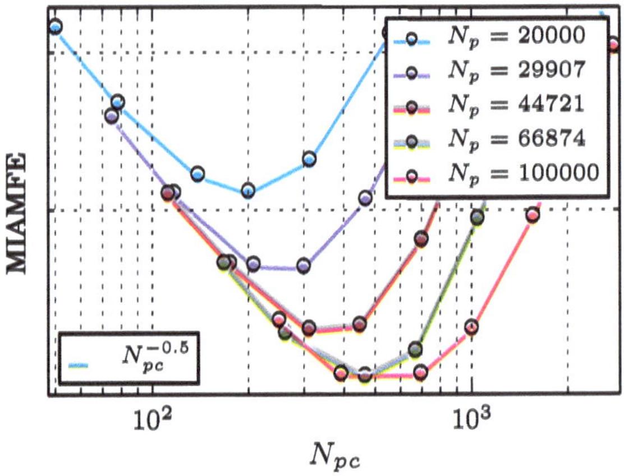

Fig. 3.4 L^1 error as a function of ε for different densities of particles $\frac{N}{\varepsilon^d}$

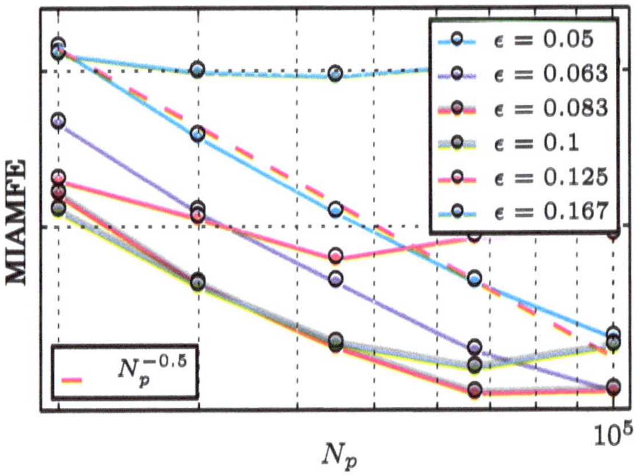

Fig. 3.5 L^1 error as a function of the total number of particles, for different value of ε

Given this bias-variance trade-off, one may be interested in finding the optimal value of ε that minimizes the expected L^1 error for a given number of particles. From the simulations we ran for different couples (ε, N) of parameters, we plot the surface of the error in Fig. 3.6 (left). We can then plot the curve of optimal ε as a function of the number of particles which is very close to $\frac{1}{\varepsilon^4}$ (for $d = 2$). This result is in-line with what we expected from Theorem 3.1 where the optimal value of window size is given by $N^{-\frac{1}{d+2}}$. Moreover, if we plot the error associated with the optimal couple of parameters as a function of ε, we can observe the optimal experimental rate of convergence of the algorithm. The theoretical optimal error (3.15) in Theorem 3.1, is of order $\mathcal{O}(N^{-\frac{1}{4p}})$, with p close to 1, while in Fig. 3.6 (right), we observe a rate of order close to $-\frac{1}{4}$ to $-\frac{1}{3}$. Theoretical and observed convergence rates are here in a very good adequacy.

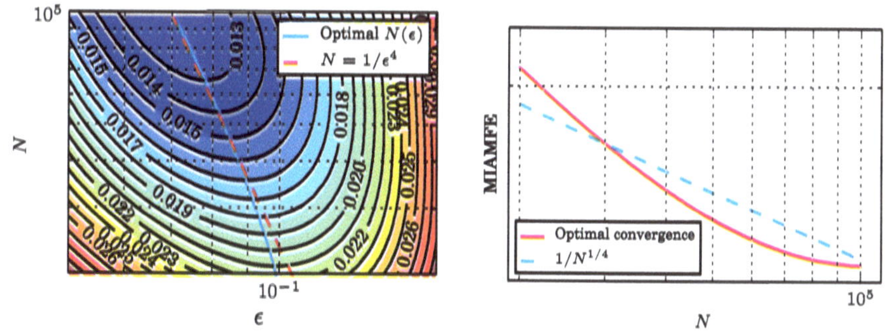

Fig. 3.6 L^1 error as a function of ε and N (left). Optimal rate of convergence for the L^1 error (right)

3.2.3 Particle in Mesh Method

We end this section with some experiments on the particle-mesh version of the algorithm. The principle of the Particle-Mesh methods is to aggregate the N scattered data points $(X^i, f(U^i))$, for $1 \leq i \leq N$ onto a regular mesh covering the simulation domain $\mathscr{D}$, thus reducing the size of the data set to the number of nodes in the mesh. The mean field is evaluated from the mesh charges at each particle position using standard regression techniques as in (3.7), (3.9). If we design the charge assignment and the force interpolation operation such that they can be performed in constant time for each particle, the Particle-Mesh algorithm has a $\mathscr{O}(M)\,\mathscr{O}(N)$ complexity, i.e. it has linear complexity with respect to the total number of particles. This is a tremendous improvements over the previous kernel regression method, and the speed-up is not only theoretical but is actually achieved in practical simulations.

The drawback of this approach is that it introduces new sources of numerical errors, and unlike classical particle computer simulations, increasing the number of nodes in the mesh does not necessarily reduce the error if the total number of particles is left unchanged. Moreover, refining the mesh increases the computational cost, so it is particularly important to be able to reduce the errors for a given mesh size in order to achieve the best compromise between quality and computational cost. In this regard, we will consider three charge assignment and interpolation functions that are designed to be optimal according to smoothness and spatial localization of errors criteria: the Nearest Grid Point (NGP), the Cloud in Cell (CIC), and the triangular Shaped Cloud (TSC) (see Fig. 3.7 for details).

Charge Assignment

Consider a mesh of cell size h (also called window size). Let x_i be the position of the ith node. Then the charge c_i and the charge density d_i assigned at node i are defined by

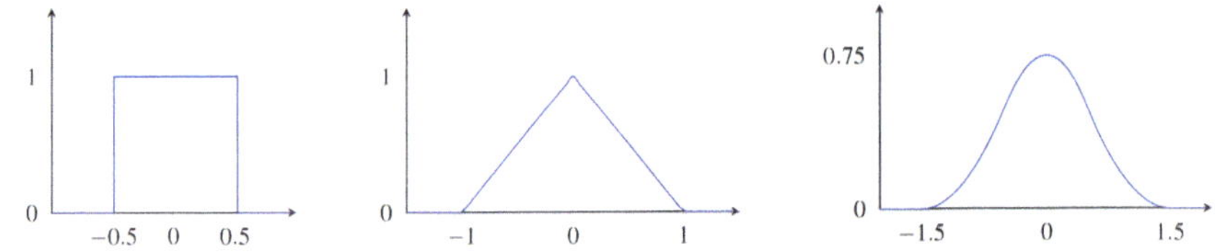

(a) Nearest Grid Point: $K_h(x) := \mathbb{1}_{\{\|x\| \leq \frac{h}{2}\}}$

(b) Cloud in Cell: $K_h(x) := (1 - \|x\|)\mathbb{1}_{\{\|x\| \leq h\}}$

(c) Triangular Shaped Cloud: $K_h(x) := (\frac{3}{4} - \frac{x^2}{h^2})\mathbb{1}_{\{\|x\| \leq \frac{h}{2}\}} + \frac{1}{2}(\frac{3}{2} - \frac{|x|}{h})^2\mathbb{1}_{\{\frac{h}{2} \leq \|x\| \leq \frac{3h}{2}\}}$

Fig. 3.7 Charge assignment functions (from left to right: NGP, CIC, TSC)

$$c_i := \frac{1}{N}\sum_{j=1}^{N} f(U^j)K_h(x_i - X^j) \quad d_i := \frac{1}{N}\sum_{j=1}^{N} K_h(x_i - X^j),$$

where K is a charge assignment function. By definition of c_i and d_i the ratio $\frac{c_i}{d_i}$ is simply the kernel regression estimate at the node point x_i as in (3.7):

$$\mathbb{E}[b(U_t)|X_t = x] \sim \frac{c_i}{d_i} = \frac{\frac{1}{N}\sum_{j=1}^{N} f(U^j)K_h(x_i - X^j)}{\frac{1}{N}\sum_{j=1}^{N} K_h(x_i - X^j)}.$$

The computation of the mesh charge values can be performed efficiently in $\mathcal{O}(N)$ with an outer loop on the particles and the use of a mesh localization procedure that makes it possible to loop only on the nodes charged by a given particle.

Of course, it is important that the localization of the particle in the mesh and the computation of the list of nodes charged by the particle be performed in constant time. In practice, the lists of neighbor cells are computed once and for all (in linear time) at the beginning of the procedure to speed up the execution of the algorithm.

In Fig. 3.8, we measured the influence of the regularity order of the charge assignment function K_h. Aside from the smoothing aspect of the obtained velocity field, we can observe a gap between the error produced by the partitioning estimates (corresponding to NGP assignment charge) and the higher order CIC or TSC functions, and CIC appears to be a good compromise between the error level and the ease of implementation.

3.3 Wind Farm Simulation Experiment with SDM

Our SDM model has been evaluated against measurements collected at a wind farm located in Bonneval, a small town 100 km Southwest of Paris, France (at 48.20 °N and 1.42 °E). The wind farm is operated by Zephyr ENR, a private company managing five other wind farms. The Bonneval wind farm, called *Parc de Bonneval*, has been

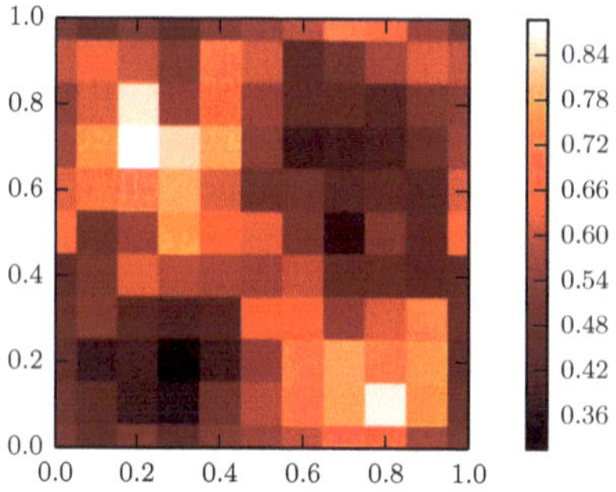

(a) Velocity norm for the NGP scheme

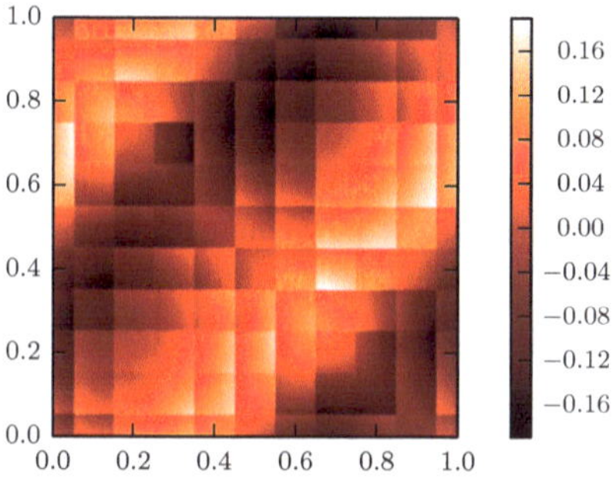

(b) Velocity error for the NGP scheme

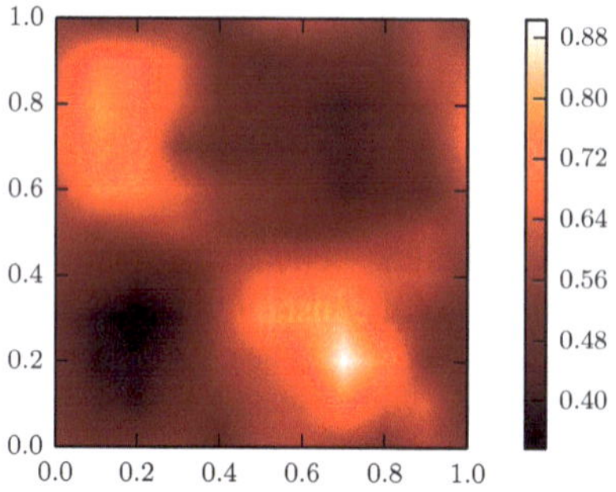

(c) Velocity norm for the CIC scheme

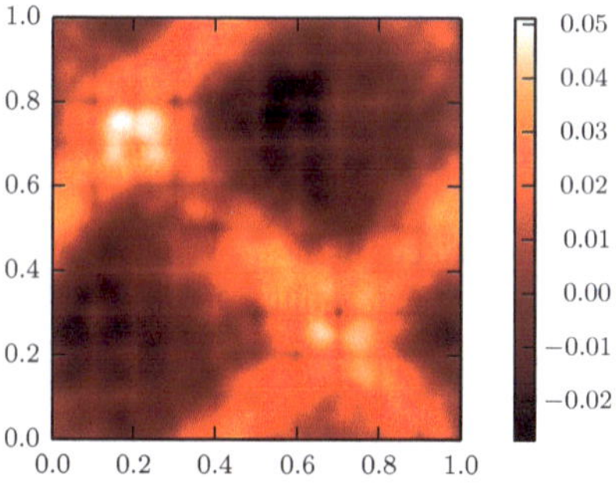

(d) Velocity error for the CIC scheme

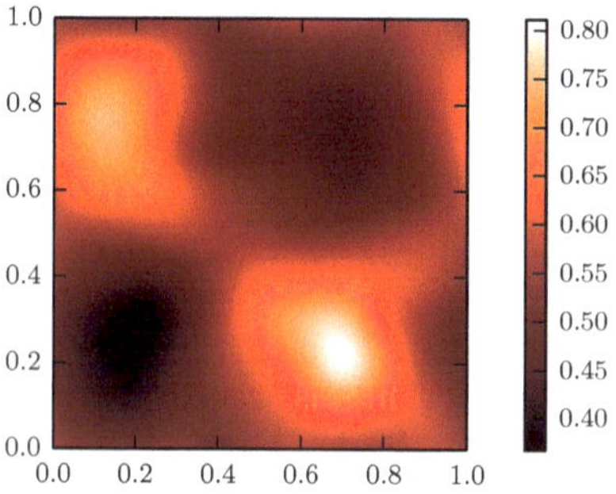

(e) Velocity norm for the TSC scheme

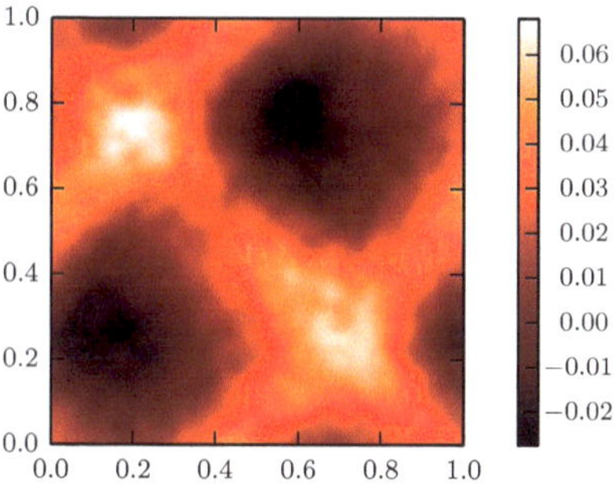

(f) Velocity error for the TSC scheme

Fig. 3.8 Velocity norm and average error for the NGP, CiC and TSC schemes

implemented in 2006 and is composed of six wind turbines, each with a power rated of 2.0 MW. In order to evaluate the SDM simulations with the data collected at *Parc de Bonneval*, wind turbines have been numerically integrated in SDM,based on an actuator disk model. This model allows the simulation of the dynamical effect of the presence of wind turbines, in the form of trailing wakes, as well as the computation of the wind energy production.

3.3.1 SDM for Atmospheric Boundary Layer Simulation

We run SDM for the winter day of December 22th 2016, with the Eq. (3.3) configured for the case of the neutral atmosphere hypothesis. Here and in the sequel we denote by

$$U_t = (u_t^{(1)}, u_t^{(2)}, u_t^{(3)}) = (u_t, v_t, w_t)$$

the velocity components (with numbering or with letters, depending on how it is convenient in the equations), and for the components of the instantaneous turbulent velocity:

$$U_t - \langle U\rangle(t, X_t) = (u_t'^{(1)}, u_t'^{(2)}, u_t'^{(3)}) = (u_t', v_t', w_t').$$

In order to elaborate the SDM model, we start from the General Langenvin model introduced by Pope [15]:

$$\begin{aligned}
dX_t &= U_t dt, \\
&\text{with } U_t = (u_t^{(i)}, i = 1, 2, 3) \text{ and } \mathbf{u}_i'(t) = u_t^{(i)} - \langle u_t^{(i)}\rangle \qquad (3.20)\\
du_t^{(i)} &= -\partial_{x_i}\langle \mathscr{P}\rangle(t, X_t)dt + \left(\sum_j G_{ij}\left(u_t^{(j)} - \langle u^{(j)}\rangle\right)\right)(t, X_t)dt + \sqrt{C_0\varepsilon(t, X_t)}dB_t^{(i)}.
\end{aligned}$$

As a stand-alone PDF method, all the Eulerian statistical means needed by the SDM model in (3.20) are computed within the simulation. In the ABL, we pay great attention to the modeling of the ground effects. We incorporate to SDM a model for the effect of the wall blocking of normal velocity component (following [19], see also [5] for details). For the wind farm simulation, we further incorporate a model for the effect of pressure reflection from the surface (by adapting the Durbin elliptic relaxation method [6]). This model refinement mainly impacts the form of the (G_{ij}) relaxation tensor we use in (3.20). We shortly describe (G_{ij}), decomposing the tensor in this common basic diagonal relaxation term $\frac{1}{2}\frac{\varepsilon}{tke}$ and the more complex γ_{ij} part, decomposed itself in its near wall part $\gamma_{ij}^{\text{wall}}$ and its internal flow part $\gamma_{ij}^{\text{homogeneous}}$:

$$G_{ij}(t,x) = -\gamma_{ij}(t,x) - \frac{1}{2}\frac{\varepsilon(t,x)}{tke(t,x)}\delta_{ij}, \qquad \text{with } C_0\varepsilon(t,x) = \frac{2}{3}\sum_{i,j}\langle\gamma_{ij}\rangle\langle u'_i u'_j\rangle(t,x)$$

$$\text{and} \quad \gamma_{ij}(t,x) = (1-\alpha(t,x)tke(t,x))\gamma_{ij}^{\text{wall}}(t,x) + \alpha(t,x)tke(t,x)\,\gamma_{ij}^{\text{homogeneous}}(t,x)$$

$$-\gamma_{ij}^{\text{homogeneous}} = -\frac{1}{2}(C_R-1)\frac{\varepsilon}{tke}\delta_{ij} + C_2\frac{\partial\langle u^{(i)}\rangle}{\partial x_j}, \qquad \text{and } -\gamma_{ij}^{\text{wall}} = -7.5\frac{\varepsilon}{tke}n_i n_j$$

where n is the wall-normal unit vector. The coefficients C_0 and C_2 have to satisfy some realizability constraints (see [7, 13]). The elliptic blending coefficient α (that balances $\gamma_{ij}^{\text{wall}}$ and $\gamma_{ij}^{\text{homogeneous}}$) solves near the ground the Poisson equation:

$$L^2\nabla^2\alpha - \alpha = -\frac{1}{tke},$$

where L is a length scale defined as a maximum of the turbulent scale and the scale connected with dissipative eddies.

Finally, we make use of the Lagrangian methodology to easily introduce complex terrain description in SDM: when a fluid-particle meets the ground during the simulation, according to the wall-boundary condition, we perform a reflection of it velocity, according to the friction velocity computed as

$$u_*(t,x) = \kappa\frac{\sqrt{\langle u\rangle^2(t,x) + \langle v\rangle^2(t,x)}}{\log\left(x^{(3)}/z_0(x)\right)}$$

where the roughness length z_0 may vary with the surface terrain.

Lagrangian Actuator Disk Model

SDM method allows some fluid and structure interaction modeling, in particular when the structure are porous objects like actuator disk models for turbine.

The SDM approach could be used with various actuator disk modelling options (see [5] and the references therein). In the actuator disc approach, each mill is represented as an immersed surface which concentrates all the forces exerted by the mill on the flow. In the SDM context, the presence of wind mills is taken into account thanks to an additional force f that represents the body forces that the blades exert on the flow. This force term is incorporated in the SDEs that govern the movement of the particles. To this end, Eq. (3.20), which governs the time evolution of the velocity $U_t = (u_t, v_t, w_t)$ of a particle, is modified as follows:

$$\begin{aligned} dU_t &= -\frac{1}{\rho}\nabla_x\langle\mathscr{P}\rangle(t,X_t)dt + f(t,X_t,U_t)\,dt \\ &\quad - G(t,X_t)\Big(U_t - \langle U\rangle(t,X_t)\Big)dt + C(t,X_t)dW_t \end{aligned} \tag{3.21}$$

where the term $f(t,x,U)$ represents the body forces of the turbine seen by the particle at point x with velocity U. We refer to [5] for a detailed discussion on the

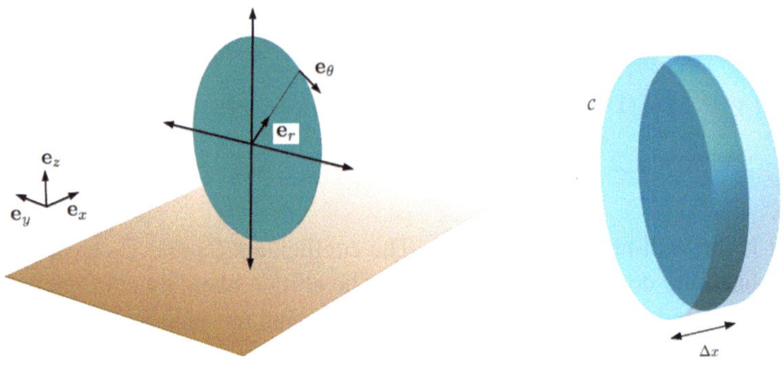

(a) Local coordinates (b) The cylinder $\mathscr{C}$

Fig. 3.9 Non rotating uniformly loaded actuator disc model. **a** The local reference frame at the actuator disc of the turbine, using cylindrical coordinates; **b** The cylinder $\mathscr{C}$ that extends the actuator disc. Mill forces are applied to particles that lie inside

turbine force terms implementation in the Lagrangian context (including nacelle and mast forces).

For the simulation of the *Parc de Bonneval* wind farm presented hereafter, we have chosen a rather basic non rotating uniformly loaded actuator disc model. Such model can be easily parametrized with the characteristic data of thrust coefficient C_T and power coefficient C_p, provided by the turbine manufacturer, and varying with the dynamics of the inflow wind at the turbine.

We describe the force f, using the local reference frame of cylindrical coordinates centered at the hub of the turbine, with basis vectors $\mathbf{e}_x$, $\mathbf{e}_r$ and $\mathbf{e}_\theta$ as shown in Fig. 3.9a. Assuming that the flow moves along the positive direction of the x axis, and that the turbine's main axis is aligned with the x axis, so that it faces the wind directly, the total thrust force exerted by the turbine is formally given by (see e.g. [17])

$$F_x = -\frac{1}{2}\rho A C_T U_\infty^2 \mathbf{e}_x,$$

where U_∞ is the unperturbed velocity far upstream from the turbine's location, A is the surface area of the turbine's disc, ρ is the density of air, and C_T is a dimensionless, flow dependent parameter called the *thrust coefficient*. As in Réthoré et al. [16], the local velocity magnitude U_D is used instead of U_∞ and the thrust force expression in SDM becomes

$$F_x = -\frac{1}{2}\rho A C_T U_D^2 \mathbf{e}_x \quad \text{with } U_D(t) = \mathbb{E}[U_t^2 | X_t \in D]. \tag{3.22}$$

In order to adapt this thrust force model to particles, the disc is extended to a cylinder $\mathscr{C}$ of length Δx and mass $\rho A \Delta x$ (see Fig. 3.9b). The force per unit mass inside region $\mathscr{C}$, and to include in (3.21), is then given by:

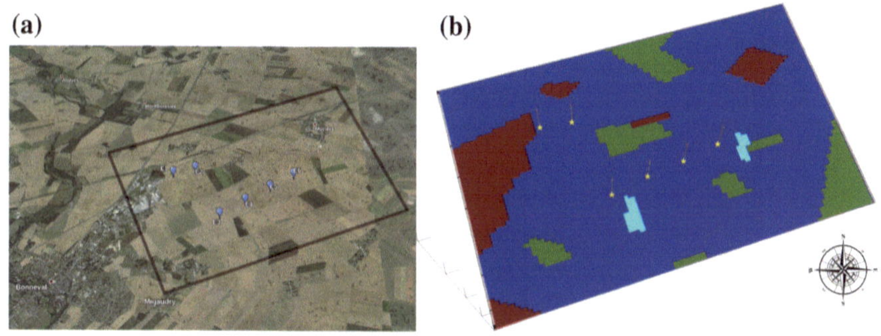

Fig. 3.10 **a** Aerial view of the *Parc de Bonneval* from Google-Earth; **b** Aerial view of the simulated wind farm. The pattern define the roughness length. Blue part represents farmland (0.04 m), red are small town (0.4 m), green are uncut grass (0.01 m), cyan are small forest (0.15 m). Yellow stars represent the turbines. From left to right: the top line are numbered 1 and 2; the turbines in the bottom line are numbered 3, 4, 5 and 6.

$$f(t, x) = -\frac{1}{\Delta x} C_T U_D^2(t) \mathbb{1}_{\{x \in \mathscr{C}\}} \mathbf{e}_x. \tag{3.23}$$

The available power is computed following the same idea:

$$P(t) = \frac{1}{2} \rho A C_p U_D^3(t).$$

Numerical Setup

The modeled domain is a 3D box, with flat ground surface and a variable roughness length inferred from Google-Earth and lookup tables of roughness lengths for typical types of land-use. Four different roughness lengths have been used with respect to the land-use pattern shown in the Fig. 3.10. The roughness length varies between 0.01 and 0.4 m. The characteristics of the numerical domain of the simulation and of the turbines are summarized in Table 3.1.

The initial and boundary conditions are generated from the MERRA reanalysis with a 3-hourly time sampling [8]. All MERRA fields are provided on the same 5/8 degree longitude by 1/2 degree latitude grid. The data used to extract initial and boundary conditions are those of the closest grid point located at 25 km Southwest of *Parc de Bonneval* (48 °N and 1.25 °E). The vertical mesh has 72 pressure levels but only the first three levels from the surface up to 970 hPa (about 400 m) are used. The pressure level coordinates are converted into altitude coordinates using the surface pressure from the MERRA reanalysis. The wind components are then interpolated onto the refined grid of SDM. The time step of the SDM simulation is 5 s. The profiles extracted from the MERRA reanalysis at the closest grid point are therefore interpolated linearly in time with a 5 s time sampling.

Table 3.1 Main parameters of the simulation

(a) Configuration of the simulation	
Domain size x	3000 m
Domain size y	4787 m
Domain size z	408 m
75 cells in x	$\Delta x = 40\,\mathrm{m}$
75 cells in y	$\Delta y = 63.83\,\mathrm{m}$
85 cells in z	$\Delta z = 4.8\,\mathrm{m}$
Particles per cell	80
Final time is 24 h	Time step is 5 s
(b) Parameters of the mill	
Hub height	100 m
Radius	40 m
Nacelle radius	4 m
Rotational speed	$1.75\,\mathrm{rad\,s^{-1}}$

Case Study Description

Parc de Bonneval is composed of six turbines of type Vestas V80-2.0 MW, each named by its number from 1 to 6 in Fig. 3.10a. The simulated study-case corresponds to the 22th December 2016, a winter day, allowing neutral atmosphere approximation, and chosen for its typical wind events, producing wake effects. Figure 3.11 displays the time evolution of the measured wind direction, wind speed and wind energy production at the 6 turbines. The wind speed and direction are measured directly at *Parc de Bonneval* by anemometers located on the hub of each turbine. The wind energy production is also provided directly from the generator. Those time series are used to evaluate SDM model performance, with a sampling period of 10 min. The chosen episode is characterized by a strong wind blowing until 5:00 local time (LT). Between 5:00 and 16:00 LT, the wind speed weakens from $10\,\mathrm{ms^{-1}}$ to $2\,\mathrm{ms^{-1}}$. It increases again up to $6\,\mathrm{ms^{-1}}$ and decreases down to less than $2\,\mathrm{ms^{-1}}$ in 2 h. As a consequence, the turbines production vary from 0 to almost the turbine nominal power of 2 MW during this day. Moreover, the wind shifts progressively from the South to the North. According to the position of the turbines (see Fig. 3.10), a wind direction around 230° lines up the turbines 3 to 6, and a direction around 250° lines up the turbines 1 and 2. We mainly chose this particular episode of December 22th, as it contains such wind event, happening between 7:00 and 9:00 LT. Indeed we can observe the wake effect in Fig. 3.11. The phenomenon decreases the production downstream by 50%.

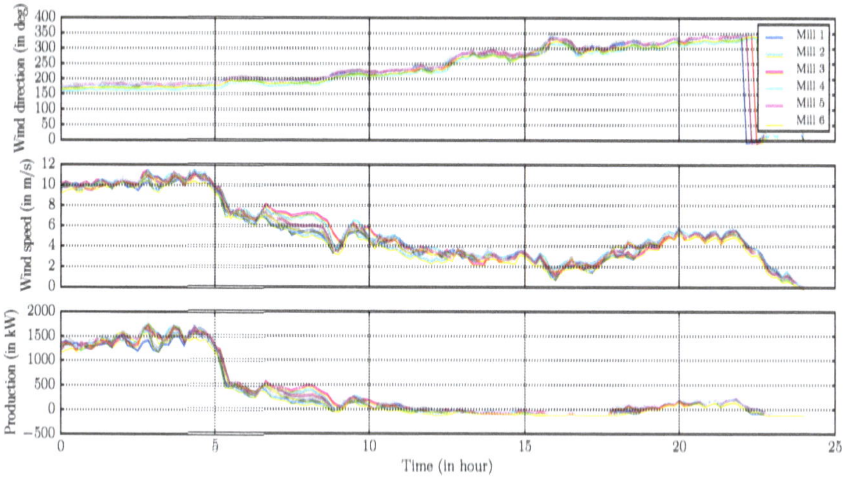

Fig. 3.11 Time evolution of *Parc de Bonneval* measurements during the 22th December 2016

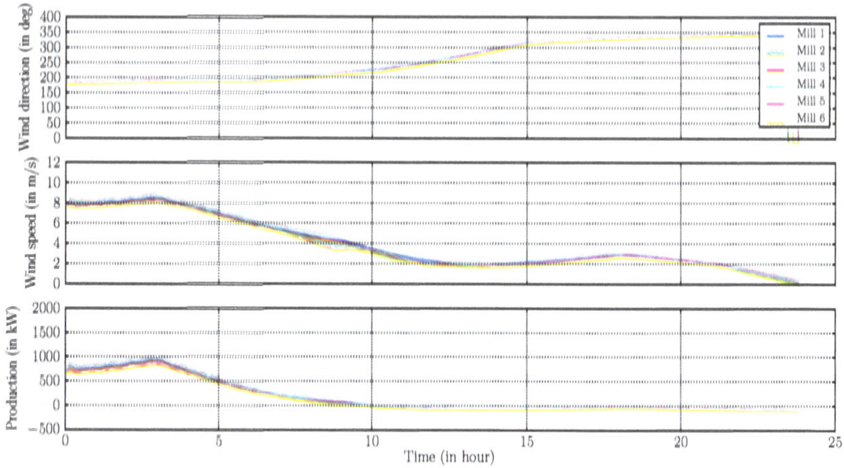

Fig. 3.12 Time evolution of SDM results for the 22th December 2016

Results

Figure 3.12 displays the time evolution of the simulated wind direction, wind speed and wind energy production at the 6 turbines. It can be directly compared to Fig. 3.11. The time variability is well reproduced with a slightly increasing wind speed between 0:00 and 3:30 LT and a constant wind direction. The wind speed increases between 8 and $9.2\,\mathrm{ms}^{-1}$. The simulated wind speed is slightly weaker than the measured wind speed which remains constant and equal to $10\,\mathrm{ms}^{-1}$ over this period of time. Such underestimation is caused by the initial and boundary conditions from MERRA reanalysis which provide a weaker wind speed at the hub height. The wind direction

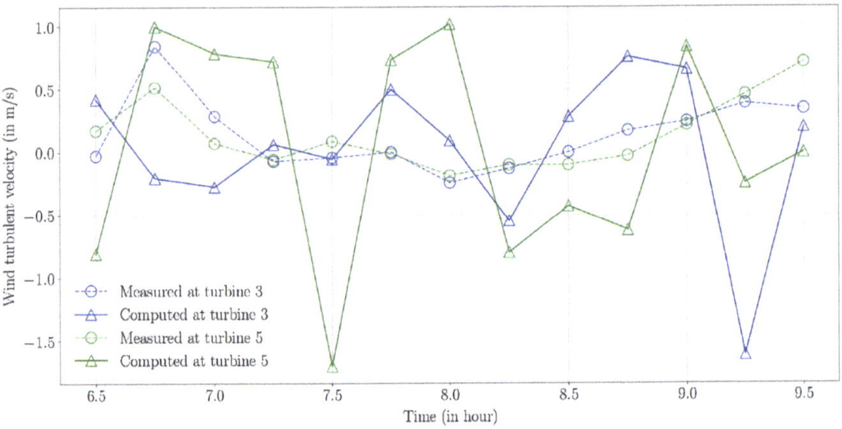

Fig. 3.13 Evolution of the wind turbulent velocity between 6:30 and 9:30 LT, when turbines 3, 4, 5 and 6 are lined up. Blue curves display the velocity for turbine 3 (upstream) and green curves display the velocity for turbine 5 (downstream). Dotted line with circles are measured at *Parc de Bonneval* and solid line with triangles are computed in SDM

is also slightly biased by about 10°. The simulated wind speed then decreases at a similar rate than the observed wind speed. The short increase of the wind speed followed by a fast decrease between 15:00 and 23:00 LT is underestimated in the simulation as the wind speed peaks at about 3.4 ms^{-1} in the simulation versus 6 ms^{-1} in the measurements. The bias in wind direction disappears after 8:00 LT. Finally, we observe that the high frequency variability is much too smooth in the simulated mean velocity. We mainly impute this phenomenon to the combination of low frequency data set for the initial and boundary conditions, with the small size of the numerical domain, that induces a strong forcing by the lateral inflow boundary conditions. However, as shown in Fig. 3.13, the intrinsic variability contains in the model is representative of the observations variability. Figure 3.13 displays the evolution of the norm of the turbulent part of the wind $U' = U - \langle U \rangle$ between 6:30 and 9:30 LT, when turbines 3, 4, 5 and 6 are lined up. During the wake alignment period, computed and measured turbulent velocity norms are displayed at a forefront turbine (turbine 3), and at a downstream turbine (turbine 5). To this end, in SDM, we have extracted a realization of the turbulent part of the velocity, by randomly picking-up every 10 min, one particle velocity at the neighborhood of the rotors. Hourly moving means are computed and subtracted to its instantaneous velocity. We proceed similarly with the measured velocity. In both cases, the variability around the downstream turbine is higher than the variability around the forefront turbine. Moreover, the variability of the turbulent velocity computed in SDM is higher than the one measured at *Parc de Bonneval*. This can be explained by the way the instantaneous velocity is retrieved. For SDM we used an instantaneous velocity at 5 s frequency picked every 10 min. For *Parc de Bonneval*, the velocity measured by anemometers is at a high frequency, but then it as been averaged over 10 min. This time averaging decrease the variability in the observations.

Wake effect. Going back to Figs. 3.11 and 3.12, we observe that the wake effect is well reproduced in the simulation between 7:00 and 12:00 LT. The magnitude is underestimated but the sheltering effect by the forefront turbines is clearly visible. The difference of wind speed between the forefront turbines and those located downstream is about 1–1.5 ms^{-1} in the simulation against 2 ms^{-1} in the measurements. Figure 3.14 displays a zoom between 6:00 and 13:00 LT of the measured and simulated wind direction, wind speed and wind energy production. In detail, the measured wind speed and energy production displays a continuously decrease

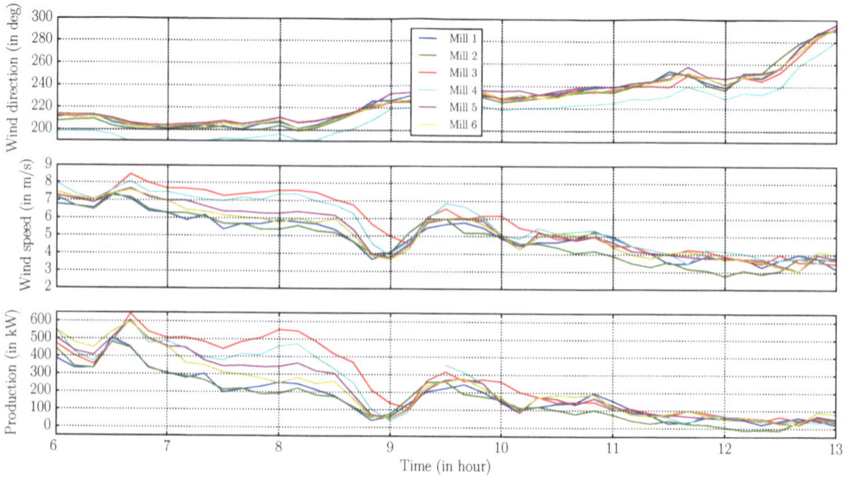

(a) Observed wake effect

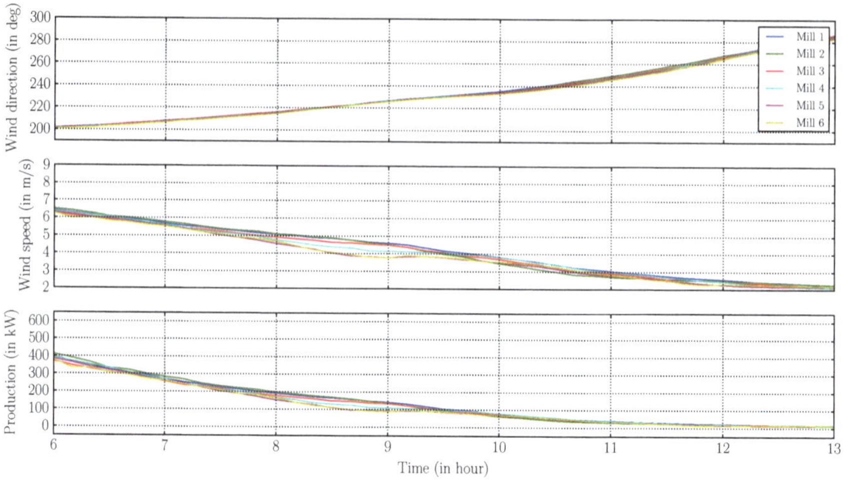

(b) Simulated wake effect

Fig. 3.14 Zoom between 6:00 and 12:00 LT

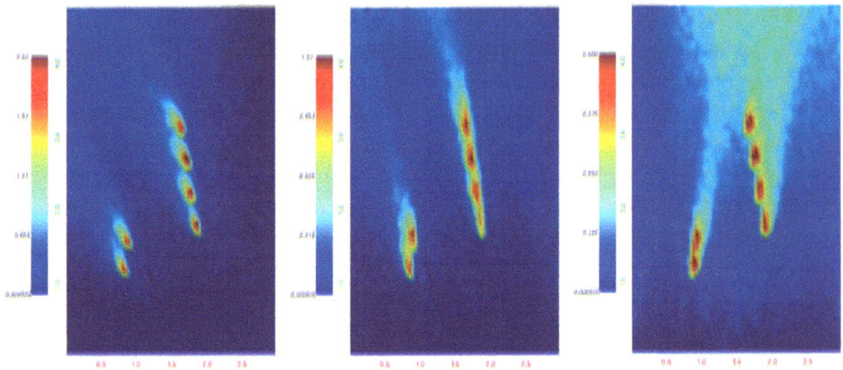

(a) at 00:20 LT (b) at 8:00 LT (c) at 11:00 LT

Fig. 3.15 Surface view at hub height (100 m) at different times. The three panels show the turbulent kinetic energy

between the forefront turbines and the most downstream turbines. At *Parc de Bonneval*, we can distinguish two groups of wind turbines. The forefront turbine 3 with turbines 4, 5 and 6 downstream in the wake between 6:30 and 9:00 LT and forefront turbine 1 with turbine 2 downstream in the wake between 10:00 and 12:00 LT. The simulation displays a similar behavior with however significant differences. Between 6:30 and 9:00 LT, wind speed and energy production at turbines 1 and 2 are similar to wind speed and energy production simulated at turbine 3, and turbines 4, 5 and 6 are in the wake of turbine 3 as observed. Between 10:00 and 12:00 LT, the simulated wind speed and energy production varies as observed at the locations of the wind turbines with however a weaker difference between the forefront and the trailing wind turbines.

Figure 3.15 shows surface views of the simulated turbulent kinetic energy at the hub height (100 m) at different times (0:20, 8:00 and 11:00 LT). At this altitude the main source of turbulence is due to the interaction with the turbines. Figure 3.15a displays the turbulent kinetic energy pattern 20 min after the beginning of the simulation at 00:20 LT. At this time the turbines are not lined up and they all produce the same energy. Figure 3.15b is similar as Fig. 3.15a at 8:00 LT. At this time, the wind direction is around 220°. Consequently, the turbines 3, 4, 5 and 6 are lined-up. Figure 3.15b displays the sheltering effect by the forefront wind turbine and the turbulence generated in its wake. At 11:00 LT (see Fig. 3.15c), the wind veers so that turbine 1 creates a wake which reaches turbine 2.

To summarize the performance of the simulation against the measurements, Table 3.2 displays skill scores: the Normalized Root Mean Square Error (NRMSE) and the MAE (Mean Absolute Error) defined by

Table 3.2 Indicator of the deviation between the simulated wind $\widehat{y}$ and the observed wind y over the six turbines

	NRMSE (in %)	MAE (in m/s)
Turbine 1	14.57	1.369
Turbine 2	14.56	1.334
Turbine 3	15.88	1.578
Turbine 4	16.83	1.681
Turbine 5	14.92	1.455
Turbine 6	14.71	1.425

$$\text{NRMSE} = \frac{\sqrt{\frac{1}{N}\sum_{i=1}^{N}(\widehat{y_i} - y_i)^2}}{y_{max} - y_{min}}, \qquad \text{MAE} = \frac{1}{N}\sum_{i=1}^{N}|\widehat{y_i} - y_i|. \tag{3.24}$$

N is the number of measurements. It is equal to 145 (one measurement every 10 min from the 22th December 2016 00:00 LT to the 23th December 2016 00:00 LT). We make use of the same number of simulated data saved at the same time. y is the measured wind speed and $\widehat{y}$ is the simulated wind speed.

Table 3.2 shows a systematic bias of 1.5 ms^{-1} between the simulation and the measurements, while the NRMSE range varies between about 14.5 to 17%. This is in part due to the initial and lateral boundary conditions from MERRA reanalysis.

Figure 3.16 shows vertical profiles of the wind at different times and locations. Both panels display one profile forefront and one profile downstream, at 8:00 (left) and at 11:00 LT (right). The profiles displaying a continuously increasing wind speed (blue curves) correspond to forefront profiles. They are taken at the same location, in front of the turbines and far from their interaction in the middle of the domain. As a consequence, it displays the upstream vertical wind. At 8:00 LT (Fig. 3.16a), the profile displaying a strong wind speed decreased between 60 and 150 m height (green curve) is extracting downstream turbine 6. This decrease is due to the forefront turbines which disrupt the flow and slowdown the wind in front of the downstream turbines. Indeed, at 8:00 LT, turbines 3, 4, 5 and 6 are lined up. At 11:00 LT (Fig. 3.16b) the green profile is extracting downstream turbine 2. At this time, turbines 1 and 2 are lined up and this is why the wind speed downstream the turbine 2 is slowed by turbine 1. In both cases, the interaction with the turbines decreases the wind speed from 2 ms^{-1} maximum at 80 and 120 m height (just under and above the hub). This figure well describes the wake effect.

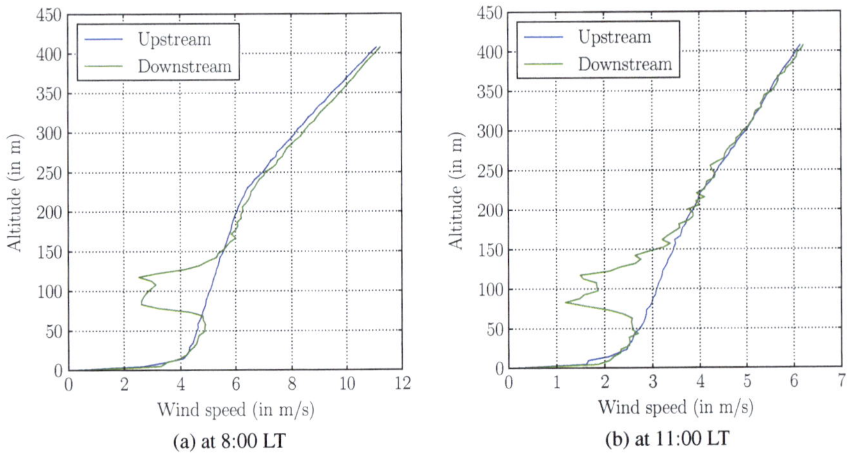

Fig. 3.16 Vertical profils taken at different time and place. **a** is taken when the turbines 3, 4, 5 and 6 are lined up; **b** is taken when the turbines 1 and 2 are lined up

Conclusion

In this paper, we have presented some first numerical experiments obtained from the SDM numerical approach, for a wind farm simulation in condition of use, and we have compared the obtained result with the reality of measures at the turbines.

We have also presented some numerical analysis and experiments that evidence the way the numerical algorithm for SDM is converging.

Some other experiments of wind farm simulation are in preparation, with improvements both in the model and in the description of initial and boundary condition. The objectives are to perform better and reduce the bias against measure, but also to illustrate the ability of SDM to compute not only the mean velocity, but also the local distribution of the turbulent wind, who takes part in the uncertainty of wind power production.

References

1. F. Bernardin, M. Bossy, C. Chauvin, P. Drobinski, A. Rousseau, T. Salameh, Stochastic downscaling methods: application to wind refinement. Stoch. Environ. Res. Risk Assess. **23**(6), 851–859 (2009)
2. F. Bernardin, M. Bossy, C. Chauvin, J.-F. Jabir, A. Rousseau, Stochastic Lagrangian method for downscaling problems in computational fluid dynamics. ESAIM: M2AN **44**(5), 885–920 (2010)
3. M. Bossy, Some stochastic particle methods for nonlinear parabolic PDEs, in *GRIP—Research Group on Particle Interactions*. ESAIM: Proceedings, vol. 15 (EDP Sciences, Les Ulis, 2005), pp. 18–57

4. M. Bossy, L. Violeau, Optimal rate of convergence of particle approximation for conditional McKean–Vlasov kinetic processes (2018)
5. M. Bossy, J. Espina, J. Morice, C. Paris, A. Rousseau, Modeling the wind circulation around mills with a Lagrangian stochastic approach. SMAI J. Comput. Math. **2**, 177–214 (2016)
6. P.A. Durbin, A Reynolds stress model for near-wall turbulence. J. Fluid Mech. **249**, 465–498 (1993)
7. P.-A. Durbin, C.-G. Speziale, Realizability of second-moment closure via stochastic analysis. J. Fluid Mech. **280**, 395–407 (1994)
8. W. McCarty, L. Coy, R. Gelano, A. Huang, D. Merkova, E.B. Smith, M. Sienkiewicz, K. Wargan, MERRA-2 input observations: summary and assessment, in *NASA Technical Report Series on Global Modeling and Data Assimilation*, vol. 46 (2016)
9. J.-P. Minier, Statistical descriptions of polydisperse turbulent two-phase flows. Phys. Rep.*665*(Supplement C), 1–122 (2016)
10. J.-P. Minier, S. Chibbaro, S.B. Pope, Guidelines for the formulation of Lagrangian stochastic models for particle simulations of single-phase and dispersed two-phase turbulent flows. Phys. Fluids **26**(11), 113303 (2014)
11. A. Niayifar, F. Porté-Agel, Analytical modeling of wind farms: a new approach for power prediction. Energies **9**(9), 741 (2016)
12. S.B. Pope, Lagrangian PDF methods for turbulent flows. Annu. Rev. Fluid Mech. **26**, 23–63. Annual Reviews, Palo Alto, CA, 1994
13. S.B. Pope, Lagrangian pdf methods for turbulent flows. Annu. Rev. Fluid Mech. **26**, 23–63 (1994)
14. S.B. Pope, Particle method for turbulent flows: tegration of stochastic model equations. J. Comput. Phys. **117**(2), 332–349 (1995)
15. S.B. Pope, *Turbulent Flows* (Cambridge University Press, Cambridge, 2000)
16. P.E. Réthoré, N.N. Sørensen, A. Bechmann, F. Zahle, Study of the atmospheric wake turbulence of a CFD actuator disc model, in *Proceedings of European Wind Energy Conference*, Marseille, France, 16–19 March 2009
17. J.N. Sørensen, Aerodynamic aspects of wind energy conversion. Annu. Rev. Fluid Mech. **43**(1), 427–448 (2011)
18. A. Stohl, Computation, accuracy and applications of trajectories. A review and bibliography. Atmos. Environ. **32**(6), 947–966 (1998)
19. M. Waclawczyk, J. Pozorski, J.-P. Minier, Probability density function computation of turbulent flows with a new near-wall model. Phys. Fluids **16**(5), 1410–1422 (2004)

Chapter 4
Day-Ahead Probabilistic Forecast of Solar Irradiance: A Stochastic Differential Equation Approach

Jordi Badosa, Emmanuel Gobet, Maxime Grangereau and Daeyoung Kim

Abstract In this work, we derive a probabilistic forecast of the solar irradiance during a day at a given location, using a stochastic differential equation (SDE for short) model. We propose a procedure that transforms a deterministic forecast into a probabilistic forecast: the input parameters of the SDE model are the AROME numerical weather predictions computed at day $D-1$ for the day D. The model also accounts for the maximal irradiance from the clear sky model. The SDE model is mean-reverting towards the deterministic forecast and the instantaneous amplitude of the noise depends on the clear sky index, so that the fluctuations vanish as the index is close to 0 (cloudy) or 1 (sunny), as observed in practice. Our tests show a good adequacy of the confidence intervals of the model with the measurement.

Keywords Solar power · Probabilistic forecast · Stochastic differential equation

4.1 Introduction

Context. As the conventional energy sources such as coal, oil, and gas are considered to be one the main factors responsible for climate change, solar energy has been recognized as a viable alternative. Solar powered plants which utilize photovoltaic (PV)

J. Badosa
LMD/IPSL, Ecole Polytechnique,
Route de Saclay, 91128 Palaiseau Cedex, France
e-mail: jordi.badosa@lmd.polytechnique.fr

E. Gobet (✉) · M. Grangereau
CMAP, Ecole Polytechnique, 91128 Palaiseau Cedex, France
e-mail: emmanuel.gobet@polytechnique.edu

M. Grangereau
e-mail: maxime.grangereau@polytechnique.edu

D. Kim
Ecole Polytechnique, 91128 Palaiseau Cedex, France
e-mail: daeyoung.kim@polytechnique.edu

P. Drobinski et al. (eds.), *Renewable Energy: Forecasting and Risk Management*, Springer Proceedings in Mathematics & Statistics 254,
https://doi.org/10.1007/978-3-319-99052-1_4

and Concentrated Solar Power (CSP) have penetrated the electricity grid, and they are contributing to meet the electricity balance between production and consumption. The need for clean energy and energy-generation independence has driven the solar farms to grow drastically in number. This is backed up by policy supports including feed-in tariffs.

One of the challenges of the electricity generation through solar panels is that it is intermittent and driven by meteorological conditions (mainly related to cloudiness), which results in high uncertainty in the final amount of production. Such uncertainty put grid operators at risk since they might have to adjust their production on very short notice and energy storage capacity is limited. Solar irradiance forecast in this context is crucial not only to predict the solar power generation amount but also to save the start and shutdown costs of conventional generators [19].

Use of day-ahead predictions. Day-ahead predictions are of particular importance for application in the energy market, where the day-ahead auction of power price plays a major role in many countries [22]. For example, in France, one may take part in the bidding by the noon on the day before the operation.[1] Besides the market participation, the day-ahead prediction can also be useful for unit commitment and energy storage dispatch [7, 11]. Simulation studies have shown that day-ahead forecasts may provide significant cost savings to grid operators and fast start and lower efficiency power plants [19]. PV-based microgrids also make use of day-ahead predictions for power planning [14].

Regarding the types of day-ahead solar irradiance forecast, we can distinguish two categories: deterministic forecasts and probabilistic forecasts. In the state of the art of irradiance forecasting, most of the literature relies on deterministic forecasts, also known as point forecasts, since their outputs are specific *values* at given times in the future. On the other hand, most probabilistic forecasts give the full *distributions* of the values of interest at the times considered. Each one of these distributions can be represented by histogram or cumulative density function for example. The representation of uncertainty information takes into account potentially extreme scenarios. It also allows the operators to gain additional trust in the forecasts [21]. Furthermore, having at hand a probabilistic forecast for the solar irradiance simultaneously at any hour (continuous-time forecast), accounting for the inter-temporal probabilistic dependency allows to properly solve some problems where the full distribution of the inputs plays an important role (see for instance [10]). For applications in energy management, see [1] and for the connection with electricity derivatives, see [2]. The continuous-time forecast may also be updated whenever new data becomes available.

Actually, methods for deriving predictions depend much on both the considered time horizon and the amount of data available at the time when the prediction is made. Moreover, the flows of data and the frequency for updated predictions are tightly related. For horizons of minutes to hours, satellite images (see *Meteosat* for Europe) are much informative on a global area, a new image is available every 15'; on the other hand, the access is not granted to anyone. For very local geographic

[1] See https://www.epexspot.com/en/market-data/dayaheadauction.

data, one can be equipped with sky cameras or simply irradiance sensors, the latter being the most common way of collecting data.

In this study, we focus on the day-ahead horizon and we consider that deterministic Numerical Weather Predictions (NWP) of solar irradiance are available at local scale as it is the case of the Meteo-France's AROME NWP data, which has a grid resolution of 1.3 km. This data is available for any individual prosumer and might for instance be used to manage batteries in order to reduce variability of the demand on the grid, as in [10].

Deterministic and probabilistic forecasts. In the deterministic forecast of day-ahead solar irradiance, NWP models are widely used, and the model value goes through Model Output Statistics (MOS) before the actual usage. MOS is a post-processing technique used to interpret empirically the outputs of a numerical model and produce site-specific forecasts [8, 12]. Statistical learning methods are often used to correct errors in the NWP model outputs and to incorporate knowledge from several models, by appropriately weighting them. These methods allow to correct biases and systematic errors in the forecasts [17]. MOS is known to improve the performance of the raw forecast of NWP by about 10–15% [27].

While the deterministic forecast has been developed for more than thirty years, the probabilistic forecast seems to be in its infancy yet. However, it seems to have a high potential of usefulness, especially in applications where risk quantification is a crucial factor, for example energy storage in connection with intermittent energy sources.

One noticeable method of probabilistic day-ahead solar irradiance forecast is the analog ensemble approach, which searches the history for similar forecasts and makes corrections to the forecast according to the error in prior forecasts. In this approach, the prior analogs become an ensemble that quantifies the uncertainty of the forecast [3]. These approach rely on non-parametric techniques and machine learning tools; on the one hand, this is a data-driven approach and therefore it is quite flexible; on the other hand, since it is not aimed at identifying a specific stochastic model, analytical or numerical methods are not applicable, which limits the resolution of some problems (like stochastic control problems). In [3] an example of ensemble approach using machine-learning based regression models – such as decision tree, K-nearest neighbors, random forests – is presented, showing that each of these models performs better than the Auto-Regressive Integrated Moving Average (ARIMA for short) model [20, 24].

Another method is the stochastic modeling, where the evolution of the system is described by a Stochastic Differential Equation (SDE). Knowledge coming from other deterministic forecasts can be incorporated in these models. As explained in [13], SDE models have several assets: we can incorporate boundedness properties, which are essential for correct modeling of the maximal solar irradiance, and the SDEs are more general than other classical time-series models, like ARIMA processes. Besides, the output of the SDE models can be both a point forecast, by simulation of a single trajectory, or a probabilistic forecast. Indeed, by simulating multiple independent Markovian evolutions of the process, it is possible to infer the full distribution of irradiance at any time (see Sect. 4.2.4). A first attempt for such a SDE modeling is proposed by [25] with a simplistic model. The authors of [13] have designed extensions and more involved models with various degrees of complexity. They assume a parametric SDE model for the solar irradiance and suppose that the observations are noisy. A convenient feature of their SDE model is to account for the maximum irradiance (also known as clear sky irradiance, [6]). For the model estimation, they cope with SDE inference with noisy data: for this, they use Kalman-type filtering techniques, which restrict their SDE model to additive Brownian noise. In the current study, we also consider a SDE model but its form is different from [13] and we assume perfect observation of the data, i.e. we assume there is no noise in the observations, see details below.

Our contribution. Our purpose is to turn deterministic forecasts into probabilistic forecasts. Our framework is to consider that a single deterministic forecast is available on the day before (i.e. on $D-1$): in our case, this comes from Météo-France (AROME NWP data). The precise description of this data set is given in Sect. 4.2.2. To model the irradiance on the day D, we use a time-dependent SDE, which models the evolutions of the Clear Sky Index (CSI for short), which is the ratio between the observed irradiance and the clear sky irradiance, that is the maximal theoretical irradiance that would be observed with no clouds. In this work we will assume that the CSI lies in the range $[0, 1]$. It will be denoted X in the equations. Some parameters of the SDE for a given day D are fixed (see forthcoming paragraph on the estimation of the parameters), while others are estimated using the Arome data at day $D-1$. Simulating the SDE gives realistic scenarios of solar irradiance, as demonstrated in Sect. 4.3. The SDE is driven by a Brownian motion W and it has the following form:

$$\mathrm{d}X_t = -a(X_t - x_t^{\texttt{forecast}})\mathrm{d}t + \sigma X_t^{\alpha}(1-X_t)^{\beta}\mathrm{d}W_t, \qquad t \in [t_0, t_1] \subset [0, 24],$$

for some parameters $a, \sigma, \alpha, \beta, (x_t^{\texttt{forecast}})_{0\le t\le 24}$ and where $[t_0, t_1]$ is the period (expressed in hours) of the day D where the sun shines at the location where irradiance is measured. The Brownian motion W allows to model the uncertainty across time. All the above parameters are identified using the Arome data at day $D-1$ and some historical data. In Sect. 4.3, we show that simulations from this model generate confidence intervals that are fully consistent with the realizations collected over a period of one year.

We now stress the similarities and differences with [13]. The form of the SDE is similar in both cases: the drift is a mean-reversion term and the diffusion term vanishes at 0 and 1. However, the form of our diffusion term is more general and allows to take into account more general form of dependence of the fluctuations, while the drift term in [13] incorporates more parameters. It would be an interesting lead to try to incorporate more parameters in the drift term of the SDE we propose as well, but one needs to make sure to have enough data to estimate them properly. Besides, in our study, the observation is assumed to be direct: by ignoring the observation noise, we do not need anymore to use filtering techniques and we can directly make use of appropriate statistical methods for SDEs [16]. This is what allows us to consider a more general diffusion coefficient, namely of the form $x^{\alpha}(1-x)^{\beta}$ with general exponents (in Sect. 4.2.3.2, we find $(\alpha, \beta) \approx (0.8, 0.7)$), while the authors of [13] take $\alpha = \beta = 1$ so that, via a Lamperti transformation, they can get back to additive noise model. Last, our SDE directly models the evolution of the clear sky index, while in [13], it models the evolution of the irradiance (although the maximal irradiance clear sky model is incorporated in the SDEs from their second model).

The estimation of the parameters of our model turns out to be fairly satisfactory, using historical data for measurements and one-day in advance predictions (Arome data). We are able to reproduce accurately enough the forecast uncertainty (see tests in Sect. 4.3); it is quite remarkable, especially because only a small amount of data is used to build the predictions (mainly AROME data).

4.2 Uncertainty Modeling from Data

4.2.1 Definitions and Notations

To be accurate, the solar irradiance (or simply irradiance in the following) that we consider refers to Global Horizontal Irradiance (GHI), that is the amount of solar radiation that reaches the surface of the Earth on a horizontal plane (its unit is W/m^2). The GHI depends on the location; in our study, it is at the latitude and longitude of 48.713 °N and 2.208 °E, at SIRTA (atmospheric observatory at Ecole Polytechnique, on the Paris-Saclay campus, http://sirta.ipsl.fr).

For each day, we consider the irradiance $I(t)$ (at the above location) as a function of time $t \in [0, 24]$ (in hours, for a given day). We assume that the irradiance evolves between two bounds, 0 and $I^{\texttt{clear sky model}}(t)$, where the second is the maximal irradiance obtained under clear sky conditions, that is, a cloudless sky. In our case, the latter has been calculated with an empirically-derived equation that takes into account the Sun-Earth geometry and the day of the year:

$$\forall t \in [t_0, t_1],\ I^{\texttt{clear sky model}}(t) = \left[83.69 \sin\left(\frac{2\pi}{365.24}(D + 82.07)\right) + 1130.44\right] \cos(\theta_z(t))^{1.2}$$

where D is the day of the year [0, 365], and $\theta_z(t)$ is the solar zenith angle. This formula was empirically obtained for the SIRTA site by Badosa et al. from 11 years of measurements (2003–2013); they selected 239 half-days (mornings or afternoons) for which the weather conditions were cloudless. One minute data was considered for this. The model they considered has the form $C_1(D)\cos(\theta_z(t))^{C_2}$, where $C_1(.)$ and C_2 are adjusted on the data, see [4]. Moreover, before the sunrise ($t \le t_0$) and after the sunset ($t \ge t_1$), $I^{\texttt{clear sky model}}(t)$ is set to 0 and there is no question about uncertainty in the irradiance. From now on, we mainly stick to the case $t \in [t_0, t_1]$ (of course, this interval evolves along the days and seasons). Our aim is to propose a stochastic model for the Clear-Sky-Index (CSI) given by

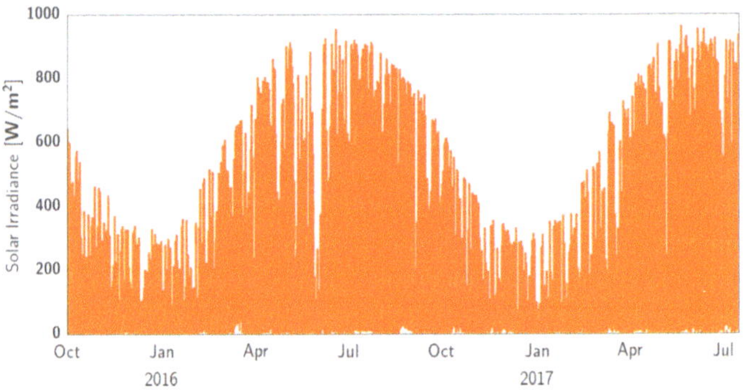

Fig. 4.1 Measurements of global horizontal irradiance from SIRTA (48.7 °N, 2.2 °E.) for the considered period

$$X_t = \frac{I(t)}{I^{\texttt{clear sky model}}(t)}. \tag{4.1}$$

X close to 1 (resp. 0) corresponds to clear-sky day (resp. dark clouds) while intermediate values relate to a variable sky (or light clouds).

To calibrate the model parameters, we also use the Accumulated CSI of the day D, defined by

$$\text{ACSI}_D = \frac{\sum_{t\in\mathscr{T}} I(t)}{\sum_{t\in\mathscr{T}} I^{\texttt{clear sky model}}(t)}, \tag{4.2}$$

where $\mathscr{T} := \{\text{measurement times in the day } D\}$. This index is an indicator of the overall clearness on day D.

4.2.2 Data: AROME Forecast and SIRTA Measurement

In our study, we consider the forecast GHI values from AROME NWP operated by Météo-France, for the closest grid point to SIRTA site. In particular, we use the run from 12:00 UTC on the previous day ($D - 1$) of the target forecast day (D), which has a time step of 1 h and covers the whole day D.

The considered period is October 1st, 2015 to July 16th, 2017 (see Fig. 4.1). Only days with both measured and forecast GHI available values were considered, which made a final dataset of 473 days. The measured data was taken at 10-min resolution. Forecast values with 10-min resolution were generated from the forecast with 1-h resolution using linear interpolation.

4.2.3 Fitting the SDE Model

4.2.3.1 Heuristic Derivation

The stochastic model we propose for the irradiance CSI is aimed at giving a precise meaning to the approximation

$$X_t \approx x_t^{\texttt{forecast}} + \text{error}(t)$$

and of its continuous-time evolution, where $x_t^{\texttt{forecast}}$ is calculated from the AROME forecast

$$x_t^{\texttt{forecast}} := \frac{I^{\texttt{forecast}}(t)}{I^{\texttt{clear sky model}}(t)} \in [0, 1].$$

We assume that $x_t^{\texttt{forecast}}$ is given in continuous time (as mentioned before, we use linear interpolation to compute values at times where no forecast is available in the data). The term error(t) stands for the unpredictable part in the prediction and reflects the uncertainty of the forecast: observe that, since the CSI lies in the range [0, 1], it is certainly not appropriate to assume that the error has a Gaussian distribution.

To begin with, the deterministic forecast $x_t^{\texttt{forecast}}$ is considered as a (time-dependent) mean-reversion level, i.e. when the realized CSI is far from the forecast CSI, it is expected that (in mean) X_t gets closer to $x_t^{\texttt{forecast}}$. In other words, a temporary prediction error is possible but it tends to vanish. We model this feature via a dissipative linear Ordinary Differential Equation that describes the time-evolution[2] $t \mapsto \mathbb{E}[X_t]$,

$$\mathrm{d}\mathbb{E}[X_t] = -a(\mathbb{E}[X_t] - x_t^{\texttt{forecast}})\mathrm{d}t$$

for some mean-reversion speed parameter $a > 0$ that will be estimated later.

We now model the stochastic fluctuations around the above relation, i.e. we write

$$\mathrm{d}X_t = -a(X_t - x_t^{\texttt{forecast}})\mathrm{d}t + \text{noise}(\mathrm{d}t).$$

First, the amplitude of the noise over an infinitesimal interval $[t, t + \mathrm{d}t]$ has to decrease as the CSI gets closer to 0 (cloudy sky) and 1 (clear sky): this is requested to maintain the model values in the unit interval. For this reason, we set

$$\text{noise}(\mathrm{d}t) = X_t^{\alpha}(1 - X_t)^{\beta}\widetilde{\text{noise}}(\mathrm{d}t)$$

for two positive parameters α, β. Once the latter will be chosen appropriately, we expect that the amplitude of the noise noise(dt) is well tuned so that the newly-renormalized noise $\widetilde{\text{noise}}(\mathrm{d}t)$ does not depend anymore on the CSI (see later Fig. 4.3).

Moreover, as explained in the next paragraph, a statistical analysis of $\widetilde{\text{noise}}(\mathrm{d}t)$ shows that it can be modeled by a Gaussian distribution $\mathcal{N}(0, \sigma^2\mathrm{d}t)$ and from now

[2] $\mathbb{E}[.]$ is the expectation attached to the forthcoming probabilistic model.

on, we take it as a Brownian increment $\sigma \mathrm{d}W_t$. This is not the only Gaussian model, but it has the advantage of a very low parametric dimension (which is a nice asset when one deals with few data).

Gathering all the previous arguments, we finally obtain a SDE of the form

$$\mathrm{d}X_t = -a(X_t - x_t^{\texttt{forecast}})\mathrm{d}t + \sigma X_t^{\alpha}(1 - X_t)^{\beta}\mathrm{d}W_t, \quad t \in [t_0, t_1], \tag{4.3}$$

with a given initial value $X_{t_0} \in [0, 1]$. In the case of $x_t^{\texttt{forecast}}$ constant, and $\alpha = \beta = 1$, we retrieve the well-known Fisher-Wright diffusion used in mathematical genetics. In the following, we will take $(\alpha, \beta) \in [\frac{1}{2}, +\infty) \times [\frac{1}{2}, +\infty)$.

From a mathematical point of view, we can justify that this SDE model is well-posed, despite that the coefficients are not globally Lipschitz and that it takes values in the unit interval [0, 1] as requested. These properties are rigorously proved in Appendix.

4.2.3.2 Specification of the Parameters

As it is well-know in the statistics for SDEs [9, 16], the estimation of the parameters entering in the drift and diffusion coefficients can be made independently, asymptotically as the frequency gets larger and larger.

Statistical estimation of a. The Itô formula gives that, for any $t \geq s$,

$$e^{at}X_t = e^{as}X_s + \int_s^t e^{au}x_u^{\texttt{forecast}}\mathrm{d}u + \int_s^t e^{au}\sigma X_u^{\alpha}(1 - X_u)^{\beta}\mathrm{d}W_u.$$

Since the first time integral is deterministic, it readily follows that

$$\mathrm{Cov}\left(\int_s^t e^{au}x_u^{\texttt{forecast}}\mathrm{d}u, X_s\right) = 0.$$

Moreover the stochastic integral is computed on $[s, t]$ and the SDE solution is adapted to the Brownian filtration, therefore

$$\mathrm{Cov}\left(\int_s^t e^{au}\sigma X_u^{\alpha}(1 - X_u)^{\beta}\mathrm{d}W_u, X_s\right) = 0.$$

To summarize, we get $\mathrm{Cov}\left(e^{at}X_t, X_s\right) = e^{as}\mathrm{Var}(X_s)$, that is

$$\mathrm{Cov}(X_t, X_s) = e^{-a(t-s)}\mathrm{Var}(X_s), \qquad t \geq s. \tag{4.4}$$

In (4.4) we retrieve a nice relation available for any SDE which drift is linear, see [5]. Therefore, it is enough to compute the correlogram of the process X from the data and to extract the parameter a using an exponential fit. The empirical correlogram

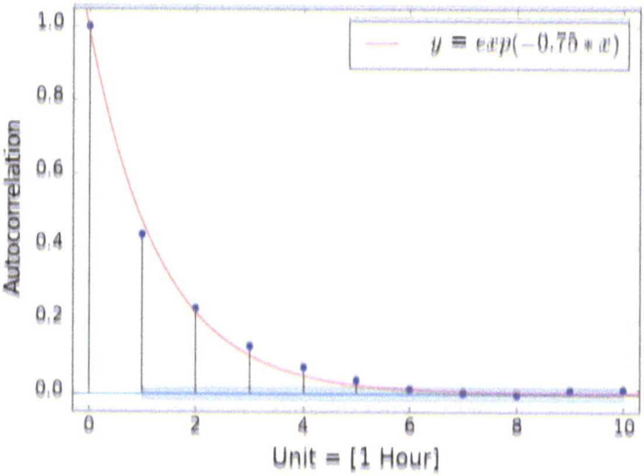

Fig. 4.2 Empirical correlogram and exponential curve fit

from Fig. 4.2 has been computed using all the data for the 473 days between sunrise and sunset. In our case, we obtain $a \approx 0.75\,\mathrm{h}^{-1}$.

Specification of σ, α, β. For high-frequency data (that is when the monitoring frequency is considered high), the increments of $X_t - X_s, t \geq s$ can be approximated by

$$X_t - X_s \underset{t-s \text{ small}}{\approx} \sigma X_s^{\alpha}(1 - X_s)^{\beta}(W_t - W_s). \tag{4.5}$$

Indeed, as $t - s$ is small, the drift term in (4.3) is of magnitude $t - s$, which can be neglected compared to the order $\sqrt{t-s}$ arising from $W_t - W_s$. This is the usual Euler approximation of SDE in small time [18]. Of course, one needs to make sure that the time step used in the simulations is small enough to guarantee that X remains in its domain $[0, 1]$.

In our model, we seek a SDE parametrization that, each day, adapts automatically to the Arome $D - 1$ forecast. Our strategy is

- first to identify a pair of exponents (α, β) that will be available in average for all the days (these parameters are somehow day-invariant),
- then to estimate σ that may be day-dependent and that will be estimated from Arome $D - 1$ forecast. In the sequel, the parameter σ for the day D will be written $\sigma_D > 0$.

Because of the relation (4.5), for measurements collected at a 10'-period ($t_{i+1} - t_i = 10' = \delta$) in the same day D, we expect to have (provided that α, β are well chosen)

$$\frac{X_{t_{i+1}} - X_{t_i}}{X_{t_i}^{\alpha}(1 - X_{t_i})^{\beta}} \overset{d}{\approx} \mathcal{N}(0, \sigma_D^2\delta), \tag{4.6}$$

where $\overset{d}{\approx}$ means *approximation in distribution*, with an independence property with respect to t_i and X_{t_i} (thanks to the properties of the Brownian increments). Therefore, it is enough to estimate empirically the variance (day by day) of the quantities (4.6) to get an estimator of $\sigma_D^2 \delta$ for that day D, and therefore of σ_D. Actually, instead of considering increments of the irradiance CSI, we take the increments of the forecast errors, i.e.

$$\text{REI} = \frac{[X_{t_{i+1}} - x_{t_{i+1}}^{\texttt{forecast}}] - [X_{t_i} - x_{t_i}^{\texttt{forecast}}]}{X_{t_i}^{\alpha}(1 - X_{t_i})^{\beta}}$$

which asymptotic behavior with respect to $\delta \to 0$ is similar to (4.6) (indeed, the curve of forecast $x^{\texttt{forecast}}$ is of finite variation). REI stands for Renormalized Error Increment.

We could identify the parameters $(\alpha, \beta, \sigma_{D_1}, \ldots, \sigma_{D_{473}})$ using a Maximum Likelihood Estimation technique (via the Gaussian approximation (4.6)), requiring an optimization procedure in dimension 475! We proceed in a simpler way, by choosing α and β so that the σ_D's computed as mentioned (day by day) are decorrelated from the average irradiance of the day (represented by the ACSI); it shows in a way that α and β are suitably tuned to make (4.6) valid. This is a way to have a single pair (α, β) independently of the day of the year. For $\alpha = 0.8$ and $\beta = 0.7$, Fig. 4.3 shows that σ_D are quite decorrelated from ACSI_D (Accumulated Clear Sky Index) defined in (4.2). Because the ACSI is computed on the measurement (see (4.2)), it makes its use not applicable as it is in the perspective of probabilistic forecast. Therefore, we seek to model σ_D in terms of the Arome forecast on day $D - 1$, using a regression analysis. Since σ_D reflects the amplitude of local variability of the irradiance, we use the Average Time Increment CSI computed on the Arome forecast as a surrogate to statistically explain σ_D:

$$\text{ATICSI}_D = \sum_{t_i \in \mathscr{T}^{\text{forecast}}} \left| \frac{I^{\texttt{forecast}}(t_{i+1})}{I^{\texttt{clear sky model}}(t_{i+1})} - \frac{I^{\texttt{forecast}}(t_i)}{I^{\texttt{clear sky model}}(t_i)} \right|, \tag{4.7}$$

where $\mathscr{T}^{\text{forecast}} := \{\text{Arome forecast times}\}$. Figure 4.4 depicts the relation between σ_D and ATICSI_D.

We exhibit a linear relation of the form

$$\sigma_D \sqrt{\delta} = 0.622 \times \text{ATICSI}_D + 0.0004 + \text{error}. \tag{4.8}$$

In the next experiments, we set the residual error in (4.8) to 0. Actually, the distribution of this error (see Fig. 4.5) is close to a Gaussian one, which could be included in our model of σ_D to possibly improve the probabilistic forecast. We haven't gone further in that direction since with the simplified version (error $= 0$), our tests reveal a good accuracy of the probabilistic forecast. Furthermore, it would be interesting to investigate the sensitivity of parameters α, β and σ with respect to the input data. This would require splitting the input data into a training set and a test set. As the data used to specify the parameters was covering a period of not much more than one year and our tests already show good accuracy of the forecast, we did not conduct this procedure.

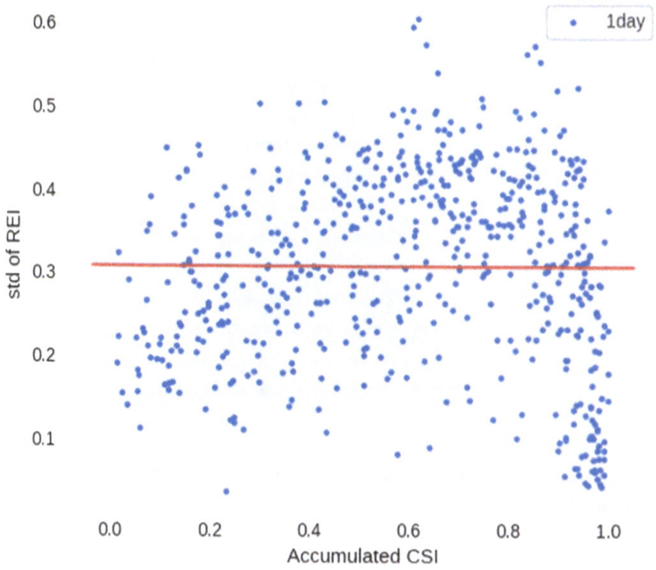

Fig. 4.3 For each day D in the data set, the standard deviation $\sigma_D\sqrt{\delta}$ of the renormalized error increment as a function of accumulated clear sky index ACSI_D, when $\delta = 10' = 1/6\ h$, $\alpha = 0.8$ and $\beta = 0.7$

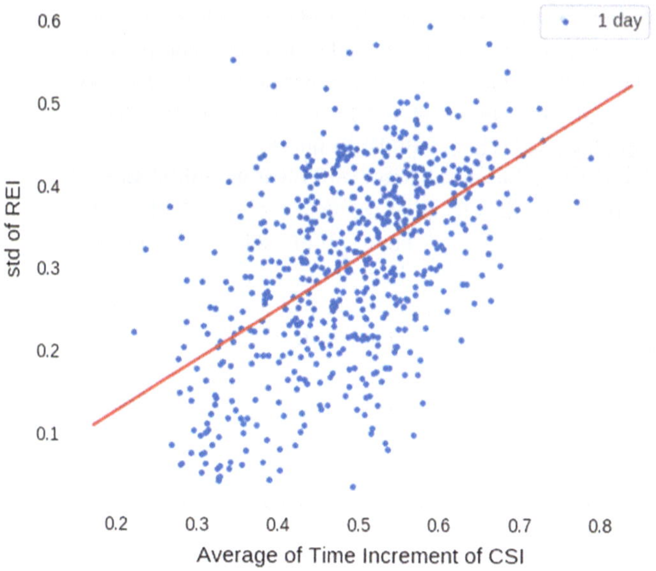

Fig. 4.4 For each day D in the data set, the standard deviation $\sigma_D\sqrt{\delta}$ as a function of the average time increment CSI defined in (4.7)

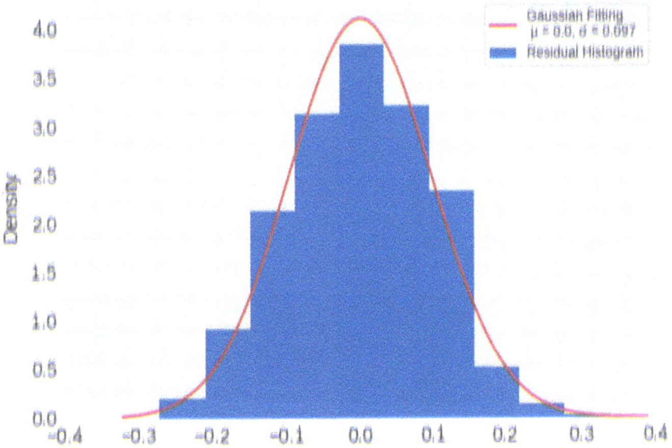

Fig. 4.5 Histogram of the residual error in (4.8), with the Gaussian density centered at 0 and with standard deviation 0.097

4.2.4 Probabilistic Forecast Computation

We present how we compute numerically the distribution of irradiance in our model on the day D, given the Arome forecast of day $D-1$.

Using the values of the parameters derived in the calibration step described earlier, from (4.3) we are now able to simulate independent trajectories of the clear sky index (and therefore of the solar irradiance accounting for the clear sky model in (4.1)). In our tests, usually we sample $M = 1000$ or $M = 10000$ paths, in order to produce accurate enough statistics and confidence intervals.

Regarding the simulation scheme itself, once the initial value X_{t_0} is chosen, we just need to simulate independent Markovian evolutions. More precisely, we use the Euler scheme (see [18] for an account on the subject of simulating SDEs). In our tests, the time step of the Euler scheme is $\Delta_t = 1'$ and when the value of the Euler scheme is outside of $[0, 1]$, the value is pushed back to $[0, 1]$ (as the exact solution). Set $t_k = t_0 + k\Delta_t$; the ith sampled path writes finally as $(1 \le i \le M)$

$$\begin{cases} X^{(i)}_{t_{k+1}} = X^{(i)}_{t_k} - a(X^{(i)}_{t_k} - x^{\texttt{forecast}}_{t_k})\Delta_t + \sigma\,(X^{(i)}_{t_k})^{\alpha}(1 - X^{(i)}_{t_k})^{\beta}(W^{(i)}_{t_{k+1}} - W^{(i)}_{t_k}), & k \ge 0, \\ X^{(i)}_{t=t_0} = X^{(i)}_{t_0}, \end{cases}$$

where $(W^{(i)})$ are M i.i.d. trajectories of the Brownian motion, and $(X^{(i)})$ are M i.i.d. trajectories of the CSI.

Initialization. A particular treatment of the independent initial value $X^{(i)}_{t_0}$ for each trajectory of the CSI needs to be implemented. Indeed, one could initialize all the processes to the same initial value of the deterministic forecast. However, this would not allow us to encompass uncertainty in the initial value of the CSI. We propose another approach, close to Markov Chain Monte-Carlo methods. The initial value $X_{t_0}^{(l)}$ is sampled according to the stationary distribution of the Euler scheme associated to the SDE

$$d\widetilde{X}_t = -a(\widetilde{X}_t - x_{t_0}^{\texttt{forecast}})dt + \sigma \widetilde{X}_t^{\alpha}(1 - \widetilde{X}_t)^{\beta} dW_t. \tag{4.9}$$

Note the term $x_{t_0}^{\texttt{forecast}}$ in the above equation. Doing so, we pick a initial point $X_{t_0}^{(l)}$ which reflects in a quite intrinsic way the uncertainty in the forecast at $t = t_0$.

In practice, we simulate the Euler scheme of (4.9) over a time interval of length T, where T is chosen long enough so that the distribution of the Euler scheme after time T is close to the target stationary distribution (and quite independent from the initial point of the SDE (4.9)); see [26] about approximation scheme for ergodic SDEs.

In practice, due to the form of the drift as a mean-reversion term and in view of the fast decorrelation (like in (4.4)), we set T as three times the characteristic time of the system, that is $T = 3/a \approx 4$ h. This procedure is repeated independently to sample each $X_{t_0}^{(l)}$.

4.3 Numerical Experiments

4.3.1 Description of the Tests

In the following graphs, we represent some features of the results we obtain for several types of days: a day with good weather, when the Clear Sky Index remains close to 1, a day with bad weather, when it remains close to 0 and a day with mitigated weather, when it takes intermediate values. For all those examples, we take days where the deterministic forecast used to establish the probabilistic forecast is fairly good, i.e. close to the real irradiance profile over the day. This is usually the case with Arome forecast.

For each day with fairly good deterministic forecasts (Figs. 4.6, 4.7, 4.8), we plot 3 graphs. The first in the top left corner features the clear sky irradiance as well as several simulated trajectories of the irradiance (at the time-scale of 1' as mentioned before). The clear sky irradiance is the irradiance we would observe if the sky were perfectly clear (this is the clear sky model as explained at the beginning of Sect. 4.2.1). The simulated trajectories of the irradiance are obtained by multiplying the simulated trajectories of the Clear Sky Index, obtained by Euler scheme of a SDE (see previous paragraph), by the clear sky irradiance (see the relation (4.1)).

The second graph in the top right corner features the clear sky irradiance, the deterministic Arome forecast irradiance, the irradiance measured on that day, a simulated trajectory of the irradiance obtained using our model, as well as confidence intervals for the irradiance forecast. This confidence region is obtained by Monte-Carlo methods: by simulating $M = 100\,000$ i.i.d. trajectories of the irradiance (using our SDE model), we can estimate accurately the 5 and 95% quantiles. If our model is accurate, we expect that the measured irradiance remains inside the

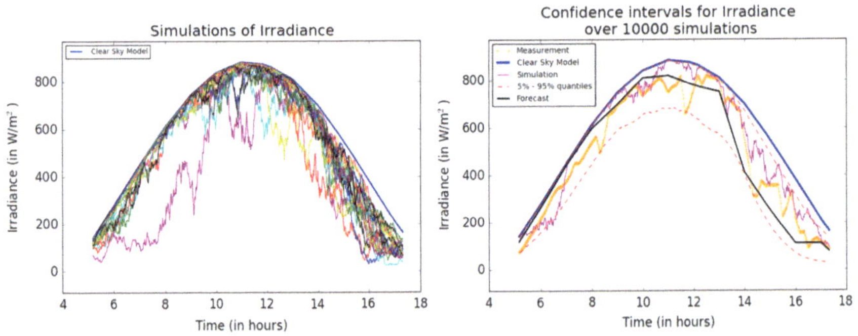

(a) Results for the probabilistic forecast

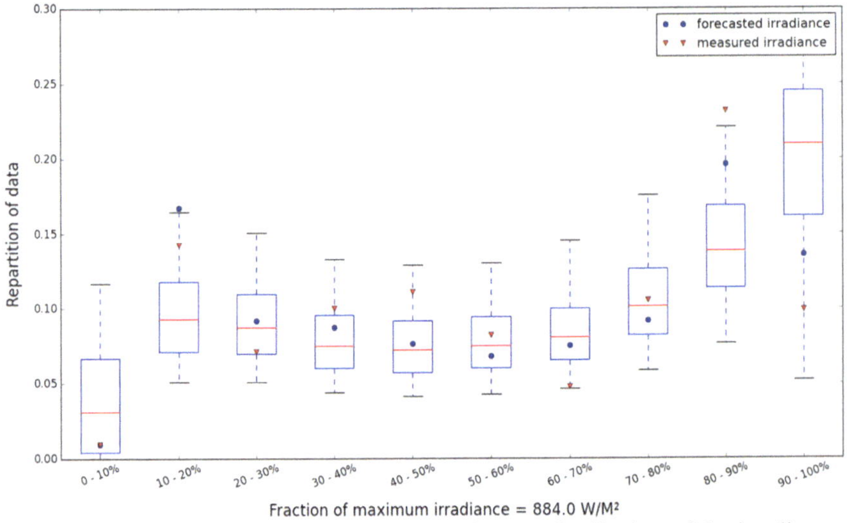

(b) Validation and accuracy of the model for the distribution of the irradiance over the day

Fig. 4.6 Results for a day with good weather (May 2nd, 2016)

confidence intervals most of the time. However, having a measurement outside this confidence area does not necessarily mean poor performance of the probabilistic forecast, because of the definition of quantiles. Of course, the deterministic forecast staying inside this confidence area is an intrinsic property of the model we have designed: we take an SDE with a drift which is a mean-reversion term, the mean being this deterministic forecast.

In the bottom graph, we represent several vertical box-plots, as well as other information. Let us explain in more details. For each day studied, we can derive the number of points of a discrete trajectory (observation, forecast or simulation) for which the irradiance value lies in each of the 10 intervals of length 10% of the maximal value of the clear sky irradiance (given by the clear sky model). Using this

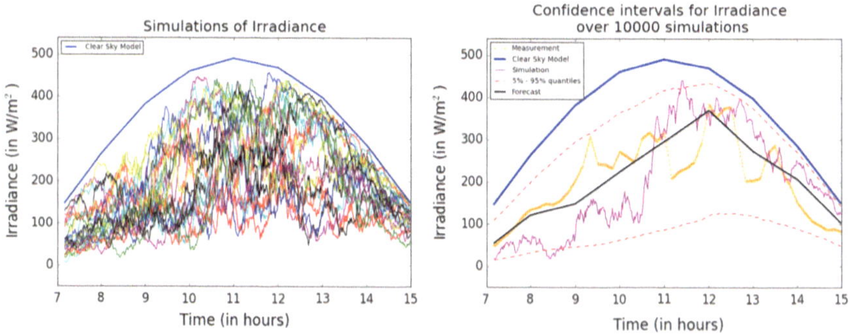

(a) Results for the probabilistic forecast

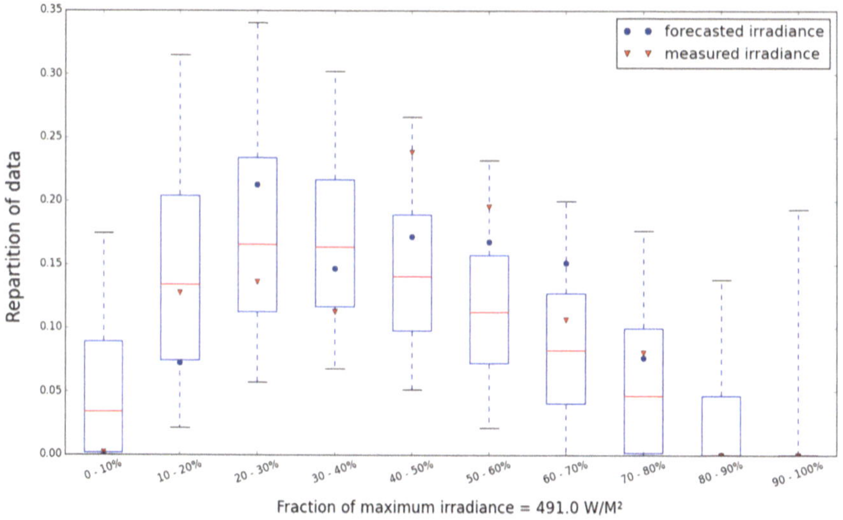

(b) Validation and accuracy of the model for the distribution of the irradiance over the day

Fig. 4.7 Results for a day with mitigated weather (October 24th, 2015)

procedure for the measured irradiance, we obtain by renormalization an estimation of the proportion of time spent by the irradiance in a given interval over the day. A possible application is to derive an estimation of the PV energy that can be produced on that specific day. For example, a red triangular point with 0–10% on the x-axis and 0.4 on the y-axis means that the measured irradiance was smaller than 10% of the maximal theoretical irradiance about 40% of the time of sun exposure over this day. We can do the same thing for the forecast irradiance and for each of the simulations. The simulations allow to estimate the distribution of these proportions of time spent in each subinterval. Indeed, using the values obtained for each simulation, we can draw box-plots which show how the irradiance was distributed over the whole day. These statistical outputs can for instance give a clear indication about the

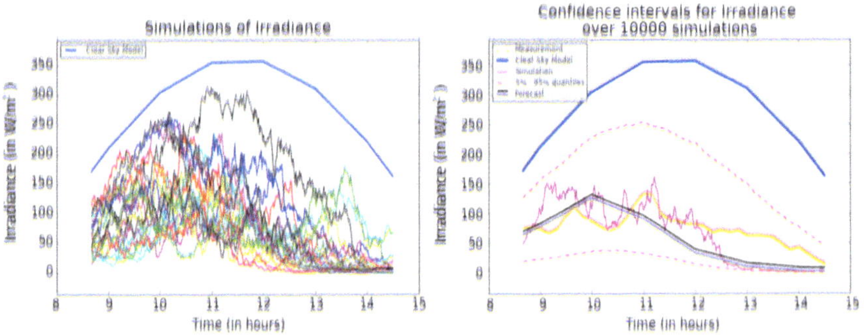

(a) Results for the probabilistic forecast

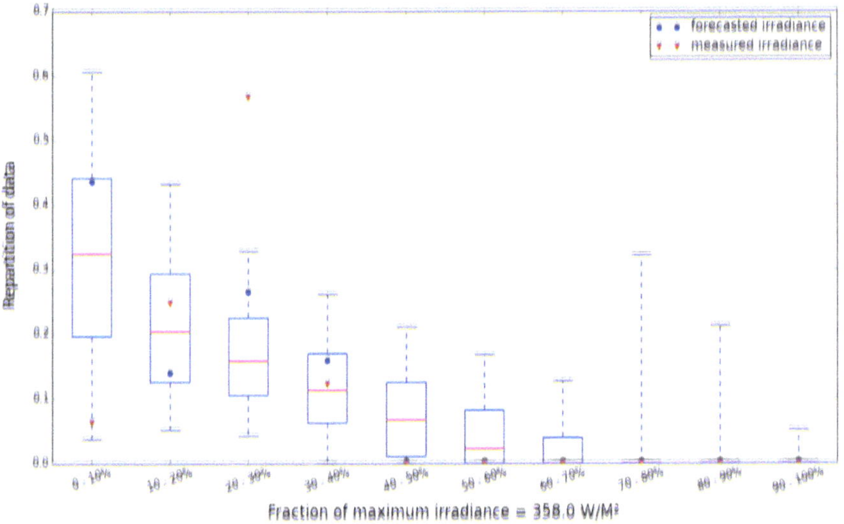

(b) Validation and accuracy of the model for the distribution of the irradiance over the day

Fig. 4.8 Results for a day with bad weather (January 22nd, 2016)

probability distribution of the energy that can be produced using PV panels. The red line represents the median, while the box extends from the first quartile to the third quartile, and the whiskers extend from the 5% to the 95% quantiles. For example, for the interval 0–10%, which regroups all data with a value corresponding to less than 10% of the theoretical maximal irradiance observable on this day, a red line with y-coordinate equal to 0.33 means that for half of the simulations, less than one third of the points of the corresponding trajectory have a value in this interval, while for the other half of the simulations, more than one third of the points of the trajectory have a value in this interval.

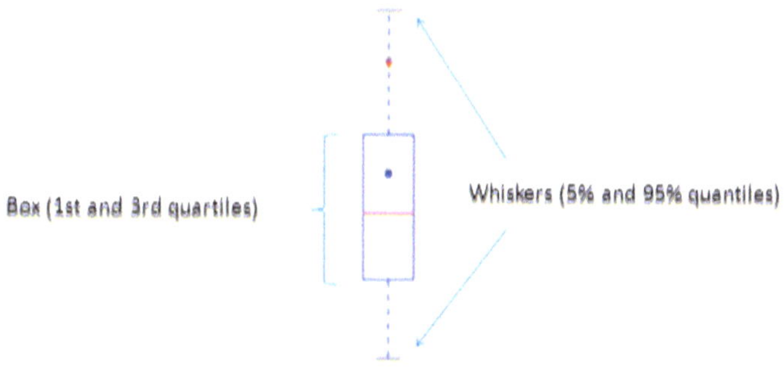

Fig. 4.9 Elements of a box-plot: explanation

A good probabilistic forecast would therefore have several characteristics:

- The trajectory of the measured irradiance would lie in the confidence area obtained by Monte-Carlo simulations.
- In the box-plots, the points for irradiance measured would lie in high density regions (between the whiskers and often inside the boxes, see Fig. 4.9).

These last points are not a guarantee that the probabilistic forecast was good. Indeed, one could imagine the following situation: if on one day, one forecasts a clear sky in the morning and a cloudy sky in the afternoon, but in practice, these events occur in the other order, the box-plots might be consistent (i.e. the measurements should lie inside the confidence area represented by the box and the whiskers), whereas the forecast and measured irradiance paths would be radically different. Even the forecast and measurements having close trajectories does not necessarily mean that all statistical properties of the irradiance are captured by our model.

4.3.2 Analysis of the Results

From these 3 figures (that are quite representative of what we have observed throughout our tests on the full history), we have general observations. First, as expected, the forecast lies in the confidence area: this is mathematically consistent with the model and in accordance with the intuition. Second, generally speaking, the points drawn using the measured irradiance data lie in the confidence areas obtained with our Monte-Carlo procedure; this can be observed in two ways.

- From the second graph, one checks that the measured irradiance (at any given time) remains most of the time in the confidence area. It helps to answer positively the question whether the profile of the irradiance observed over the day is in accordance with the forecast distribution that is induced by our SDE model.
- From the third graph, we analyse whether the repartition of irradiance over the day is correctly predicted by our model, regardless of when phenomenon occurred. The

answer is yes. This criterion is of course less severe than the previous one, but it has the merit to accept events in the days which were predicted but occurred at another time that what the forecast predicted. For example, if the forecast predicts a sunny morning and a cloudy afternoon, the time frame when the transition happens may not be accurately predicted by Arome models.

These results constitute empirical evidence of the good performance of our model: in a way, we are able to correctly reproduce the distribution of the irradiance over the day (either by time interval – second graph – or by repartition of irradiance – third graph). Further measures of performance will be investigated in future works.

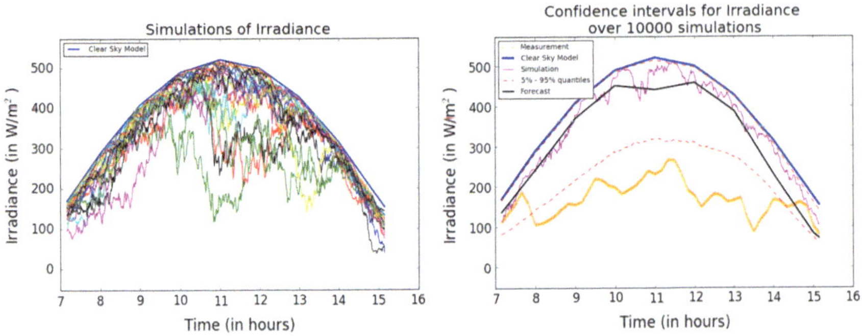

(a) Results for the probabilistic forecast

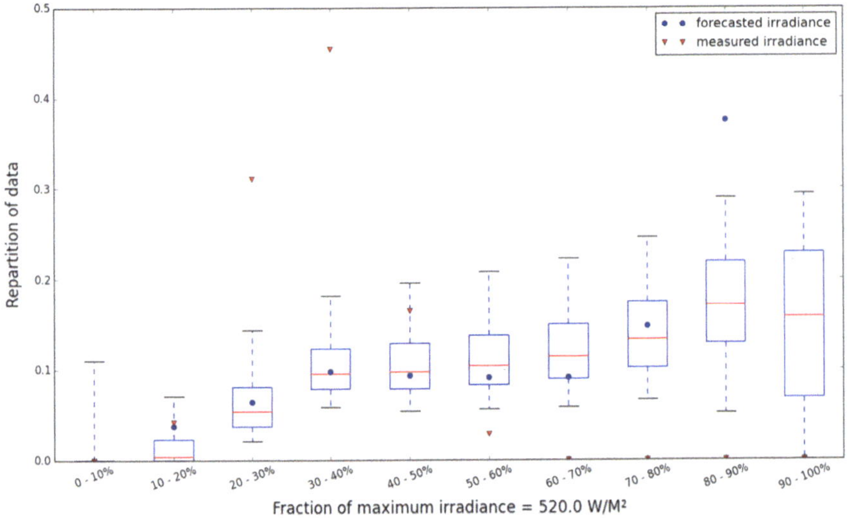

(b) Validation and accuracy of the model for the distribution of the irradiance over the day

Fig. 4.10 Results for a day with bad forecast (October, 19th, 2015)

4.3.3 Limits

We show in the next Fig. 4.10 that if the deterministic forecast is completely erroneous, i.e. if the irradiance profile observed is completely different from the predicted one, then the probabilistic forecast performs poorly as well. It is not surprising at all, since the probabilistic forecast is built upon the deterministic one.

In the graph at the top right corner, we see that due to a poor forecast, the measured irradiance fails to be inside the confidence area over the whole day. In the box-plots at the bottom, it is also clear that the distribution of the proportions of time spent in each subinterval of irradiance is not estimated correctly.

4.4 Conclusion

We have designed a stochastic differential equation that models the solar irradiance for a given day D. This gives rise to a probabilistic forecast. The parameters of the model change from day to day and can be tuned automatically: they depend only the irradiance from clear sky model on the day D and the deterministic Arome forecast computed on day $D-1$ for the day D. By simulating the SDE in a Monte-Carlo framework, we obtain the distribution of the forecast and related statistics. Despite its apparent simplicity, the model is able to produce quite accurate confidence intervals for the irradiance at a given time, and for the repartition of irradiance during the day.

Acknowledgements This research is part of the Chair *Financial Risks* of the *Risk Foundation*, the *Finance for Energy Market Research Centre* and the ANR project *CAESARS* (ANR-15-CE05-0024). The work benefits from the support of the Siebel Energy Institute and it was conducted in the frame of the TREND-X research program of Ecole Polytechnique, supported by Fondation de l'Ecole Polytechnique. The authors acknowledge Météo-France and the Cosy project for the numerical weather prediction data used in the study.

Appendix: Proof of Well-Posedness of the SDE Model (4.3), When $\frac{1}{2} \leq \alpha$ and $\frac{1}{2} \leq \beta$

Because the exponents α and β are possibly non integers in the definition of (4.3), the signs of X_t and $1 - X_t$ may be an issue. Therefore, we start with a modification of the SDE model (4.3) avoiding the sign problems:

$$\mathrm{d}X_t = -a(X_t - x_t^{\texttt{forecast}})\mathrm{d}t + \sigma \mathbf{1}_{X_t \in [0,1]} X_t^{\alpha}(1 - X_t)^{\beta}\mathrm{d}W_t, \tag{4.10}$$

where $X_0 \in [0, 1]$ is a given deterministic initial value.

Existence/uniqueness. A direct application of [23, Chapter IX, Theorem 3.5-(ii), p. 390 and Theorem 1.7 p. 368] shows that the model (4.10) is well-posed, in the sense that there is an unique strong solution on the probability space $(\Omega, \mathcal{F}, \mathbb{P})$ where the filtration is the natural filtration of the Brownian motion completed as usually with the $\mathbb{P}$-null sets. In [23, Chapter IX, Theorem 3.5-(ii), p. 390] we have used $\frac{1}{2} \leq \alpha$ and $\frac{1}{2} \leq \beta$.

The solution (4.10) *takes values in* $[0, 1]$. We invoke a comparison theorem for SDEs. Denote $b^X(t, x) = -a(x - x_t^{\texttt{forecast}})$ the drift coefficient of X and now, consider the solution to

$$dY_t = -aY_t dt + \sigma \mathbf{1}_{Y_t \in [0,1]} Y_t^{\alpha} (1 - Y_t)^{\beta} dW_t, \qquad Y_0 = 0. \tag{4.11}$$

Its initial condition fulfills $X_0 \geq Y_0$, its drift $b^Y(t, y) = -ay$ is globally Lipschitz in space (and b^X too) and last, we have $b^Y(t, x) - b^X(t, x) = -ax_t^{\texttt{forecast}} \leq 0$. Therefore, [15, Chapter V, Proposition 2.18, p. 293] shows that $X_t \geq Y_t$ for any t with probability 1. But since *the* solution to (4.11) is 0, the above proves that X remains positive.

Similarly, set

$$dY_t = -a(Y_t - 1)dt + \sigma \mathbf{1}_{Y_t \in [0,1]} Y_t^{\alpha} (1 - Y_t)^{\beta} dW_t, \qquad Y_0 = 1.$$

Clearly, $Y_0 \geq X_0$, $b^X(t, x) - b^Y(t, x) = -a(1 - x_t^{\texttt{forecast}}) \leq 0$, $Y_t = 1$ and we conclude that $X_t \leq 1$. This justifies why we can remove the indicator function in (4.10) to get (4.3). □

References

1. R. Aïd, L. Campi, N. Langrené, H. Pham, A probabilistic numerical method for optimal multiple switching problem in high dimension. SIAM J. Financ. Math. **5**(1), 191–231 (2014)
2. R. Aid, *Electricity Derivatives*. Springer Briefs in Quantitative Finance (Springer, Berlin, 2015)
3. S. Alessandrini, L. Delle Monache, S. Sperati, G. Cervone, G. Cervone, An analog ensemble for short-term probabilistic solar power forecast. Appl. Energy **157**, 95–110 (2015)
4. J. Badosa, M. Haeffelin, H. Chepfer, Scales of spatial and temporal variation of solar irradiance on Reunion tropical Island. Solar Energy **88**, 42–56 (2013)
5. B.M. Bibby, I.M. Skovgaard, M. Sorensen, Diffusion-type models with given marginal distribution and autocorrelation function. Bernoulli **11**(2), 191–220 (2005)
6. R.E. Bird, R.L. Hulstrom, Simplified clear sky model for direct and diffuse insolation on horizontal surfaces. Technical report, Solar Energy Research Institute, Golden, CO (USA), 1981
7. M. Delfanti, D. Falabretti, M. Merlo, Energy storage for PV power plant dispatching. Renew. Energy **80**, 61–72 (2015)
8. H.M. Diagne, P. Lauret, M. David, Solar irradiation forecasting: state-of-the-art and proposition for future developments for small-scale insular grids, in *WREF 2012-World Renewable Energy Forum* (2012)
9. E. Gobet, LAN property for ergodic diffusion with discrete observations. Ann. Inst. H. Poincaré Probab. Statist. **38**(5), 711–737 (2002)
10. E. Gobet, M. Grangereau, McKean optimal stochastic control of microgrid equipped with photo-voltaic panels and battery. Preprint (2018)

11. R. Hanna, J. Kleissl, A. Nottrott, M. Ferry, Energy dispatch schedule optimization for demand charge reduction using a photovoltaic-battery storage system with solar forecasting. Solar Energy **103**, 269–287 (2014)
12. D. Heinemann, E. Lorenz, M. Girodo, Solar irradiance forecasting for the management of solar energy systems. Energy Meteorology Group, Oldenburg University, Energy and Semiconductor Research Laboratory, 2006
13. E.B. Iversen, J.M. Morales, J.K. Møller, H. Madsen, Probabilistic forecasts of solar irradiance using stochastic differential equations. Environmetrics **25**(3), 152–164 (2014)
14. H. Kanchev, D. Lu, F. Colas, V. Lazarov, B. François, Energy management and operational planning of a microgrid with a PV-based active generator for smart grid applications. IEEE Trans. Sustain. Energy **58**(10) (2011)
15. I. Karatzas, S.E. Shreve, *Brownian Motion and Stochastic Calculus*, 2nd Edn. (Springer, Berlin, 1991)
16. M. Kessler, A. Lindner, M. Sorensen. *Statistical Methods for Stochastic Differential Equations* (CRC Press, Boca Raton, 2012)
17. J. Kleissl, *Solar Energy Forecasting and Resource Assessment* (Academic Press, London, 2013)
18. P.E. Kloeden, E. Platen, *Numerical Solution of Stochastic Differential Equations*. Applications of Mathematics, vol. 23 (Springer, Berlin, 2010). 4th Corrected Printing
19. C.B. Martinez-Anido, B. Botor, A.R. Florita, C. Draxl, S. Lu, H.F. Hamann, B.M. Hodge, The value of day-ahead solar power forecasting improvement. Solar Energy **129**, 192–203 (2016)
20. A.A. Mohammed, Z. Aung, Ensemble learning approach for probabilistic forecasting of solar power generation. Energies **9**(12), 1017 (2016)
21. K.D. Orwig, M.L. Ahlstrom, V. Banunarayanan, J. Sharp, J.M. Wilczak, J. Freedman, S.E. Haupt, J. Cline, O. Bartholomy, H.F. Hamann, B.M. Hodge, C. Finley, D. Nakafuji, J.L. Peterson, D. Maggio, M. Marquis, Recent trends in variable generation forecasting and its value to the power system. IEEE Trans. Sustain. Energy **6**(3), 924–933 (2015)
22. R. Perez, E. Lorenz, S. Pelland, M. Beauharnois, G. Van Knowe, K. Hemker Jr., D. Heinemann, J. Remund, S.C. Müller, W. Traunmüller, G. Steinmauer, D. Prozo, J.A. Ruiz-Arias, V. Lara-Fanego, L. Ramirez-Santigosa, M. Gastin-Romero, L.M. Pomares, Comparison of numerical weather prediction solar irradiance forecasts in the US, Canada and Europe. Solar Energy **94**, 305–326 (2013)
23. D. Revuz, M. Yor, *Continuous Martingales and Brownian Motion*. Comprehensive Studies in Mathematics, 3rd Edn. (Springer, Berlin, 1999)
24. C.V.A. Silva, L. Lim, D. Stevens, D Nakafuji, Probabilistic models for one-day ahead solar irradiance forecasting in renewable energy applications, in *2015 IEEE 14th International Conference on Machine Learning and Applications (ICMLA)* (IEEE, 2015), pp. 1163–1168
25. T. Soubdhan, R. Emilion, Stochastic differential equation for modeling global solar radiation sequences, in *Proceedings of the IASTED International Conference: Modelling, Identification, and Control (AsiaMIC 2010)* (2010), pp. 14–17
26. D. Talay, Second-order discretization schemes of stochastic differential systems for the computation of the invariant law. Stoch. Stoch. Rep. **29**, 13–36 (1990)
27. A. Tuohy, J. Zack, S.E. Haupt, J. Sharp, M. Ahlstrom, S. Dise, E. Grimit, C. Mohrlen, M. Lange, M.G. Casado, J. Black, M. Marquis, C. Collier, Solar forecasting: methods, challenges, and performance. IEEE Power Energy Mag. **13**(6), 50–59 (2015)

Chapter 5
Homogeneous Climate Regions Using Learning Algorithms

Mathilde Mougeot, Dominique Picard, Vincent Lefieux and Miranda Marchand

Abstract Climate analysis is extremely useful to understand better the differences of electricity consumption within the French territory and to help electricity consumption forecasts. Using a large historical data base of 14 years of meteorological observations, this work aims to study a segmentation of the French territory based on functional time series of temperature and wind. In a first step, 14 clustering instances, one for each year, have been performed using, for each instance, one year of data. Each year, the clustering exhibits several homogeneous and spatially connected regions. Benefits of this approach let to study the stability of the previous regions over the years and to highlight the inter-annual variability of the French climate. A final aggregation of all clustering instances shows a segmentation map in easily interpretable, geographically connected climate zones over the last years. Especially, we observe that the number of clusters remains extremely stable through the years. Exhibiting stable homogeneous regions bring then some valuable knowledge for potentially installing new wind or solar farms on the French territory.

Keywords Climate segmentation · Graph partitioning · Clustering

M. Mougeot (✉) · D. Picard
Université Paris Diderot, LPSM UMR 8001, Sorbonne Paris Cité, 75013 Paris, France
e-mail: mougeot@epsm.paris

D. Picard
e-mail: picard@epsm.paris

V. Lefieux
RTE-EPT & UPMC-ISUP, 92919 La Défense Cedex, France
e-mail: vincent.lefieux@rte-france.com

M. Marchand
RTE-R&DI, 92919 La Défense Cedex, France
e-mail: miranda.marchand@rte-france.com

P. Drobinski et al. (eds.), *Renewable Energy: Forecasting and Risk Management*,
Springer Proceedings in Mathematics & Statistics 254,
https://doi.org/10.1007/978-3-319-99052-1_5

5.1 Introduction

The United Nations Conference on Climate Change COP21 has set a goal of 30% renewable energy in French overall energy supply by 2020. Nowadays, wind energy represents 4% of the national electricity production and should double by 2020 [25]. Since electricity can hardly be stored, forecasting tools are essential to properly balance electricity consumption and generation. In France, heating represents about 10% of the annual electricity consumption and depends strongly on the felt temperature which is directly linked to meteorological conditions. Climate understanding is extremely useful to better understand the differences of electricity consumption on the French territory and to help electricity consumption forecasts. RTE,[1] the French electricity transmission system operator, is responsible for operating, maintaining and developing the high and extra high voltage network.

In France, Météo France company owns a large historical data base of meteorological observations of the French territory, most of them recorded at specific places. Based on these observations and on a numerical model of the atmosphere, the Arpège program provides meteorological data for all the French territory, up to a resolution of 10 km. Wind and temperature time series extracted from the Arpège model are in particular available for 259 points covering all the French territory, for a period of 14 years, at an hourly sample rate (see Fig. 5.1).

This joint work between RTE and LPSM[2] investigates the segmentation of the French territory based on this set of data. Our main goal is to provide a segmentation of the French territory in homogeneous regions based on these data.

A delicate problem with meteorological data is the scale: how to find a best scale to describe a phenomena? In this study our aim was not purely meteorological: it was to find homogeneous regions in terms of climate but also convenient in an energy planning perspective. With this objective, producing algorithms for the whole dataset would not have allowed to provide indices of confidence for this classification. Instead, we preferred to part the dataset into 14 years. It seemed a reasonable scale to study the variability along time as well as the robustness of the algorithms - smaller scales obviously contained too much variability.

Hence, our goal is to provide segmentations of the French territory in homogeneous regions based on yearly temperature and wind information, and to exhibit, at a same time, the regions which tend not to be classified in the same group over years.

In this application, each grid point is considered as an observation. The total number of observations for the French territory is then $n = 259$. Each temperature (resp. wind) at each datetime is considered as a variable and let p define the number of studied variables. The global data set may be then characterized by a (n, p) matrix with $n = 259$ and p equals to the number of variables i.e. the length of the studied time series.

[1]Réseau de Transport d'Electricité.

[2]Laboratoire de Probabilités, Statistique, Modélisation.

Fig. 5.1 The 259 Meteorological grid points of the Arpège model

Data segmentation is commonly achieved with clustering algorithms. Different methods may be used for this task such as, among others, hierarchical clustering, k-means, model based clustering or spectral clustering [5]. Most of the clustering algorithms use different notions of closeness of the inputs, generally Euclidean distances coupled with a first representation of the data.

Clustering approaches have been frequently used to understand meteorological data. Atmospheric profiles have been computed with the help of clustering by mixture decomposition of copulas [30, 31] and many works have been conducted in order to understand the repartition of temperature [2, 24] and wind [15, 27] for different regions of the earth.

In statistical machine learning, a major issue is the selection of an appropriate feature space where inputs show nice properties for solving the problem. This space can either be specified explicitly by hand-coded features or be automatically learned. Choosing the feature space is not an easy task, and without theoretical assumptions, it is quite hard to propose a first choice. Moreover, it is well known that the choice of this feature space has a strong impact on the final results.

In our application, the time series of temperature or wind are characterized by high dimensional data with more than $p = 132\ 860$ dimensions for each point of the grid, observed over the period 2001–2014. A clustering method provides groups of homogeneous observations. In this case, appropriate representations are especially required before using segmentation algorithms, to avoid the curse of dimensionality and to speed up time computation.

A natural pre-processing, to reduce the size of the time series frequently used consists in averaging raw data to a given scale, for example to a day, a week, a month or a year scale. Averaging leads also to a natural *smoothing* of the initial raw data.

An alternative and smart method to reduce the raw inputs is to model the time series by using a nonparametric regression model associated with a dictionary of atoms [18, 19]. In this case, the choice of an appropriate dictionary may provide a sparse representation allowing to represent the original signals using only few coefficients

[1, 13]. The atoms of the dictionary can be chosen using transforms as the Fourier or the Wavelets transforms or directly learned from the data using for example the well-know Karhunen–Loeve, leading to so the called Principal Component Analysis decomposition (PCA) -see for instance [17]- or the Non negative Matrix Factorization (NMF) decomposition [11].

Representing the initial signals using fixed or data-driven atoms induces, as for temporal aggregation, a natural *smoothing* of the data which may also lead to different results of clustering. This work will especially study the impact of smoothing on the segmentation of the meteorological time series: different methods of smoothing are investigated to extract, at each point of the grid, salient features and/or to model the temporal signal.

It is well known that climatic conditions fluctuate from one year to another. In order to highlight the inter-annual variability of homogeneous meteorological regions, we study in this paper successive segmentations, using each time one year of data. Based on our dataset, 14 segmentation instances are consequently computed, one for each year. This allows us to evaluate the stability and robustness of the segmentations over the years.

Finally, in order to provide a global map of homogeneous regions over the years, we propose to aggregate this set of clustering instances using a graph partitioning approach based on spectral clustering. This method allows us to take into account the inter-annual variability of the homogeneous regions in the final segmentation.

This document is organized as followed. The following section presents the initial meteorological data. Section 5.3 presents the different methodologies for extracting the features. The methods for clustering and for aggregating the set of clusters are presented in Sect. 5.4. Numerical results computed on the French meteorological time series are discussed in the last section.

5.2 The Meteorological Data

Meteorological data are displayed on a regular grid with one single point every 50 km, both in latitude and longitude. The original grid covers all Europe with 6035 grid points. In this study, the 259 grid points which are located in metropolitan France and represented in Fig. 5.1 are used. There are no meteorological stations at the regular locations of the grid points, so our data is in fact an output of a meteorological model. The model combines real meteorological observations at geographically irregular locations to obtain what would have been the meteorology at a set of geographically regular locations as a grid. This operation is called “assimilation”. The model used here is Arpège Climat from Météo France. Temperature data correspond to temperature at 2 m high, in Celsius. Wind data correspond to wind at 100 m high, and is composed of two components (South–North and West–East), from which the wind speed is computed in m/s. The interest of such meteorological model is to work with coherent and regularly spaced data for both wind and temperature.

Over the period 2001–2014, each temperature or wind signal is characterized by 132 860 hourly observations. On an interval of one year, each time series is then

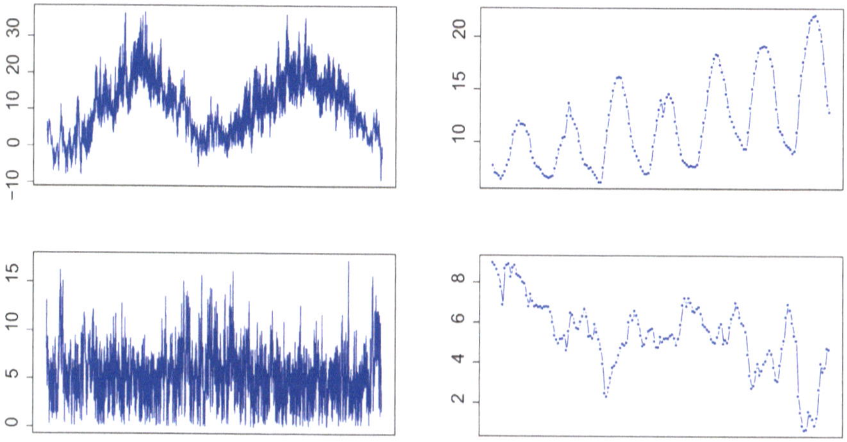

Fig. 5.2 Temperature (top) and wind (bottom) for 2 years of data (left) and for one week extracted from the Arpège model

characterized by a vector of $p = 8760$ observations (for 365 days). For illustration, Fig. 5.2 shows 2 years of hourly data from 2013 to 2014 and a zoom of a week in May 2014. Year and day periodicity can clearly be observed on the temperature time series. For wind data, it should be stressed that no evident periodicity can be observed at any scale.

5.3 Smoothing and Feature Extraction

Statistical modeling brings tools for generating descriptions of the observed data and our first work focuses on extracting salient features from the raw data. In our first approach, we model the temporal signals using a functional approximation based on a dictionary of functions. Two kinds of dictionaries are studied here: dictionaries with generic atoms and dictionaries with data-driven atoms. Theoretically, a comparison between these two approaches could have been provided with the minimum description length framework which discriminates between competitive models based on the complexity of each descriptor, see for instance [20]. Here, we focus on the practical comparison of both methods using the meteorological time series.

5.3.1 Functional Approximation

We consider that the original data are function of time observed at regular intervals. For each time series at location i, we observe $(Y_j^i, j/p)$, $1 \leq j \leq p$ where

$$Y_j^i = f^i(j/p) + \varepsilon_j^i.$$

The functions f^i are considered as unknown and the ε_j^i's are Gaussian random variables $\mathcal{N}(0, (\sigma^i)^2)$, where σ^i is the standard deviation of the noise of each signal at location i, $1 \leq i \leq n = 259$. We propose to estimate each f^i in a nonparametric way using a dictionary of d functions $\mathcal{D} = \{g_1, \ldots, g_d\}$ and write:

$$f^i = \sum_{l=1}^{d} \beta_\ell^i g_\ell + h^i$$

The Ordinary Least Square (OLS) allows to estimate the coefficients β^i of the model by minimizing $||Y^i - G\beta^i||^2$ where G is the design matrix, defined by the column vectors of the functions of the dictionary evaluated at times j/p and Y^i is the $(p \times 1)$ vector of meteorological observations for the spot number i.

Adaptive Choice of the Meaningful Coefficients

When the matrix $(G^T G)$ is invertible, the estimated coefficients $\hat{\beta}$ computed with OLS are then simply $\hat{\beta}^i = (G^T G)^{-1} G^T Y$. Without precisely knowing the level $(\sigma^i)^2$ of the noise, it is a bit hard to select the main basis functions needed to estimate f^i. In this work, we chose to select these functions with the help of a criterion which evaluates the quality of the approximated signal. This criterion is defined by the ratio between the ℓ_2 norms of the approximated and initial signal. A classical forward variable selection helps select the smallest set of functions of the dictionary which lets a ratio up to a threshold T. During an iterative procedure, the unitary functions of the dictionary are added one after another in the different models, testing at each step whether the added function significantly improves the fit. If we note $\hat{Y}^i_{j_0} = \sum_{j=1}^{j_0} \hat{\beta}^i_{(j)} g_j$ and $\hat{\beta}^i_{(1)}, \ldots, \hat{\beta}^i_{(j_0)}$ the j_0 largest coefficients in absolute value $|\hat{\beta}^i_{(1)}| \geq \cdots \geq |\hat{\beta}^i_{(j_0)}|$, the procedure stops when the reconstruction is considered as "satisfactory" i.e. if $||\hat{Y}^i_{(j_0)}||^2/||Y^i||^2 \geq T_{NP}$ where T_{NP} denotes the threshold for this nonparametric approach.

It should be underlined that the choice of T_{NP} and of the dictionary of functions $\mathcal{D}$ has a strong impact on the number of selected coefficients and consequently on the sparsity of the representation and on the nature of the approximation. For a same value of the threshold T_{NP}, different values of sparsity (and estimated coefficients) may emerge depending on the chosen functions of the dictionary. This leads to distinct reconstructions. Considering two different dictionaries, even if the number of non zero selected coefficients is similar, the approximated signals computed with each dictionary are different (or slightly different depending on the threshold value T_{NP}). It should be stressed that the approximations of the signals are non linear and computed independently at each localization. There is therefore no evidence that the non zero coefficients for all approximated signals (for the different spots) share the same support and consequently the same subspace.

Restrained Support Choice for the Meaningful Coefficients

In order to provide at the same time a sparse representation for all the signals and a common support for the coefficients, we introduce a second method hereafter described. The first step consists in estimating the coefficients β^i of each signal i, $1 \leq i \leq n$ using the standard OLS procedure. For each atom of the dictionary, we compute an index, $\kappa_j = \sum_{i=1}^{n} (\hat{\beta}_j^i)^2$ to quantify the amount of energy brought by this atom over all the signals of the data base. The κ_j indices are then ordered by decreasing order $\kappa_{(1)} \geq \cdots \geq \kappa_{(d)}$. We select the size, J_0 of the common supports by taking the minimum number of coefficients κ_j for which the following ratio exceeds a quantity: $\frac{\sum_i \|\hat{Y}^i_{(j)}\|^2}{\sum_i \|Y^i\|^2} \geq T_{all}$. This ratio quantifies, in average, the amount of energy kept in the approximation with the constraint of a common support between all the approximations. T_{all} denotes the threshold for this common support approach. Section 5.5 studies the impact of these different representations, using a free decomposition or a decomposition with a common constrained support on the segmentation of the French meteorological data.

5.3.2 *Data Driven Dictionary*

Principal Component Analysis (PCA) is a method widely used for feature extraction before curve segmentations. For example, in [9], principal components are introduced to model sparse functional data in biology. Let us recall that the PCA algorithm consists in producing for a data set where each data point is in $\mathbb{R}^p$, some axes on which to project the data points to best capture their variability. It becomes rather intuitively clear that if the data set stores the observations of different populations, the first PCA axes will probably be adapted to principally describe the differences between these populations. PCA operates a linear smoothing when retaining a smaller number of axes which may be interpreted as adaptive dictionary vectors and is also a feature selection method. PCA clustering consists in processing the data with a PCA algorithm, retaining the number of principal axes in such a way that for the threshold $T_{PCA} = 95\%$ of the variance is kept.

5.3.3 *Kernel-PCA*

However one major disadvantage of PCA is to consider only linear transformations of the data. Obviously, linear methods are not accurate enough to handle typical situations for instance where the data are sitting on a regular but not linear manifold of small dimension. In this case, the idea consists in 'featuring' the data i.e. transforming the data set with the help of 'feature'-functions $\varphi_l(Y^i)$, $l = 1, \ldots k$ [12]. Finding these transformations is not always an easy task. A useful technique is the 'kernel-trick' due to Vapnik [28], consisting in introducing a new kernel $K(Y^i, Y^j)$ measuring the disparities between signals. The theory of Mercer kernels [6] shows that if K is precisely a Mercer kernel, then a space and a function φ (called feature space and function) exist such that K is the matrix of scalar products $\langle \varphi(Y^i), \varphi(Y^j) \rangle$. A standard PCA on this matrix $K(Y^i, Y^j)$, transforms the implicit features φ into 'principal features' $\varphi_1(Y^i), \ldots, \varphi_k(Y^i)$, which are explicit.

Kernel clustering consists hence to process the data with a PCA starting from the matrix $K(Y^i, Y^j)$. Examples of kernels currently producing very accurate results are the following (see [22]).

$K(x, x') = \langle x, x' \rangle^d$
$K(x, x') = tanh(\kappa \langle x, x' \rangle + \theta)$
$K(x, x') = \exp(-\|x - x'\|2/\mu)$.

Each kernel definition involves regularisation parameters as d, θ and μ. In the class of kernels presented above, the last one $K_L(x, x') = \exp(-\|x - x'\|2/\mu)$ plays a special role because of its meaningful mathematical interpretations, and the so called heat kernel clustering consists in adding a smoothing and a normalizing step to the steps mentioned above (see the Spectral Clustering in the next section).

5.3.4 Temporal Aggregation

Alternatively, a simple and natural pre-processing to reduce the size of the time series inputs is to average the raw data at a given scale. Most of the times, averaging erases the noise in the data but may also erase specific patterns at the same time. Daily average, for example, which decreases the size of the raw inputs of a factor of 24, erases also potential day/night meteorological fluctuations. Yearly average gets also rid of annual fluctuations and compresses the data of a factor 365. Weekly and monthly averages are also scales commonly used to study the weather but are not directly linked to natural meteorological fluctuations. Four levels of temporal aggregation are studied in this project corresponding to raw hourly, daily, weekly or monthly data. Considering one year of data with 365 days, each point of the grid is respectively represented with $p = 8760$ (hourly), $p = 365$ (daily), $p = 52$ (weekly) or $p = 12$ points (monthly).

5.4 Segmentation

Many methods of clustering are available for segmentation: among them, hierarchical clustering [32], k-means [7], spectral clustering [29] or model based clustering [3]. All these algorithms are often used for many applications and we briefly recall the advantages and disadvantages of each of the first three methods which are used in our application. More details of each method can be easily found in the literature as for example in [5, 8, 14].

5.4.1 Clustering Algorithms

Hierarchical clustering (HC) is a very common algorithm for clustering data [5]. HC produces a nested series of partitions successively created by a bottom-up or a top-down approach. The bottom-up approach begins with each observation in a distinct singleton cluster and successively merges clusters together until a stoping criterion is satisfied. Many variants of HC clustering exist and differ in the way they characterize the similarity between a pair of clusters. The choice of the merging criterion induces different results of clustering. In the single-link method, the distance between two clusters is the minimum of the distances between all pairs of elements drawn independently from the two clusters. In the complete-link algorithm, the distance between two clusters is the maximum of all pairwise distances between elements in the two clusters. The clusters obtained by the complete-link algorithm are more compact than those obtained by the single-link algorithm. The Ward criterion is frequently used as merging criterion and let to obtain balanced clusters. The HC algorithm produces a dendrogram and does not need any assumption on the number of groups before computation. The number of groups is chosen afterwards (see Sect. 5.4.3). Due to the computation of the distance matrices for all observations and for a large number of nested partitions, the HC algorithm is numerically costly, especially for high dimensional data.

The **k-means** algorithm is the simplest and most commonly used algorithm to partition data. In contrast with the HC which produces a clustering structure, **k-means** partitioning has the advantage of being less complex, particularly in applications involving large data set or in high dimension data set for which the construction of a dendrogram is computationally prohibitive. The k-means algorithm produces clusters by optimizing a criterion function, which is frequently chosen as the squared error criterion. It is an easily implementable and interpretable (as MLE estimation) procedure which aims at producing a local minimum to the following problem: Given the number k of clusters (we'll discuss this point later), one seeks for a partition $C_1, \ldots, C_k$ of $\{1, \ldots, n\}$ minimizing:

$$min_{C_1,\ldots,C_k} \sum_{l=1}^{k} \sum_{i,j \in C_l} \|Z^i - Z^j\|^2.$$

In our application, Z^i may be the raw data Y^i, or the projection coordinates after PCA or the adaptive coefficients computed after the nonparametric estimation procedure, at a given time scale (hour, day...) as well as other results of pre-processing feature extractions.

As so defined, the minimization problem is exponentially difficult. However, seeking for a local minimum to this problem is much easier. First, if for each group l, $m(C_l)$ is the centroid of the candidate cluster C_l, the minimization problem reduces to find the optimal partition $C_1, \ldots, C_k$, for the following problem:

$$min_{C_1,\ldots,C_k} \sum_{l=1}^{k} \sum_{i \in C_l} \|Z^i - m(C_l)\|^2.$$

The algorithm starts with a random affectation of the points into classes $C_1, \ldots, C_k$ and then returns at each new step a centroid to each 'cluster' and re-affects each data point according to its proximity to the centroid, then re-adjusts the centroid, and repeats the operation until stabilisation. This search naturally reduces at any steps the quantity $\sum_{l=1}^{k} \sum_{i \in C_l} \|Z^i - m(C_l)\|^2$, producing a local minimum.

In practice, the algorithm is therefore, typically run multiple times with different starting states, and the best configuration obtained from all the runs is used as the output clustering.

In our case, we chose to run the k-means algorithms to compute the clustering of each segmentation instance 10 000 steps for 100 initial random configurations in order to attempt to reach the global minimum of the error criterion. The squared error criterion, mostly used in the k-means, might not be appropriate for high dimensional data. We compensate this by a pre-processing of pattern extraction or sparse representation of the data as described in the previous section. As mentioned, in the k-means algorithm, the number of clusters k has to be chosen in advance before running the algorithm. The choice of the optimal number of clusters is detailed in Sect. 5.4.3.

Spectral clustering (SC) considers the initial observations, $Z^1, \ldots, Z^n$ ($n = 259$), as the nodes of a graph [16, 29] where Z^i represents a feature vector for location i in $\mathbb{R}^p$. An affinity matrix A is associated to this graph, and SC algorithm uses the spectrum of this matrix to perform the clustering. Many variants of this algorithm may be found in the literature [23]. The affinity matrix, A, is a $n \times n$ matrix which both defines the architecture of the graph (absence or presence of a link between two nodes) and the strength of the relation between two nodes i and j, represented by a weight w_{ij}, $1 \leq i, j \leq n, i \neq j$. In the graph literature, different methods are introduced to put an *edge* between two nodes i and j of the graph [29].

For the ε *neighborhood* method, two nodes i and j are connected by an edge if the Euclidean distance $||Z_i - Z_j||$ is lower than a chosen threshold ε.

In the B *nearest neighbors* method, two nodes i and j are connected by an edge if i is among the B nearest neighbors of j and j is among the B nearest neighbors of i. The case $B = n$ provides a fully connected network. In this case, the presence/absence of an edge depends on the relative proximity between nodes i and j and not on the units of Z_i and Z_j, as in the previous case.

Both previous methods lead to symmetric graphs and depend on the values of the tuning parameters (ε or B). After defining the architecture of the graph (existence of a link or not), the affinity matrix A is constructed by attributing weight values to each edge.

For the *Simple-minded* graphs, $w_{ij} = 1$ if and only if nodes i and j are connected by an edge.

For the *Heat Kernel* graphs, the connection between two nodes i and j is evaluated by a weight:

$$w_{i,j} = e^{\frac{-\|z_i - z_j\|_2^2}{\mu}}$$

with $\mu > 0$. Again, the choice of the value of μ may have a strong impact on the results of the algorithm. Using previous notations, the normalized Laplacian graph is computed: $L = I - D^{-1/2}AD^{-1/2}$ where D is the diagonal degree matrix with $D_{i,i} = \sum_j w_{i,j}$. Assuming that the number k of clusters is known, the k largest eigenvectors of the Laplacian matrix are used to group the data into classes [16]. Let A be the adjacency matrix, $A \in \mathbb{R}^{n\times n}$, k the given number of clusters, the following steps defines the SC algorithm:

1. Compute the normalized Laplacian L.
2. Compute the first k eigenvectors $u_1, \ldots u_k$ of L corresponding to the k largest eigenvalues, $\lambda_1, \ldots, \lambda_k$ of L,
3. Extract the matrix $T \in \mathbb{R}^{n\times k}$ from U, where $U \in \mathbb{R}^{n\times k}$ is the matrix containing the vectors $u_1, \ldots, u_k$ as columns, by normalizing the rows to unitary norm such that $t_{i,j} = u_{i,j}/(\sqrt{\sum_{l=1}^{k} u_{il}^2})$
4. Cluster the points (t_i), $1 \leq i \leq n$ with the k-means algorithm into clusters $A_1, \ldots, A_k$ where $t_i \in R^k$ is the vector corresponding to the ith row of T.

Regarding a given set of observations, the SC algorithm needs to first compute the affinity matrix, and to specify the number k of clusters. The affinity matrix depends both on the architecture and the weights of the graph.

The parameters of the architecture (ε neighborhood, nearest neighbors) and of the graph (simple minded or heat kernel) are critical choices which have a strong impact on the computation of the clusters. However, when the affinity matrix is well fixed, the SC algorithm is a very powerful tool to cluster the different nodes of the graph. In this application, the SC algorithm is used for the aggregation of year by year clustering results (see the following section).

5.4.2 *Aggregation of Clustering Instances*

As already mentioned in the introduction, a natural study interval for partitioning the climate of the French territory is the year. On the basis of the available information in our data base, this study provides 14 clustering instances for the entire period of analysis. Aggregation of clustering instances may be in this case particularly welcome to differentiate the couples of spots which tend to be always classified in the same group from those which tend to be systematically classified in different groups. In order to aggregate the clusters, we construct for each instance l of clustering, $1 \leq l \leq 14$, a co-cluster indicator matrix D^l defined by $D^l_{ij} = 1$ if the nodes i and j are in the same cluster for year l and $D^l_{ij} = 0$ otherwise as already introduced in [33], $1 \leq i, j \leq n$. Hence in this application, D^l is a $n \times n$ symmetric matrix with $n = 259$ spots.

$D = \frac{1}{14}\sum_l D^l$ corresponds to the average of the co-cluster indicators over the complete period for 14 years of data. $D_{ij} = 1$ (resp. $D_{ij} = 0$) means that both nodes i and j tend to be classified always (resp. never) in the same cluster over 14 years.

An affinity matrix is naturally defined by $A = I_n - D$ where I_n is the $n \times n$ identity matrix. The SC algorithm is, in this case, very appropriate to compute the final clustering based on A.

5.4.3 Calibration of the Number of Clusters

A major challenge in cluster analysis is the estimation of the optimal number of "clusters". During these last years, different methods have been introduced, in the literature, to identify this optimal number. Most of these methods are based on the evolution of the within W_k or the between B_k cluster dispersion, function of the number of clusters k. To compute the optimal number of clusters, [4] proposes to maximize the between over within variance ratio depending on the number of clusters. In [7, 10], the number of appropriate clusters is identified when the increase of the number of groups does not correspond to an appropriate decrease of the within variance. In [26], the gap statistics based on the within sum of squares function is introduced to compute the optimal number of clusters. A visual method, called "silhouette" is also proposed in [21] to identify the optimal number of clusters. After implementing and testing all the previous indicators on simulated data (where the number of groups is exactly known), the conclusion we have drawn is that most of these indices provide the same results in cases when there is no doubt on the "right" number of clusters, that is when the number of clusters can be easily and visually identified. In other cases, these indices may provide different results but, in these situations, it is also hardly obvious to know the exact answer.

In this work, we decided to take a modeling point of view introduced in [7] and to estimate the optimal number of clusters, k_0 by studying the evolution of the relative decrease of the within variance function of the number of clusters. The optimal number of clusters k_0 corresponds to the smallest number such that the ratio $\frac{W_{k+1}-W_k}{V}$ is smaller than 5% where V is the total variance of the set of data. The choice of the value 5% corresponds to a relative decrease of the total variance provided by the increase of the number of groups from k to $k+1$. The value of 5% is here chosen accordingly to the previous threshold values already introduced to approximate the signals in Sect. 5.3.

5.5 Numerical Results

This section presents the results of the French climate segmentation carried on the available 14 years of hourly Arpège data of temperature and wind.

5.5.1 *Smoothing and Sparsity*

Tables 5.1 and 5.2 provide the number of non zero estimated coefficients (sparsity) obtained using a PCA data-driven dictionary and a nonparametric signal regression model with a generic dictionary (Fourier and Haar basis) respectively for temperature and wind. The sparsity is studied for adaptive or constrained support (mentioned with a (*) in the tables). The thresholds (T_{NP}, T_{all}, T_{PCA}) chosen to compute the approximated signals take the same value $T = 0.95$ for all approaches. The interval of study equals one year with different temporal scales (hour, day, week, month). The sparsity results are averaged over the 14 years and the standard deviation is mentioned.

Both for temperature and wind data, the PCA decomposition provides the highest sparsity for all scales. For temperature data, the compression rate obtained with PCA is very high with a ratio equal to 381 for hourly data, 36 for daily data, 13 for week data and 5 for month data (8760/23; 365/10.21; 52/4.07, 12/2.36). For wind data, the compression rate observed for the PCA decomposition is much smaller with ratio equal to 134 for hourly data, 14 for daily data, 5 for week data and 3 for monthly data (8760/65.35; 365/26.4; 52/10.64; 12/4.21). At a month scale, the compression ratio is similar for temperature and wind data.

For temperature data at the hour scale, the non-parametric model with a Fourier dictionary shows as expected a higher sparsity compared to the Haar decomposition.

Table 5.1 Temperature. Sparsity average (and standard deviation) computed for an interval study of one year for different smoothing methods from 2001 to 2014

Temporal scale	Hour	Day	Week	Month
p	8760	365	52	12
PCA	23 (2.5)	10.21 (1.05)	4.07 (0.26)	2.36 (0.49)
Fourier	637 (93)	126 (3.4)	18.5 (0.78)	3.58 (0.33)
Haar	976 (87)	133 (3.9)	18.34 (0.89)	4.76 (0.33)
Fourier*	761 (62)	161.21 (9.61)	23.42 (2.44)	3.35 (0.63)
Haar*	1172 (67)	149.85 (11.85)	17.3 (2.37)	4.14 (0.53)

Table 5.2 Wind. Sparsity average (and standard deviation) computed for an interval study of one year for different smoothing methods from 2001 to 2014

Temporal scale	Hour	Day	Week	Month
p	8760	365 (0)	52 (0)	12 (0)
PCA	65.35 (25.6)	26.4 (5.9)	10.64 (1.08)	4.21 (0.57)
Fourier	937 (257)	176 (2.9)	23.72 (0.9)	4.49 (0.35)
Haar	1233 (218)	197 (4.2)	23.82 (1.02)	5.01 (0.36)
Fourier*	1303 (342)	259 (0)	40 (0)	7.78 (0.57)
Haar*	1878 (359)	259 (0)	37.35 (1.82)	6.85 (0.36)

This is a consequence of the periodic temporal oscillation of the temperature: night/day and seasonal oscillations. For wind data, the generic dictionary does not bring any interesting sparsity for all scales. From a sparsity point of view, PCA appears to be very interesting to represent temperature or wind data. At small scales, from a sparsity point of view, the use of a data-driven dictionary or the non-parametric model seems to be more or less equivalent, especially for wind data.

Comparing temperature and wind approximation for all implemented methods shows that temperature time series let to the highest level of sparsity.

It is important to notice also that if PCA gives the highest sparsity (it is not surprising since it is the aim of the method), it lacks stability over the years. Hence it gives a very powerful tool for insight on the data, but a much less interesting instrument in a forecast perspective.

5.5.2 *Number of Clusters*

Tables 5.3 and 5.4 provide the number of clusters computed using the criterion on the decrease of the within variance (function of the number of clusters) at a threshold of 5% (see Sect. 5.4.3), for temperature and wind data, for different scales and for an interval study of one year. For the k-means or Hierarchical clustering method, the clustering is performed using the features (coefficients) computed with the previous decompositions (PCA, non-parametric models with Fourier and Haar dictionaries using an adaptive or constrained (*) support). The results are averaged over the 14 years and the standard deviation is mentioned.

For both temperature and wind data, we surprisingly observe that, for each scale, the average number of clusters obtained with the different smoothing methods is globally the same. It means that the various representations do not induce an important variation of the final number of clusters on the studied periods. The number of clusters decreases slightly with the scale. For temperature, more clusters are obtained with hourly data (5 to 6 clusters regarding all methods) than with monthly data (less than 5 groups in general).

Table 5.3 Temperature. Average number of clusters computed with the Kmeans algorithm for different smoothing methods from 2001 to 2014

Scale	k means				Hierarchical clustering			
	Hour	Day	Week	Month	Hour	Day	Week	Month
Raw data	5 (0)	5 (0)	4.9 (0.3)	4.8 (0.4)	5.4 (1.7)	5.9 (1.1)	6 (1.0)	5.2 (0.8)
PCA	5.1 (0.3)	5 (0)	4.9 (0.3)	4.8 (0.4)	5.8 (1.3)	6.3 (1.8)	5.9(1.0)	5.2 (0.9)
Fourier	5.0 (0)	5 (0)	4.8 (0.4)	4.8 (0.4)	6.1 (1.4)	5.8 (0.9)	5.6 (1.0)	5.8 (0.8)
Haar	4.8 (0.4)	5 (0.4)	4.8 (0.4)	5.0 (0.4)	5.6 (1.4)	6.6 (0.9)	5.7(0.8)	5.5 (1.0)
Fourier*	5 (0)	5 (0)	4.8 (0.4)	4.8 (0.4)	6.0 (1.9)	5.5 (1.0)	5.6 (0.8)	6.6 (1.3)
Haar*	5 (0)	5 (0)	4.8 (0.4)	4.6 (0.5)	6.0 (1.8)	5.3 (1.0)	5.5 (0.5)	5.4 (0.9)

Table 5.4 Wind data. Average number of clusters computed with the k-means and the hierarchical clustering algorithm for different smoothing methods from 2001 to 2014

Scale	k means				Hierarchical clustering			
	hour	day	week	month	hour	day	week	month
Raw data	4 (0)	4.1 (0.4)	4.2 (0.4)	4.2 (0.4)	4.3 (0.8)	4.2 (0.6)	4.5 (0.8)	4.9 (1.2)
PCA	4 (0)	4.2 (0.4)	4.3 (0.5)	4.3 (0.5)	4.2 (0.4)	4.5 (0.8)	5.1 (1.4)	4.7 (1.0)
Fourier	4 (0)	4.1 (0.3)	4.1 (0.4)	4.2 (0.3)	4.1 (0.3)	4.1 (0.3)	4.9 (1.4)	4.6 (0.8)
Haar	4 (0)	4.1 (0.3)	4.1 (0.3)	4 (0)	4.2 (0.4)	4.4 (0.7)	4.6 (0.7)	4.5 (0.7)
Fourier*	4.1 (0.4)	4.1 (0.4)	4.2 (0.4)	4.3 (0.5)	4.0 (0.5)	4.2 (0.6)	4.7 (1.0)	3.9 (0.6)
Haar*	4.0 (0)	4.2 (0.4)	4.3 (0.5)	4.1 (0.4)	4.1 (0.3)	4.4 (0.7)	4.6 (0.7)	4.4 (0.6)

For wind data, the number of clusters is the same for all methods. The level of sparsity i.e. the number of non zero coefficients used to represent the data, does not have any influence on the final results. However, it should be stressed that the number of coefficients used to represent the data has a strong influence on the computation time needed to provide all the results.

5.5.3 *Meteorological Segmentation Maps*

Figures 5.3 and 5.5 (resp. Figures 5.4 and 5.6) illustrate the segmentation maps of the 14 years using the k-means algorithm (resp. HC) drawn for temperature and wind data, using an interval study of one year. For these figures, the selected features are, in both cases, computed with PCA using a threshold $T_{PCA} = 0.95$. It should be underlined that the two-step methods previously described, first step of basis expansion followed by step of clustering, does not introduce any topological constraints on the $n = 259$ spots for the clustering: for instance the GPS coordinates of the grid points do not appear at any stage of the procedure. However, we observe that neighboring points of the grid tend to be classified in the same cluster. For temperature data, the number of clusters found is around $k = 5$ for all the 14 years (Figs. 5.3 and 5.4). The computed maps highlight principally four main and connected regions of homogeneous temperature localized in the four quadrant of France respectively the north-west, the north-east, the south-west and the south-east. The last cluster corresponds to the mountains areas (Vosges, Massif Central, Alps and Pyrenees) localized in different disconnected regions.

For wind data, the number of clusters found is slightly lower than for temperature data and varies from 4 (mainly) to 5. As for temperature, the four computed cluster maps correspond to connected and oriented area, with a north-east/south-west orientation (Figs. 5.5 and 5.6).

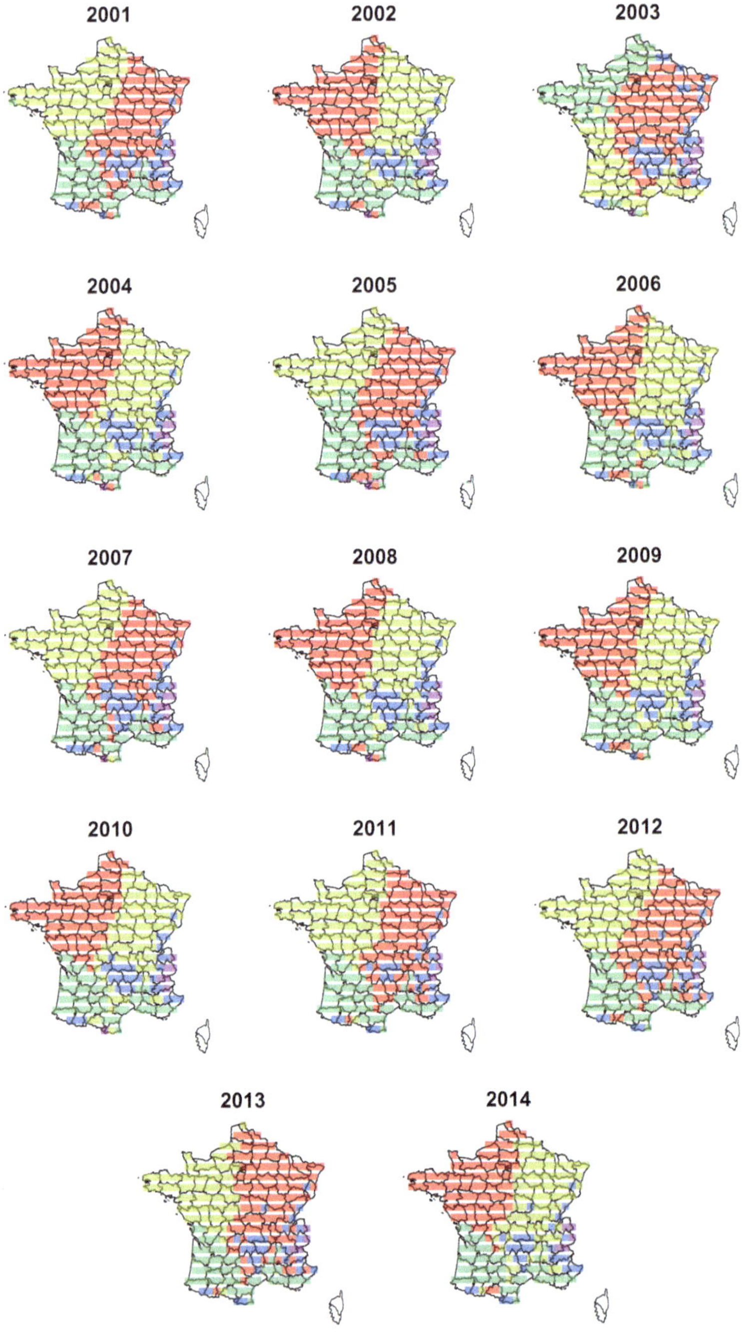

Fig. 5.3 Temperature segmentation maps (kmeans) with daily data 2001–2014

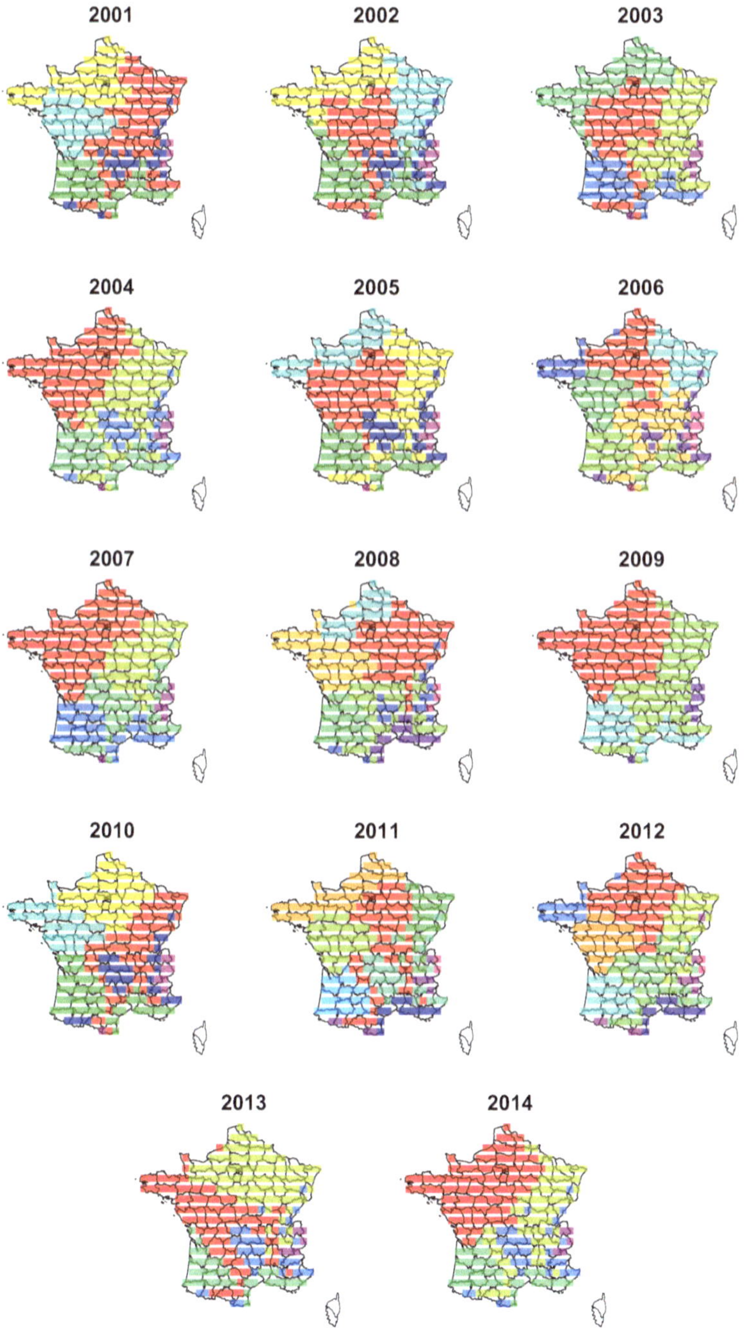

Fig. 5.4 Temperature segmentation maps (HC) with daily data 2001–2014

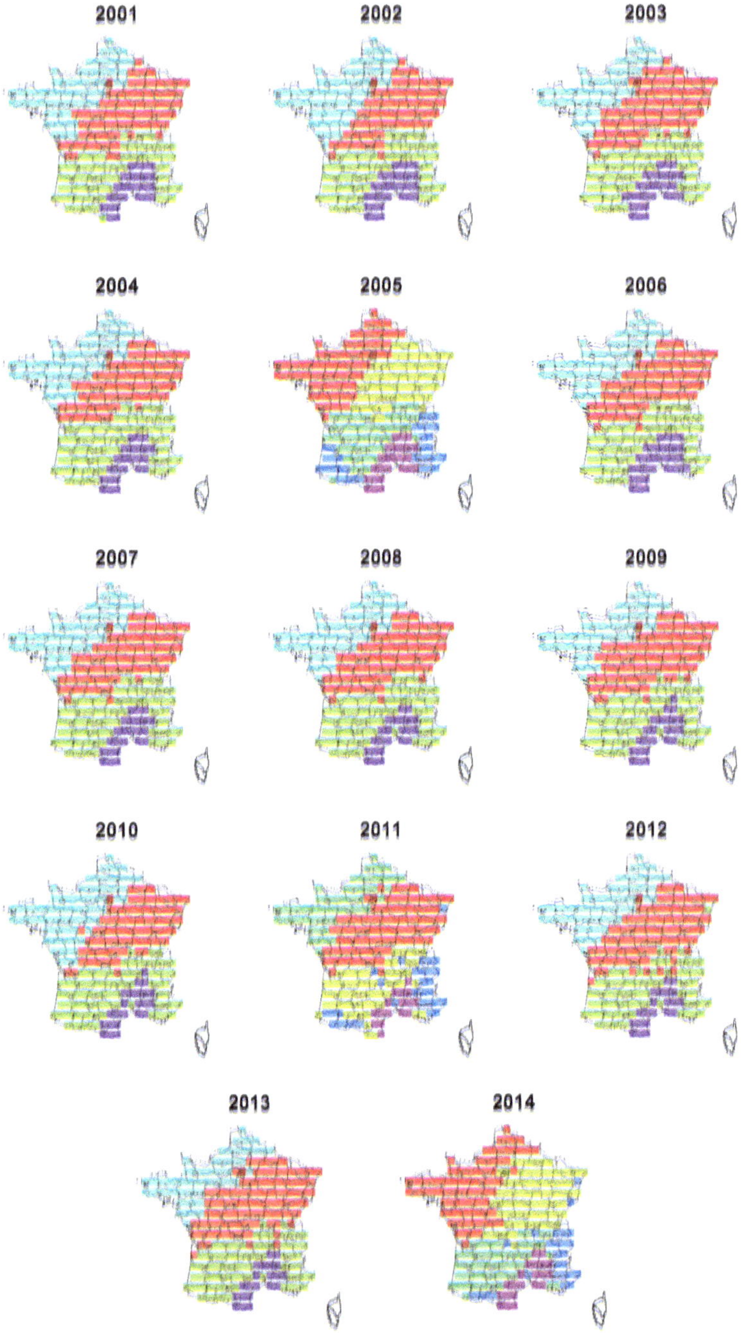

Fig. 5.5 Wind segmentation maps (k-means) with daily data 2001–2014

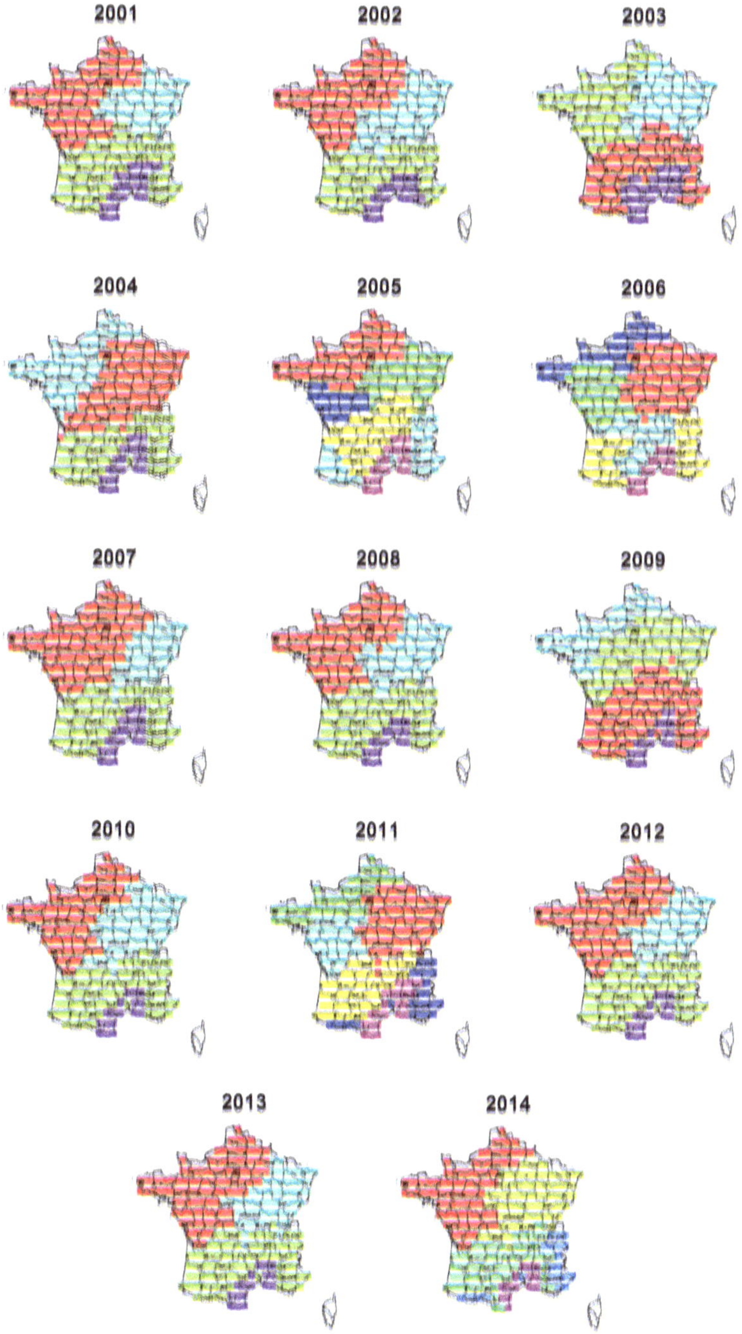

Fig. 5.6 Wind segmentation maps (HC) with daily data 2001–2014

5.5.4 Clustering Aggregation

Even if the number of clusters seems to be stable over the whole period 2001-2014 for temperature and wind data, the repartition of the points in the different groups varies from one year to another. These variations can be quantified by studying the co-cluster indicator matrix D already described in Sect. 5.4. As an illustration, Fig. 5.7 shows the values of D for three specific points of the grid, Brest (West of France, $i = 46$), Dijon (East of France $i = 110$) and Toulouse (South of France $i = 238$), localized with a red star on the different maps. In this example, the co-cluster indicator matrix is computed for temperature and wind using the k-means algorithm with PCA and daily time series. We decided to take a very conservative point of view by representing only the points j such that $D_{ij} = 1$ (with a red cross) and $D_{ij} = 0$ (with a blue dash). Recall that $D_{ij} = 1$ means that both points i and j are always in the same cluster during all 14 years and $D_{ij} = 0$, the points i and j are never in the same cluster, $1 \leq j \leq 259$.

We observe, in Fig. 5.7, that for these three specific cities, neighborhood points of the grid tend to be in the same cluster, throughout all the studied period (2001–2014). This is also true for all points of the grids which are spatially localized in the 'center' of a cluster region. The points of the grid which are localized in the frontiers between two regions tend to belong to different clusters from one year to another. As an example, Toulouse is a such an unstable point for wind classification.

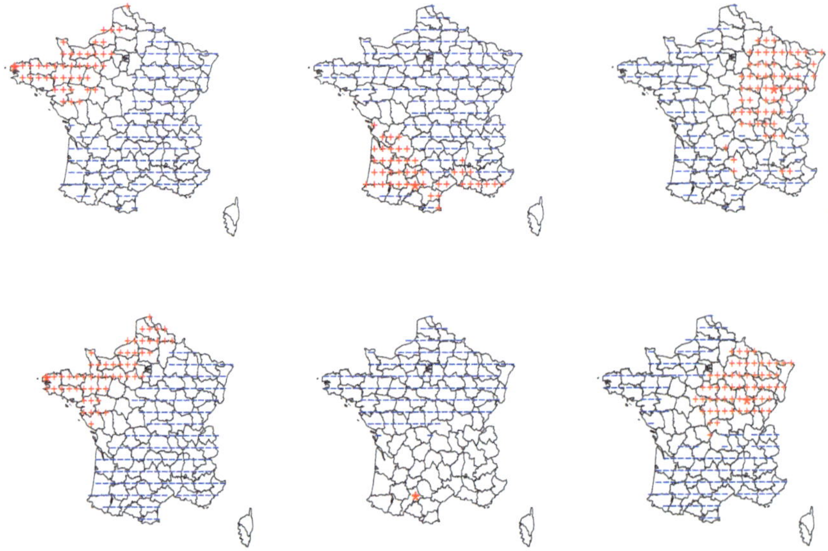

Fig. 5.7 Visualization of the 1 and 0 values of the co-cluster indicator matrix for Brest (left), Toulouse (middle), Dijon (right). $C_{ij} = 1$: red '+'; $C_{ij} = 0$: blue '−' for temperature (top) and wind (bottom)

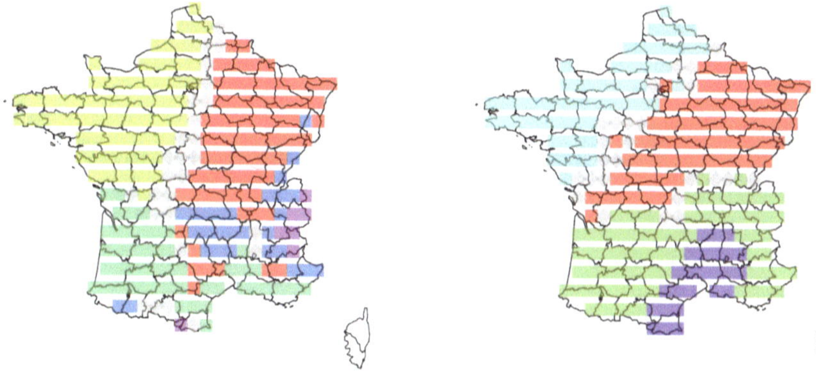

Fig. 5.8 Aggregation of 14 clustering instances using the spectral clustering for temperature (left) and wind (right) for daily data

The spectral clustering algorithm is in this case very appropriate to perform an analysis on the indicator matrix D. The nodes of the graph are defined by the points of the grid. To define the architecture of the graph, we decided to put an edge between two nodes i and j if they tend to be classified most always in the same way: either systematically in the same cluster or in different clusters, throughout the 2001–2014 period, i.e. $D_{ij} \geq (1 - \varepsilon)$ or $D_{ij} \leq \varepsilon$, ε is a tuning parameter which allows a weak variability throughout the period. As for heat kernel graphs, the connection between two nodes is defined with a weight function of the co-cluster indicator: $w_{ij} = e^{\frac{1-D_{ij}}{\mu}}$.

Figure 5.8 shows the results computed with SC for temperature and wind data on the previous data. The results obtained with the SC clustering are, in this case, very robust and do not depend on a particular choice of the μ parameter; here $\mu = 0.20$ and $\varepsilon = 0.20$. As expected, 5 main homogeneous regions appear on the map for temperature data and 4 main regions for wind data. The points colored in gray correspond to spots with a majority of unstable connections along the years where an unstable connection between two spots i and j is characterized here by $\varepsilon \leq D_{ij} \leq (1 - \varepsilon)$ with $\varepsilon = 0.20$.

This approach appears to be particularly interesting because it does not only show the regions with homogeneous temperature or wind but also the frontiers of variability between those regions corresponding to the inter-annual climate variations. In this case, the computed homogeneous regions can be interpreted as confidence intervals of "robust regions" with climatical similarities (for temperature or wind).

Notice also that computing only one clustering instance using the entire time series would have been, in this case, much less informative for inter-variability highlighting.

Comparing the final results computed with unsupervised methods as for k-means and hierarchical clustering is quite difficult. In order to characterize the different solutions, we draw the histogram of the values of the co-clustering matrix nD, with $D_{ij} > 0, i > j$ and $n = 14$. To enhance the representation and to focus on the relative

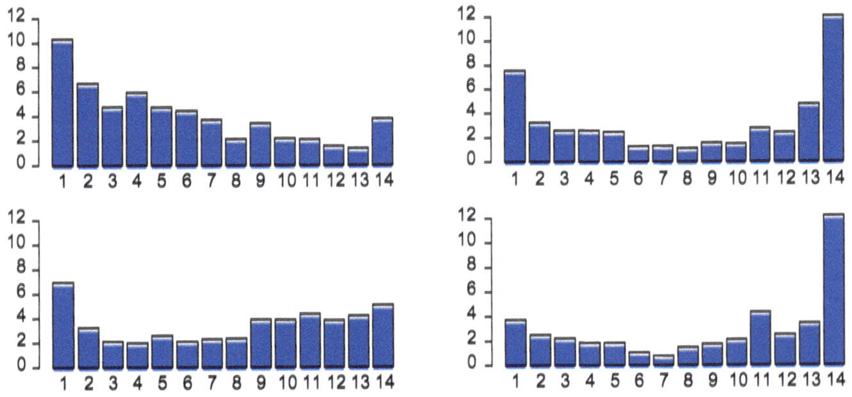

Fig. 5.9 Histogram of co-clustering indicators (x14) computed for HC (left) and k-means (right) for temperature (top) and wind (bottom) daily data

proportion of spots which tend to be classified always in the same cluster ($nD_{ij} = n$), the zero modality in not represented in the graphs for k-means or HC instances.

We observe in Fig. 5.9 that the number of spots which tend to be classified in the same cluster along the years are much higher for k-means than for HC algorithm for temperature or wind data. For the k-means clustering, 12% of the co-clustering matrix (nD) equals $n = 14$ for temperature and wind. This clearly indicates that the HC algorithm provides more variable clusters along the years as k-means does.

5.6 Conclusion

This work provides a segmentation of the French territory in homogeneous climate regions using learning methods, and more specifically clustering algorithms. Several instances of clustering (one per year) are performed to highlight the inter-annual variability of the French climate. For both temperature and wind data, we observe that the average number of clusters remains extremely stable over years. It means that, the various representations do not induce an important variation on the final number of clusters on the studied periods. Using the variability study along the different years, we clearly find that HC algorithm is more variable, much less robust than k means. A final aggregation of this set of clustering instances is provided using a graph partitioning approach based on spectral clustering. This approach allows us to give as a result a global instance of classification regions which is cautious and stable, and even highlights confidence zones.

It is remarkable that the learning techniques investigated in this study, give as segmentation results climate zones which are easily interpretable, geographically connected (except mountains regions for obvious reasons), without introduction of any hints which could lead the algorithms in this direction.

It is important to notice also that among the smoothing methods, if PCA gives the highest sparsity (it is not surprising since it is the aim of the method), it undoubtedly lacks stability. Hence it gives a very powerful tool for insight on the data, but a much less interesting instrument in a forecast perspective. Obviously more elaborated techniques such as functional representations of the data could lead to more stable and predictive classifications. A balance between sparsity and reproducibility should be investigated in this domain, in a future perspective.

Acknowledgements The authors thank RTE for the financial support through one industrial contract, LPSM for hosting our researches and Agence Nationale de la Recherche (ANR-14-CE05-0028) through the FOREWER project.

References

1. A. Antoniadis, X. Brossat, J. Cugliari, J.-M. Poggi, Une approche fonctionnelle pour la prevision non-parametrique de la consommation d'electricite. Journal de la Société Française de Statistique **155**(2), 202–219 (2014)
2. M. Bador, P. Naveau, E. Gilleland, M. Castellà, T. Arivelo, Spatial clustering of summer temperature maxima from the CNRM-CM5 climate model ensembles & E-OBS over europe. Weather Clim. Extrem. **9**, 17–24 (2015)
3. J.D. Banfield, A.E. Raftery, Model-based Gaussian and non-Gaussian clustering. Biometrics 803–821 (1993)
4. T. Caliński, J. Harabasz, A dendrite method for cluster analysis. Commun. Stat. Theory Methods **3**(1), 1–27 (1974)
5. J. Friedman, T. Hastie, R. Tibshirani, *The Elements of Statistical Learning*. Springer Series in Statistics, vol. 1 (Springer, New York, 2001)
6. M. Girolami, Mercer kernel based clustering in feature space. IEEE Trans. Neural Netw. (2001)
7. J.A. Hartigan, M.A. Wong, Algorithm as 136: a k-means clustering algorithm. J. R. Stat. Soc. Ser. C (Appl. Stat.) **28**(1), 100–108 (1979)
8. A.K. Jain, M. Narasimha Murty, P.J. Flynn, Data clustering: a review. ACM Comput. Surv. (CSUR) **31**(3), 264–323 (1999)
9. G.M. James, T.J. Hastie, C.A. Sugar, Principal component models for sparse functional data. Biometrika **87**(3), 587–602 (2000)
10. W.J. Krzanowski, Y.T. Lai, A criterion for determining the number of groups in a data set using sum-of-squares clustering. Biometrics 23–34 (1988)
11. J. Mairal, F. Bach, J. Ponce, Sparse modeling for image and vision processing. Found. Trends® Comput. Graph. Vis. **8**(2–3), 85–283 (2014)
12. S. Mika, B. Scholkopf, A. Smola, K.-R. Müller, M. Scholz, G. Ratsch, Kernel PCA and denoising in feature spaces, in *Advances in Neural Information Processing Systems 11* (MIT Press, Cambridge, 1999), pp. 536–542
13. M. Mougeot, D. Picard, V. Lefieux, L. Maillard-Teyssier, Forecasting intra day load curves using sparse functional regression, in *Modeling and Stochastic Learning for Forecasting in High Dimensions* (Springer, Berlin, 2015), pp. 161–181
14. K.P. Murphy, *Machine Learning: A Probabilistic Perspective* (MIT Press, Cambridge, 2012)
15. J. Najac, J. Boé, L. Terray, A multi-model ensemble approach for assessment of climate change impact on surface winds in france. Clim. Dyn. **32**(5), 615–634 (2009)
16. A.Y. Ng, M.I. Jordan, Y. Weiss, On spectral clustering. Adv. Neural Inf. Process. Syst. **2**, 849–856 (2002)
17. C. Radhakrishna Rao, The use and interpretation of principal component analysis in applied research. Sankhyā Indian J. Stat. Ser. A (1961–2002) **26**(4), 329–358 (1964)

18. J.O. Ramsay, B.W. Silverman, *Applied Functional Data Analysis: Methods and Case Studies*, vol. 77 (Springer, New York, 2002)
19. J.O. Ramsay, B.W. Silverman, *Springer Series in Statistics* (Springer, Berlin, 2005)
20. J. Rissanen, *Minimum Description Length Principle* (Wiley Online Library, London, 1985)
21. P.J. Rousseeuw, Silhouettes: a graphical aid to the interpretation and validation of cluster analysis. J. Comput. Appl. Math. **20**, 53–65 (1987)
22. B. Schölkopf, A.J. Smola, *Learning with Kernels: Support Vector Machines, Regularization, Optimization, and Beyond* (MIT Press, Cambridge, 2002)
23. J. Shi, J. Malik, Normalized cuts and image segmentation. IEEE Trans. Pattern Anal. Mach. Intell. **22**(8), 888–905 (2000)
24. M. Stefanon, F. D'Andrea, P. Drobinski, Heatwave classification over Europe and the Mediterranean region. Environ. Res. Lett. **7**(1), 014023 (2012)
25. Syndicat des Énergies Renouvelables, Panorama de l'électricité renouvelable en 2015. Technical report, Syndicat des Énergies Renouvelables, 2015
26. R. Tibshirani, G. Walther, T. Hastie, Estimating the number of clusters in a data set via the gap statistic. J. R. Stat. Soc. Ser. B (Stat. Methodol.) **63**(2), 411–423 (2001)
27. P.-J. Trombe, P. Pinson, H. Madsen, Automatic classification of offshore wind regimes with weather radar observations. IEEE J. Sel. Top. Appl. Earth Obs. Remote Sens. **7**(1), 116–125 (2014)
28. V.N. Vapnik, *The Nature of Statistical Learning Theory* (Springer, New York, 1995)
29. U. Von Luxburg, A tutorial on spectral clustering. Stat. Comput. **17**(4), 395–416 (2007)
30. M. Vrac, Modélisions statistiques à différentes échelles climatiques et environnementales. HDR, Université Orsay, 2012
31. M. Vrac, A. Chédin, E. Diday, Clustering a global field of atmospheric profiles by mixture decomposition of copulas. J. Atmos. Ocean. Technol. **22**(10), 1445–1459 (2005)
32. J.H. Ward Jr., Hierarchical grouping to optimize an objective function. J. Am. Stat. Assoc. **58**(301), 236–244 (1963)
33. D. Yan, A. Chen, M.I. Jordan, Cluster forests. Comput. Stat. Data Anal. **66**, 178–192 (2013)

Chapter 6
Electricity Demand Forecasting: The Uruguayan Case

Andrés Castrillejo, Jairo Cugliari, Fernando Massa and Ignacio Ramirez

Abstract The development of new electricity generation technologies has given new opportunities to developing economies. These economies are often highly dependent on fossil sources and so on the price of petrol. Uruguay has finished the transformation of its energetic mix, presenting today a very large participation of renewable sources among its production mix. This rapid change has demanded new mathematical and computing methods for the administration and monitoring of the system load. In this work we present enercast, a R package that contains prediction models that can be used by the network operator. The prediction models are divided in two groups, exogenous and endogenous models, that respectively uses external covariates or not. Each model is used to produce daily prediction which are then combined using a sequential aggregation algorithm. We show by numerical experiments the appropriateness of our end-to-end procedure applied to real data from the Uruguayan electrical system.

Keywords Electricity demand forecast · Time series · Sequential aggregation

---

A. Castrillejo · F. Massa
IESTA, Fac. CCEE, Universidad de la Republica, Eduardo Acevedo, 1138 Montevideo, Uruguay
e-mail: andres@iesta.edu.uy

F. Massa
e-mail: massa@iesta.edu.uy

J. Cugliari (✉)
Université de Lyon, 2 ERIC EA, 3083 Lyon, France
e-mail: jairo.cugliari@univ-lyon2.fr

I. Ramirez
IEE, Fac. de Ingeniería, Universidad de la Republica, Av. Julio Herrera y Reissig, 565
Montevideo, Uruguay
e-mail: nacho@fing.edu.uy

P. Drobinski et al. (eds.), *Renewable Energy: Forecasting and Risk Management*,
Springer Proceedings in Mathematics & Statistics 254,
https://doi.org/10.1007/978-3-319-99052-1_6

6.1 Introduction

When the balance between electrical supply and demand is broken, the resulting grid damage or supply outages may affect the quality of the service. Electricity load anticipation is therefore an important task for the transmission system operator of an electrical grid as it helps to reduce the risk of this happening; larger prediction horizons (greater than one year) help to anticipate the needs on production means and distribution, while shorter ones (hours, weeks) are employed to decide the production and distribution plans. In general, more accurate predictions result in lower production costs.

Recent years have seen a diverse array of new technologies which allow generation of electrical energy from new renewable sources (*renewables* for short). This diversity increases the complexity of the scenario to be managed by the agent or agents responsible of the equilibrium balance on the electrical grid. Such is the case of Uruguay, whose plan for minimizing its dependence on fossil fuel has led to a very rapid increase in power generation from a wide array of renewables [12]. Such complex scenario calls for new, sophisticated methods for power management.

A comprehensive survey on traditional electricity load forecasting methods can be found in [32]; the reader is further referred to [10, 31] for an extensions and updates of the methods described therein. Broadly speaking, forecasting methods employ statistical techniques for capturing the salient features of the load demand (we discuss them in Sect. 6.2). The main difference is in the underlying hypothesis assumed in each case for justifying and providing validity to the proposed models. (Clearly, any systematic departure of the data from the assumed hypothesis results in a degradation in prediction performance.) We classify these methods into four large groups: time series analysis, pattern methods, regression analysis, and machine learning. We elaborate on such methods below.

Time series analysis methods form their prediction by combining past load information using linear models. Typical approaches are for example ARIMA (Auto Regressive Integrated Moving Average, see for example [10, 22] or [31]. An important drawback of such methods is their rigidity (that is, non-adaptability), leading to large prediction errors during highly unstable load demand periods. Such issues are alleviated for example by the use of heavy tailed error models (instead of the traditional Gaussian noise assumption) [33], or allowing for adaptive linear parameters and *state space models* as in [13].

Pattern methods are based on self-similarity within the data, that is, they assume that future load curves are well modeled as a combination of previously observed patterns in the data. A typical embodiment of this idea is given in [28], where predictions result from the weighted mean of past loads, the weights given by a similarity measure between the last observed time frame (e.g., the last observed daily curve) and all past daily curves in the recorded past. In [2, 3], daily curves are modeled as continuous functions, and tools from functional theory are used to determine their similarity to past daily curves. A variant of the previous idea is to look for similarities not in the whole past but within a small *dictionary* of typical patterns which combine past curves and exogenous covariates [25].

Regression analysis methods predict the future load one sample at a time by decomposing the load curves into various explicative effects, usually driven by exogenous factors. For instance, the predictions obtained in [9, 10, 30] are the result of a non-linear regression involving two components: one due to meteorological conditions and the other due to calendar structure. Predictions obtained with this approach can be very accurate but suffer from the same rigidity issues as those observed in time series methods. Various strategies have been proposed to alleviate such limitations. Examples include adaptive regression parameters [24], generalized additive models [26], and Bayesian models [23].

Machine learning techniques usually employ models which are generally more flexible, albeit less interpretable, than the previously described ones. Examples include [24] which uses *kernel methods* combined with Gaussian processes, and [5] which employs *gradient boosting*. See the aforementioned references for the description of such techniques.

A very successful strategy derived from this field is the idea of combining the output of several predictors, often of different type (i.e., different algorithms may be used to obtain the different predictions), to produce the *final* prediction of the system [11]. This is known as *experts aggregation*; this is one of the methods we adopt in our framework, as will be seen later.

The goal of this work is to obtain a forecasting framework that can be used to anticipate the load needs of the Uruguayan grid. In particular, we need to account for the fact that meteorological data is scarce in the Uruguayan scenario. Our work is thus focused on three models, two of which do not rely on meteorological data.

The reminder of this document is structured as follows. Section 6.2 describes both the electrical and the meteorological data from the Uruguayan system. Section 6.3 discusses the models we adopt. A series of experiments illustrate the resulting prediction framework in Sect. 6.4. A short discussion in Sect. 6.5 concludes the work.

6.2 Data

This section explores the typical features encountered in Uruguayan demand data. The electricity demand data was kindly provided by *Adminstración Nacional de Usinas y Trasmisiones Eléctricas* (UTE)[1]; the dataset in question contains hourly electrical load measurements between January 1st 2007 and 31 December 31st 2014. Meteorological data covering the same time period, with the same sampling frequency, was kindly provided by the *Instituto Uruguayo de Meteorología* (INUMET).[2]

[1] `portal.ute.com.uy/`.

[2] `inumet.com.uy/`.

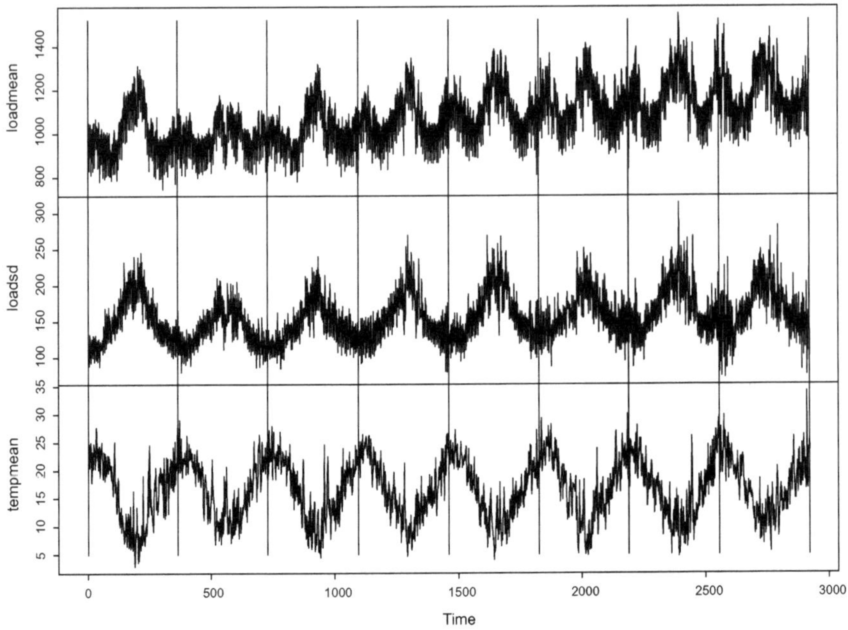

Fig. 6.1 Mean (above) and standard deviation (middle) of daily load, and mean temperature (bottom) for the Uruguayan electrical system; vertical lines separate years

Figure 6.1 aggregates the three available curves (daily load mean, daily load standard deviation and daily mean temperature). We first focus on load demand (upper panel). As residential electricity consumption represents a major portion of the total Uruguayan demand, most salient features in the curves are derived from domestic human activity patterns. First of all, a clear upwards long-term trend can be observed which is linked to population increase and increasing use of electrical household devices, in particular high powered ones. Them a number of cyclical components can also be observed. The annual cycle reflects the seasonality induced by both economical activity and meteorological phenomena. It is important to note that this annual pattern has gradually evolved across the years. While the first years are characterized by one strong mode in winter, later years show another important mode in summer, which is due largely to the recent widespread adoption of electrical cooling systems. Note that this change has a major impact in electricity supply planning, for instance on maintenance schedules. The latter feature shows the importance of having models capable of adapting themselves to major dynamic changes in the structure of the demand.

Being the domestic demand a major portion of the total, and given that cooling and heating devices usually require high power to operate, it is to be expected that the temperature (bottom curve of Fig. 6.1) has a strong effect in the load curves. This dependency is however more complex than what one might observe at a first glance.

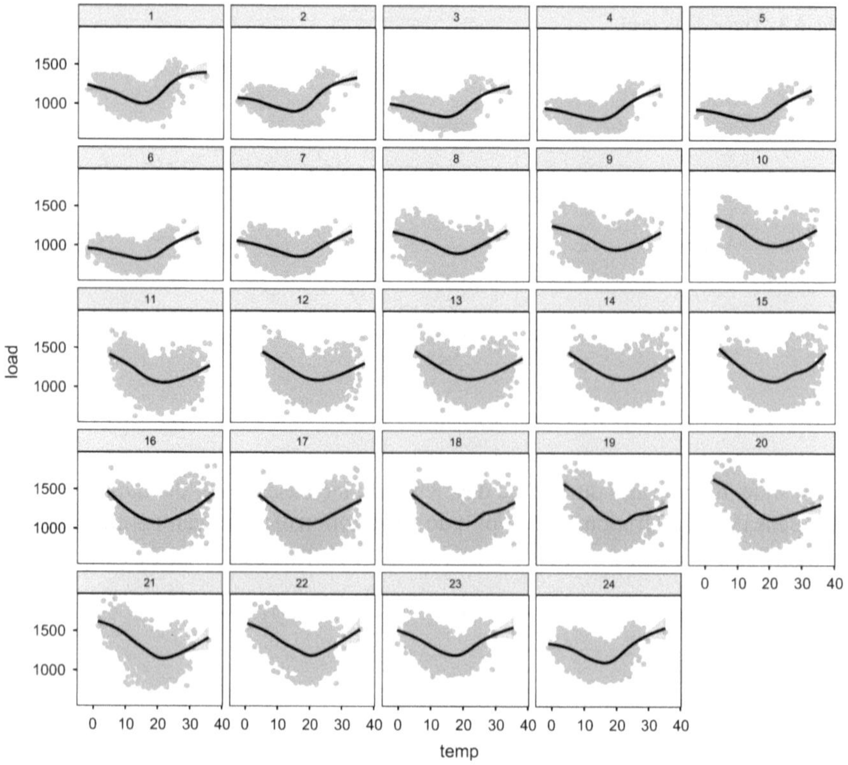

Fig. 6.2 Temperature gradients estimated at different hours. Each panel represents the scatter of the load (in Mwh) as a function of the temperature (in Celsius degrees) at one specific hour of the day. A smooth non parametric estimation of the link is added as a black line

For instance, air conditioning devices are triggered on by extreme temperatures, either low or high; this is more evident at the daily demand scale. However, this is not a simple matter of thresholds, and this is evidenced by the marked increase in standard deviation (middle curve of Fig. 6.1) during cold seasons.

Further insight into the complex dependency between temperature and demand can be observed when one plots the load curve as a function of the temperature. This representation allows one to estimate how much the electricity demand changes when the temperature changes and thus it is usually called the *temperature gradient*. Graphics shown in Fig. 6.2 represent a way to estimate the temperature gradient (obtained actually as a nonparametric fit of the curves not explained here). Each curve fit corresponds to a specific hour of the day, which allows one to see how this dependency evolves inside the day presenting in some situation very important slopes on both high and low temperatures.

Figure 6.3 exemplifies the variability found in daily load patterns. Despite such variabilities, common features can still be identified, such as lower demands during

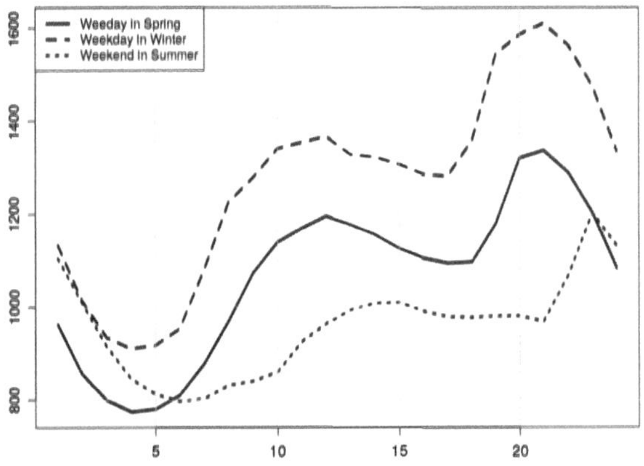

Fig. 6.3 Three daily load curves

night time, higher ones during the day, a steady increase during the morning with a first peak in mid afternoon. The amplitude and position of these features are important landmarks of the load curve.

6.3 Methods

As mentioned in Sect. 6.1, the proposed framework combines the output of various models through an experts aggregation strategy.

We now describe the individual models which make up the experts to be aggregated. For clarity of presentation we separate them into *exogenous models* (using meteorological data) and *endogenous models* (relying only on past demand patterns). We then present and discuss the expert aggregation strategies applied to obtain the final prediction.

6.3.1 Exogenous Models

The following models use external information that must be available at the moment of the prediction.

Hong's Vanilla Benchmark (HVB)

This model was introduced in [19]. This is a multiple linear regression model that incorporates the effect of temperature on the load as a third order polynomial; the parameters of the polynomial are indexed by calendar features such as day, month, and hour, resulting in the following formulation

$$\begin{aligned}\mathbb{E}[\text{load}_t] = \beta_0 + \beta_1 t + \beta_2\, \texttt{Day}_t \times \texttt{Hour}_t + \beta_3\, \texttt{Month}_t + \beta_4\, \texttt{Month}_t \times \texttt{Temp}_t \\ + \beta_5\, \texttt{Month}_t \times \texttt{Temp}_t^2 + \beta_6\, \texttt{Month}_t \times \texttt{Temp}_t^3 + \beta_7\, \texttt{Hour}_t \times \texttt{Temp}_t \\ + \beta_8\, \texttt{Hour}_t \times \texttt{Temp}_t^2 + \beta_9\, \texttt{Hour}_t \times \texttt{Temp}_t^3. \end{aligned} \tag{6.1}$$

The form of (6.1) is simple and easy to interpret, its main feature being the thermo-sensitivity being allowed to vary according to a given calendar hour, day and month. This model was used as a benchmark model in the GEFCOM 2012 competition [20], hence its name and popularity. In that competition, gains of up to 40% were obtained using more flexible models. Notice that Hong's Benchmark is a purely mid-term model, as it does not incorporate past load records; this may be useful such data is not available. It does need, however, enough temperature and load data as to adjust its parameters, something which is done using plain least squares regression. Point and interval predictions are then obtained by simple evaluation of the model using the current temperature and temperature forecasts respectively.

State Space Models (SSM)

This family of models includes an inertial term in the form of an intrinsic state α of the whole system being modeled [15]. The observed output (the demand) is a function of the state and external variables represented in ε, and the state evolution over time is modeled as a linear equation involving the previous state and other observed variables summarized in a vector η. The general formulation is given by,

$$\begin{cases} y_t &= x_t \alpha_t + \varepsilon_t \\ \alpha_{t+1} &= T_t \alpha_t + R_t \eta_t \end{cases},$$

where y_t is the target variable observed at time t, $x_t \in \mathbb{R}^{m+1}$ is a vector of *predictors*, the state at time t is represented as $\alpha_t \in \mathbb{R}^{m+1}$, T_t and R_t are known matrices, and ε_t and η_t are the noise and disturbance processes usually assumed to be independent Gaussian with zero-mean and with *unknown* covariance matrices.

One particular case is detailed in [13] for describing evolving parameters on the load forecast task. The first equation, which provides the current prediction in terms of the state and exogenous variables, is given by

$$\text{load}_t = X_t^f \beta + X_t^e \gamma_t + w_t, \tag{6.2}$$

where the time dependent parameters are included in the state γ_t and the fixed ones are in β. The regressors X_t^e associated to the time evolving parameters are meteorological variables such as temperature which is the only meteorological information we use. To represent the inertia of the weather, temperature is represented as 3 variables. First, the observed temperature at each moment of the day. Second, the lowest temperature record in the last 24 h. Third, the highest temperature record in the last 24 h. The last two series are smoothed using splines. Behind this construction we follow the strategy of capturing effects due to extreme weather which are well know by the electrical engineers in Uruguay. Extreme observations usually arrive in clusters called respectively hot and cold waves, that is sequences of days where weather is extreme. Keeping the information of the extreme is important to detect these days. During winter, low highest temperatures are an evidence of the presence of a cold wave. Reciprocally, high lowest records evidences the presence of a hot wave.

The regressors X_t^f are associated to the fixed parameters, in this case calendar variables such as day, month, etc.

In our implementation of this method we choose two components: a *trend*, which is a linear function of the discrete time index $t = 1, 2, \ldots, n$, and a seasonal component which is modeled as a superposition of three pairs of sines and cosines,

$$\sum_{j=1}^{j=3} a_j \sin\left(\frac{2\pi t}{s_j}\right) + b_j \cos\left(\frac{2\pi t}{s_j}\right),$$

where the coefficients $\{a_j, b_j\}_{j=1,2,3}$ are to be estimated. With respect to the model in [13] this is a difference since in that work calendar effects are time-varying. The second equation of the model, which describes the state transition function, is given by

$$\alpha_t = \alpha_{t-1} + \eta_t, \tag{6.3}$$

that is, a multivariate random walk; the noise magnitude (the variance of η_t) is to be estimated.

The parameters of the model are estimated using Kalman filter theory [21]. Given an initial state, the following states are obtained using closed-form equations that involve the inversion of possibly large matrices. The dimension of such matrices depends on the respective dimensions of the state space and the observation vector. Since this is an effective computational bottleneck, we follow [10] and break the state space into 24 separate SSMs which can be treated in parallel, providing a huge decrease on the computation time.

Random Forests (RF)

Originally proposed in [7], a random forest is comprised of a set of *maximal decision trees* which involve randomness in their construction two ways. First, each maximal tree is estimated using a bootstrap sample of the data. Second, at each training step of a tree, which involves splitting one of its nodes, the splitting decision is made on a random subset of the feature variables. These random subsamplings help in reducing statistical dependency between the output of each tree in the ensemble. Finally, whereas each maximal tree may exhibit a large variance, the aggregation of many such hopefully independent trees yields a predictor with significantly smaller variance. This method is applied to load demand prediction in, for example, [14], using past loads and past temperatures as predictors for the current load. Another example is [27] where the difference between the current temperature and the one observed some hours ago (that is, with a given lag) is used for prediction. In our work, we adopt the latter method using a lag of 72 h to take into account important changes in the meteorological conditions.

6.3.2 Endogenous Models

The following models predict future loads based solely on past load data and calendar information; no exogenous information is used. For these methods to work, however, very recent load data must be readily available.

Time Series Benchmark (TSB)

Well known in econometrics, Autoregressive Moving Average (ARMA) models are a family of stationary linear stochastic processes given by the following general form [8],

$$y_t = \sum_{k \le t} \theta_k \varepsilon_k, \tag{6.4}$$

where ε_k is a sequence of i.i.d Gaussian variables of zero mean and variance σ^2. The linear coefficients $\{\theta_k\}_k$ are fixed parameters which are a priori unknown.

ARMA processes can be viewed as the output of finite response filters (FIR) where the input is white Gaussian noise. The parameters are usually estimating using the Box–Jenkins method [6]. First the empirical autocorrelation function is defined as $\hat{\rho}_\ell = \hat{\gamma}_\ell / \hat{\gamma}_0$, where ℓ identifies the lag and

$$\hat{\gamma}_\ell = \sum_t (y_t - \hat{y})(y_{t-\ell} - \hat{y}). \tag{6.5}$$

As many real-life time series cannot be assumed to be stationary, the Box–Jenkins method proposes a number of possible transformations for obtaining time series which fall closer to that hypothesis. In our case, we first apply a discrete differentiation, $y'_t = y_t - y_{t-\ell}$, with ℓ typically 1 or 2. The differentiated series y'_t is then modeled as an ARMA process; we call this process an integrated ARMA (ARIMA for short) of order ℓ. A second differentiation is then performed. The second transformation is a *seasonal* differentiation, $y''_t = y_t - y'_{t-\ell'}$, which is identical in form as the first one but with a lag ℓ' corresponding to the period of a given season in the data (e.g., hours in a day, days in a week, months in a year). This two step transformation is known as a SARIMA (Season Auto-Regressive Integrated Moving Average). The process of identification, estimation and validation of such SARIMA model is usually done by an expert and can be quite time consuming, especially in the case of load demand as it typically exhibits two seasonal components. Instead, we propose a new estimation scheme based on mimicking the empirical autocorrelation function $\hat{\rho}_\ell$ to the theoretical one, ρ_ℓ, as it is done by econometrics.

For a particular SARIMA model one can write

$$S(\theta) = \sum_{\ell=0}^{L} (\rho_\ell(\theta) - \hat{\rho}_\ell)^2, \tag{6.6}$$

where θ is a vector of unknown parameters and L is a sufficiently large lag. The estimator $\hat{\theta}$ is obtained as $\arg\min_\theta S(\theta)$. In [18], a similar idea is used to estimate autoregressive models. The above estimator can be viewed as a generalized method of moments for time dependent observations. While the theoretical properties of the proposed estimator are not as good as those of the maximum likelihood estimator (MLE), its computation is significantly less demanding than the latter, which makes it our choice of estimator within our framework.

Point predictions are obtained using the last observed values of load_t and the last prediction errors. Prediction intervals can be obtained using either a normal approximation or bootstrap procedures.

Kernel Wavelet Functional (KWF)

As discussed in Sect. 6.2, the shape of the load curves carries valuable information about the context of the observed load (e.g. position of the year, type of the day, meteorological conditions). Assuming that similar past conditions induce similar future conditions, one can construct an easy-to-interpret predictor [28]. A modern version of this method, which exploits the functional nature of time series curves, is used in [1]. There, the predictor is written as

$$\widehat{\text{Load}}_{n+1}(\tau) = \sum_{m=1}^{n-1} w_{m,n}\text{Load}_{m+1}(\tau), \qquad \tau \in [0, 1]$$

where $\text{Load}_n(\tau)$ is the load curve for day n at instant $\tau \in [0, 1]$ and the weight $w_{m,n}$ is proportional to the similarity between the load curves Load_m and Load_n (more on the similarity criterion is mentioned below). Note that the curves Load_n are not observed directly, but rather estimated from the corresponding (discrete) hourly values, For instance, Load_n is estimated from load_t, $t = (n-1) \times 24 + j$, $j = 1, \ldots, 24$. The preceding estimation is performed using *wavelets*, which also serve for computing the similarity weights $w_{m,n}$ in terms of the wavelet coefficients of both curves. Non stationary patterns are treated by means of corrections applied on the wavelet coefficients (see [2] for details).

Prediction intervals can be obtained using a bootstrap strategy where the bootstrap sampling is determined by $\{w_{m,n}\}$ the weight vector (see [4]).

6.3.3 Online Mixture of Experts

The general idea of mixing experts is to combine the outputs of different individual predictors in order to produce a better overall prediction. If this mixture is properly implemented, the mixed output is theoretically guaranteed to be at least as good as the best of all the individual predictors [11]. This carries on to practice, where significant improvements are consistently obtained with this strategy in a wide variety of settings beyond time series analysis. The idea is simple, and can be summarized as follows. Consider the observations up to time T, $y_{1:T} = \{y_t, t = 1, \ldots, T\}$, and a set of K individual predictors (experts). For each $k = 1, \ldots, K$, the *performance* of the kth predictor up to time T is computed in terms of some loss function, in our case squared loss, between its output $e_{k,1:T} = \{e_{k,t}\}$ and the observed output,

$$l_{k,T}(y_{1:T}, e_{k,1:T}) = \sum_{t=1}^{T}(y_t - e_{k,t})^2.$$

Given $l_{k,T}$, $k = 1, \ldots, K$, the aggregated prediction at time $T + 1$, $\hat{y}^{\text{mix}}_{T+1}$, is computed as a weighted sum of the output of all K predictors at time $T + 1$,

$$\hat{y}^{\text{mix}}_{T+1} = \sum_{k=1}^{K} w_{k,T+1} e_{k,T+1}$$

where the weights $w_{k,T+1} = \omega(l_{k,T})$ with $\omega(\cdot)$ some decreasing function. In our case we use polynomial potential aggregation rules (ML-Poly) which computes the mixture as a weighted average of experts using polynomial weights and allowing different learning rates on each expert.

6.4 Experiments

In this section we report on a number of experiments performed in conditions which are close to the operational ones. The goal of these experiments is to evaluate the performance of each of the implemented predictors, as well as that of the aggregation strategy. We use the open source R environment [29] and the package `enercast`[3] which was developed by the authors. This package implements the proposed individual models, whereas the on-line mixing is done with the `opera` R package [17].

The performance of each model is measured through the one-day-ahead daily prediction of load demand curves during the whole last year. That is at the end of day J we use all the records of that day to predict the following 24 records. The only exception is the data used as temperature predictions which is the effectively observed values of day $J + 1$ as it is usual to estimate the model. Forecasting performance is measured using two metrics to compare predicted against actual values. One is the daily Mean Absolute Proportional Error (MAPE), which is given as a percentage,

$$\text{MAPE}(y, \widehat{y}) = \frac{100}{24}\sum_{h=1}^{24}\left|\frac{y_h - \widehat{y}_h}{y_h}\right| \tag{6.7}$$

[3] `github.com/cugliari/enercast`.

where $y = (y_1, \ldots, y_{24})$ is the effective load, and $\widehat{y} = (\widehat{y}_1, \ldots, \widehat{y}_{24})$ is the daily prediction. The other measure is the Root Mean Square Error (RMSE),

$$\mathrm{RMSE}(y, \widehat{y}) = \sqrt{\frac{1}{24}\sum_{h=1}^{24}(y_h - \widehat{y}_h)^2}, \tag{6.8}$$

which is expressed in the same units as the target variable, Megawatts per hour (MWh). Note that this measure is more consistent with the criterion used by the sequential aggregation strategy (square loss) for weighting the contribution of each expert.

Tables 6.1 and 6.2 show summary yearly statistics for the MAPE and RMSE prediction performances for each of the individual methods (experts) as well as that of the mixture model which combines all of them. To begin with HVB, an exogenous model, comes out clearly as the worst method in all the statistics, in both tables. Interestingly, the other purely exogenous model, the Random Forest, comes out as the best individual expert in almost all cases, the only exceptions being the minimum and maximum MAPE, and the minimum RMSE. We also note that the performance of TSB, which is purely endogenous, is very close to that of SSM, which also employs exogenous (temperature) information.

Finally, as can be clearly seen in both Tables 6.1 and 6.2, the mixture strategy significantly outperforms the rest of the models in all but one of the statistics, the minimum, which is also arguably the less relevant one. In particular, we obtain a

Table 6.1 Summary statistics of daily MAPE (%) over year 2014. Best results are shown in blue, while worst are shown in red

Model	Min.	1st Qu.	Median	Mean	3rd Qu.	Max.
1. HVB	2.12	4.26	5.49	5.93	6.87	30.85
2. KWF	1.83	3.40	4.30	4.61	5.31	20.17
3. TSB	0.65	2.23	3.14	4.13	5.11	22.03
4. SSM	0.58	2.29	3.42	4.12	4.98	25.84
5. RF	0.76	2.09	2.97	3.66	4.34	20.30
Mixture	0.76	1.76	2.40	3.06	3.58	17.60

Table 6.2 Summary statistics of daily RMSE (in MWh) over year 2014. Best results are shown in blue, while worst are shown in red

Model	Min.	1st Qu.	Median	Mean	3rd Qu.	Max.
1. HVB	31.83	61.05	73.38	81.27	92.32	305.30
2. KWF	25.66	48.69	60.11	65.96	74.71	278.50
3. TSB	7.75	32.94	46.37	58.92	72.94	287.60
4. SSM	11.88	32.96	46.97	57.34	68.42	288.70
5. RF	9.87	29.41	42.32	51.36	62.82	206.30
Mixture	11.19	25.19	36.08	44.35	53.08	196.04

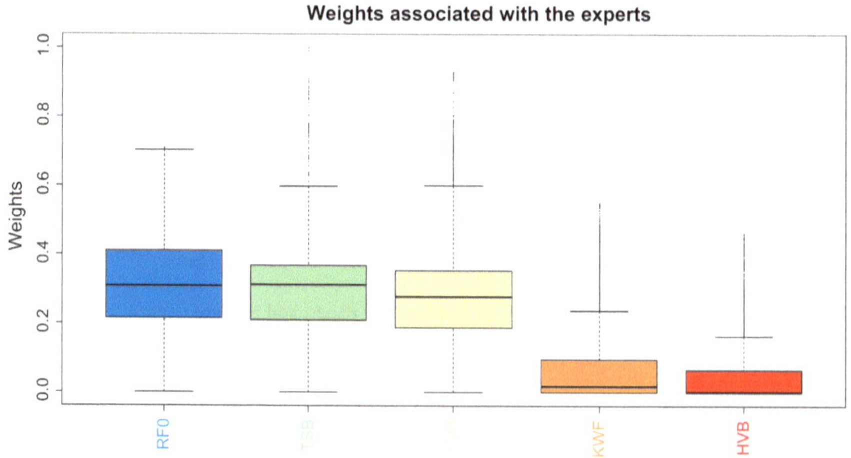

Fig. 6.4 Box plot of the empirical distribution of the weights assigned during the year 2014 to each expert

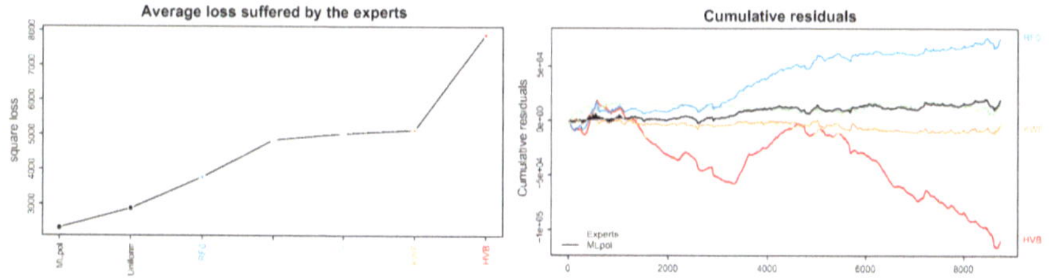

Fig. 6.5 Average square loss (left) and cumulative residuals (right) for the individual experts and the online mixture

15% reduction in RMSE w.r.t the best individual model. The following discussion aims at providing more insight into these results.

The box plot shown in Fig. 6.4 depicts the empirical distribution of the weights assigned to each individual expert along the 365 days of 2014. The figure shows two clear groups among the experts. On one side, RF, TSB and SSM exhibit a similar median weight of about 33%. On the other side, the median weight of both KWF and HVB are sensibly smaller, barely above 10%, and yet they both peak at values over 40%.

Figure 6.5 provides more insight into the benefits of the weighting mechanism. On the left, we show the average squared loss by predictor in increasing order, whereas the right plot shows the cumulative prediction error across the year for each method. Here we can see that both endogenous models, HVB and RF, exhibit strong biases during long periods, even though RF appears to be the best single expert among the ones tested. The mixture model however does not exhibit such bias, which shows

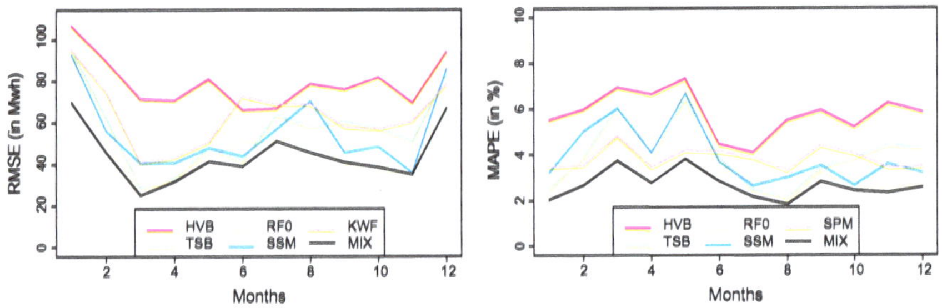

Fig. 6.6 Mean RMSE (left) and MAPE (right) for each expert and mixture (right)

its ability to temporarily shift its weight to other predictors, even if they are overall worse, when such systematic deviations occur.

6.4.1 A Closer Look at the Results

Being a yearly summary, the performance statistics shown in Tables 6.1 and 6.2 are insufficient to characterize the appropriateness of any given individual model, as they do not reflect the relative importance of different seasons, days (such as holidays), or specific hours of the day, in what respects to energy generation planning. We now focus on two of these aspects, namely, the behavior across the year (monthly performance) and across the day (hourly performance). The first is shown in Fig. 6.6 for the individual predictors as well as for the mixture. Although the mean monthly RMSE and MAPEs evolve differently for each expert, some common elements can be observed. First we look at RMSE. Here the worst performances occur during (Austral) summer (December, January and February); winter (June, July and August) is generally well predicted; autumn is better predicted than spring by all the methods except HVB. These conclusions however do not carry on to the MAPE case, that is, when the relative error is evaluated. Here the picture is quite different. The only similarity is winter, where again the best performance is obtained in general. The worst season in this case appears to be autumn, peaking in May and then March (this can be explained by the fact that May, for example, is a month with relatively low load, thus amplifying relative errors).

The hourly MAPE for all predictors is shown in Fig. 6.7. Note that the performance of endogenous hour-by-hour forecast methods, which can only rely on the previous day loads, becomes worse for larger prediction horizons; this effect is superimposed to the particular difficulties of the load pattern itself. We note that the RF and SSM curves are similar in shape to the mixture one. The shape of these predictors reflect the higher difficulty of predicting the *morning ramp* and the *afternoon plateau* (see Fig. 6.3). As expected, the performance of TSB worsens for larger horizons and yet, it is the best predictor for the first hours of the day, surpassing the performance of

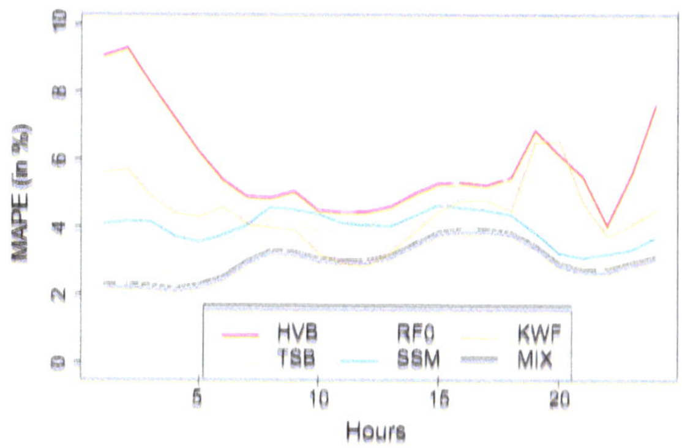

Fig. 6.7 Mean MAPE for each expert and mixture by hours

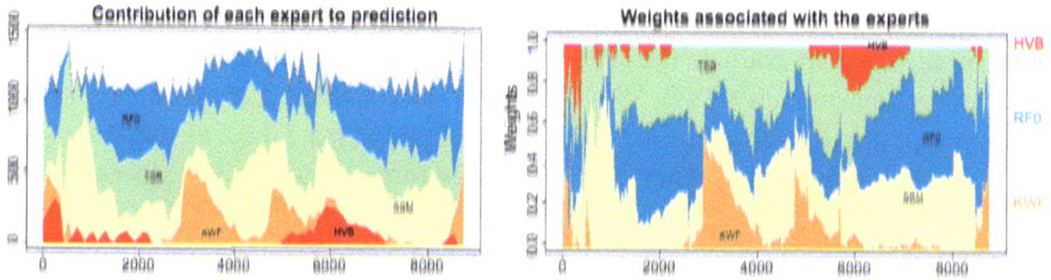

Fig. 6.8 Contribution of the individual experts to the mixture prediction (left) and weight distribution per expert by time (right)

the mixture on the first two hours of the day. The afternoon peak (between 18h00 and 20h00) is difficult for KWF and HVB, but KWF has a good performance during midday, matching the mixture during those hours.

Figure 6.8 shows the contribution of the individual experts to the global prediction as well as the weights distribution. The time axis are the same as the one on the left panel of Fig. 6.5. Clearly, the best three experts (RF, TSB and SSM) share the contributions most of the time, that is, when one of the main three experts reduces its contribution, the other two take on larger weights (cf. February, around time step 1000 where the RF contribution is relatively small and SSM and TSB take over). However, exceptions to this behavior are of interest. Such is the case of three periods of the year: Winter, the month of May and Spring. During these periods, HVB and KWF have point-wise larger participation, meaning that the other three experts are not able to follow the consumption structure during these periods. The month of May is of particular interest because the Uruguayan weather in this period is highly erratic and difficult to anticipate. In other terms, the additional information provided by the temperature in the exogenous models is of less quality. The online mixing algorithm quickly detects this phenomenon, thereby adjusting the weights in favour of the endogenous experts during these periods.

6.5 Discussion

Anticipation of load demand is a crucial element to help decision taking on dispatch. The introduction of new challenging contexts such as renewables, changes and intensification on electronic devices usages or the increasing availability of new data sources, pushes the domain towards the use of new automatic approaches to produce forecast. In this work we show the construction of some models very different in their nature. Instead of choosing the best of them to produce the final forecast, we use a sequential prediction, that is an ensemble method adapted to time series in order to enhance the prediction quality.

While we tried to be as close as possible to operational conditions some important issues are to be raised. Unfortunately, forecast measurements were not available during this work, as such information would have provided a more realistic operational scenario in which to evaluate the different predictors. For instance, the performance of the exogenous models (HVB, SSM, RF) has to be considered an optimistic estimation, since predictions from these experts were obtained from the *true* measurements of the day being predicted rather than from daily forecast, which introduces its own (significant) errors and biases. Note however that he conclusions on what respects to the purely endogenous experts as well as the benefits of using mixtures are expected to hold in the presence of such information. Also, the prediction obtained in our experiments are computed at midnight with all the data of the that day available. However, in real conditions one would have to predict many hours in advance in order to be able to use this information in the decision taking procedure for the dispatch.

A natural perspective would be to add more experts on the prediction mix. One inexpensive way to obtain this is to create specialized experts as in [16], where one trains the individual model on one specific context (e.g. during cold and hot seasons separately). Using the individual models presented here, one could combine them to produce slightly different versions. For instance, models such as KWF or TSB could be used to predict temperature curve for the next day, and then these temperature predictions can be used as input for the exogenous models. Finally, the bias correction phenomena observed in the experiments and commented in Fig. 6.5 can be exploited at the level of each expert using some bias correction technique. Then the biased corrected models can also be included as experts for the mixture.

Acknowledgements This work was partially funded by *Agencia Nacional de* Investigación e Innovación (ANII, Uruguay), grant FSE 2013/1/10886, "Modelos de previsión de demanda de corto plazo."

References

1. A. Antoniadis, E. Paparoditis, T. Sapatinas, A functional wavelet-kernel approach for time series prediction. J. R. Stat. Soc. Ser. B Stat. Methodol. **68**(5), 837 (2006)

2. A. Antoniadis, X. Brosat, J. Cugliari, J.-M. Poggi, Prévision d'un processus à valeurs fonctionnelles en presence de non stationnarites. Application à la consommation d'électricité. Journal de la Société Française de Statistique **153**(2), 52–78 (2012)
3. A. Antoniadis, X. Brossat, J. Cugliari, J.-M. Poggi, Une approche fonctionnelle pour la prévision non-paramétrique de la consommation d'électricité. Journal de la Société Française de Statistique **155**(2), 202–219 (2014)
4. A. Antoniadis, X. Brossat, J. Cugliari, J.-M. Poggi, A prediction interval for a function-valued forecast model. Int. J. Forecast. **32**(2), 939–947 (2016)
5. S. Ben Taieb, R.J. Hyndman, A gradient boosting approach to the kaggle load forecasting competition. Int. J. Forecast. **30**(2), 382–394 (2014)
6. G.E.P. Box, G.M. Jenkins, G.C. Reinsel, G.M. Ljung, *Time Series Analysis: Forecasting and Control*, Wiley Series in Probability and Statistics (Wiley, New York, 2015)
7. L. Breiman, Random forests. Mach. Learn. **45**(1), 5–32 (2001)
8. P.J. Brockwell, R.A. Davis, *Time Series: Theory and Methods*, 2nd edn., Springer Series in Statistics (Springer, Berlin, 1991)
9. A. Bruhns, G. Deurveilher, J.S. Roy, A non linear regression model for mid-term load forecasting and improvements in seasonality, in *Proceedings of the 15th Power Systems Computation Conference* (2005), pp. 22–26
10. J. Cancelo, A. Espasa, R. Grafe, Forecasting the electricity load from one day to one week ahead for the spanish system operator. Int. J. Forecast. **24**, 588–602 (2008)
11. M. Devaine, P. Gaillard, Y. Goude, G. Stoltz, Forecasting electricity consumption by aggregating specialized experts. Mach. Learn. **90**(2), 231–260 (2013)
12. DNETN, Politica energética 2005–2030. Technical report, Ministerio de Industria, Energía y Minería, Uruguay, 2008
13. V. Dordonnat, S.J. Koopman, M. Ooms, A. Dessertaine, J. Collet, An hourly periodic state space model for modelling French national electricity load. Int. J. Forecast. **24**(4), 566–587 (2008)
14. G. Dudek, Short-term load forecasting using random forests, in *Intelligent Systems'2014* ed. by D. Filev, J. Jablkowski, J. Kacprzyk, M. Krawczak, I. Popchev, L. Rutkowski, V. Sgurev, E. Sotirova, P. Szynkarczyk, S. Zadrozny. Advances in Intelligent Systems and Computing, vol. 323 (Springer, Berlin, 2015), pp. 821–828
15. J. Durbin, S.J. Koopman, *Time Series Analysis by State Space Methods*, Oxford Statistical Science Series (Clarendon Press, Oxford, 2001)
16. P. Gaillard, Y. Goude, Forecasting electricity consumption by aggregating experts; how to design a good set of experts, in *Modeling and Stochastic Learning for Forecasting in High Dimensions* ed. by A. Antoniadis, X. Brossat, J.-M. Poggi. Lecture Notes in Statistics, vol. 217 (Springer, Berlin, 2015), pp. 95–115
17. P. Gaillard, Y. Goude, Opera: Online Prediction by Expert Aggregation, 2016. R package version 1.0
18. W. Guang-Te, V.P. Singh, An autocorrelation function method for estimation of parameters of autoregressive models. Water Resour. Manag. **8**(1), 33–55 (1994)
19. T. Hong, P. Wang, H.L. Willis, A naïve multiple linear regression benchmark for short term load forecasting, in *2011 IEEE Power and Energy Society General Meeting* (IEEE, 2011), pp. 1–6
20. T. Hong, P. Pinson, S. Fan, Global energy forecasting competition 2012. Int. J. Forecast. **30**(2), 357–363 (2014)
21. R. Kalman, A new approach to linear filtering and prediction problems. J. Basic Eng. **82**(1), 35–45 (1960)
22. B. Lanzilotta, S. Rodriguez, Modelos de predicción de energía eléctrica con datos horarios para Uruguay. Informe de investigación, Fondo Sectorial de Energía. ANII, marzo, 2014
23. T. Launay, A. Philippe, S. Lamarche, Construction of an informative hierarchical prior distribution. Appl. Electr. Load Forecast. arXiv:1109.4533 (2012)
24. J.R. Lloyd, GEFCom2012 hierarchical load forecasting: gradient boosting machines and Gaussian processes. Int. J. Forecast. **30**(2), 369–374 (2014)

25. M. Mougeot, D. Picard, V. Lefieux, L. Maillard-Teyssier, Forecasting intra day load curves using sparse functional regression, in *Modeling and Stochastic Learning for Forecasting in High Dimensions* ed. by A. Antoniadis, J.M. Poggi, X. Brossat. Lecture Notes in Statistics, vol 217 (Springer, Cham) (2015)
26. R. Nedellec, J. Cugliari, Y. Goude, GEFCom2012: electric load forecasting and backcasting with semi-parametric models. Int. J. Forecast. **30**(2), 375–381 (2014)
27. S. Papadopoulos, I. Karakatsanis, Short-term electricity load forecasting using time series and ensemble learning methods, in *2015 IEEE Power and Energy Conference at Illinois (PECI)* (2015), pp. 1–6
28. J.-M. Poggi, Prévision nonprametrique de la consommation électrique. Rev. Statistique Appliquée **XLII**(4), 93–98 (1994)
29. R Core Team, R: a language and environment for statistical computing, R Foundation for Statistical Computing, Vienna, Austria, 2017
30. L.J. Soares, M.C. Medeiros, Modeling and forecasting short-term electricity load: A comparison of methods with an application to Brazilian data. Int. J. Forecast. **24**(4), 630–644 (2008)
31. J.W. Taylor, L.M. de Menezes, P.E. McSharry, A comparison of univariate methods for forecasting electricity demand up to a day ahead. Int. J. Forecast. **22**(1), 1–16 (2006)
32. R. Weron, *Modeling and Forecasting Electricity Loads and Prices: A Statistical Approach*, Wiley Finance Series, vol. 396 (Wiley, New York, 2006)
33. R. Weron, B. Kozlowska, J. Nowicka-Zagrajek, Modeling electricity loads in California: a continuous-time approach. Phys. A Stat. Mech. Appl. **299**(1), 344–350 (2001)

Chapter 7
A Flexible Mixed Additive-Multiplicative Model for Load Forecasting in a Smart Grid Setting

Eugene A. Feinberg and Jun Fei

Abstract This paper presents a mixed additive-multiplicative model for load forecasting that can be flexibly adapted to accommodate various forecasting needs in a Smart Grid setting. The flexibility of the model allows forecasting the load at different levels: system level, transform substation level, and feeder level. It also enables us to conduct short-term, medium and long-term load forecasting. The model decomposes load into two additive parts. One is independent of weather but dependent on the day of the week (d) and hour of the day (h), denoted as $L_0(d, h)$. The other is the product of a weather-independent normal load, $L_1(d, h)$, and weather-dependent factor, $f(w)$. Weather information (w) includes the ambient temperature, relative humidity and their lagged versions. This method has been evaluated on real data for system level, transformer level and feeder level in the Northeastern part of the USA. Unlike many other forecasting methods, this method does not suffer from the accumulation and propagation of errors from prior hours.

Keywords Load forecasting · Additive-multiplicative model · Smart grid

7.1 Introduction

Electric load forecasting is a useful tool needed and used by most electric utilities to make important decisions including decisions on capital expenditures, purchasing and generation of electric power, area planning, and load switching. By the forecasting horizon, load forecasting can be divided into three types: short-term (a few minutes up to a week), medium-term (a week up to a year), and long-term (over one year) [1, 2]. Depending on the aggregation scale, load forecasting for a utility

E. A. Feinberg (✉) · J. Fei
Department of Applied Mathematics & Statistics and Advanced Energy Center, Stony Brook University, Stony Brook, NY 11794, USA
e-mail: eugene.feinberg@stonybrook.edu

J. Fei
e-mail: jun.fei@stonybrook.edu

P. Drobinski et al. (eds.), *Renewable Energy: Forecasting and Risk Management*, Springer Proceedings in Mathematics & Statistics 254,
https://doi.org/10.1007/978-3-319-99052-1_7

can be performed at different levels: at the system level for the entire utility, at the load pocket level (aggregation of several closely located distribution substations), at the substation/transformer level, at the feeder level, and even at the customer level. Long-term load forecasting is important for long-term capital budgeting and planning decisions, medium-term level load forecasting is important for forward purchasing and adjusting the system capacity and reliability, and short-term load forecasting is important for energy trading, cogeneration, and load switching decisions including demand response. Customer level load forecasting is a new feature in a Smart Grid setting, which is made possible by the availability of Advanced Meter Infrastructure (AMI) and can be used to enhance the modeling of power system up to the customer level.

In the literature majority of the works on short-term load forecasting can be classified into four categories by the modeling and forecasting method used, namely statistical, intelligent systems, neural networks, and fuzzy logic [3]. They include a variety of methods: the so-called similar day approach [4], various regression models [5–8], time series [8–11], statistical learning [2], neural networks [12–14], expert systems [15, 16], fuzzy logic [17–19], and Support Vector Machine (SVM) [14, 20, 21]. Combination of several of such techniques are reported to produce more accurate forecasts [4, 9, 11, 12, 14, 15, 18]. Two methods, the end-use and econometric approach are broadly used for medium- and long-term forecasting. In particular, a multiplicative model was proposed in [2]. This model corresponds to expression (1) with $L_0 = 0$. Unlike many studies, e.g., [8], we did not use the standard mathematical solutions and software. Instead we have developed iterative computational methods to dynamically estimate the model parameters and developed the proprietary software [22] which was used by a utility.

In this paper we present a more accurate model (1) with the additional term L_0. This models is called mixed additive-multiplicative. The iterative algorithms for computing the model parameters and software for additive model were adjusted to the mixed additive-multiplicative models.

The model operates with three groups of inputs: (i) time parameters that consist of day of the week and hour during the day, (ii) weather parameters, and (iii) historical loads. Weather parameters include the ambient temperature, relative humidity, and their lagged values. We have designed and tested six variants of recursive algorithms to quickly estimate the model parameters and recommend one that worked best in our practice.

Our initial motivation for the model development was designing and coding the planning software for a utility in Northeast of US to predict the system next-year peak loads for local areas called load pockets [23]. Following the success of this application, the model was adjusted to be used on the daily basis for next-day system load forecasts described in Sect. 7.3. It was also used to predict potential overloads for substation transformers and feeders and to model loads for large customers and groups of customers based on the AMI readings.

The peak load forecasting software for load pockets conducts calculations once a year typically in the fall because annual peak loads take place in summers in the most areas of the USA. The application uses the weather and load data in a batch mode. The data in the batch consist of hourly loads and weather observations (ambient temperature and relative humidity) for the summer season. In the applications dealing with next-day forecasts, the calculations are conducted every morning also in the batch mode with the batch consisting of the last day loads, weather data, and weather forecasts. For next-day forecasts, the system recalculates model parameters on the daily basis.

This paper is organized as follows. In Sect. 7.2 we give the model description for hourly load forecasts. In Sect. 7.3 the model evaluation is presented. Section 7.4 presents some applications of the model in a smart grid setting. Conclusions are given in Sect. 7.5.

7.2 Model Description

The load is decomposed in the model into two additive parts. One depends on day of the week (d) and hour of the day (h) only, and it is denoted as $L_0(d, h)$. The other is the product of a weather-independent normal load, $L_1(d, h)$, and weather-dependent factor, $f(w)$. The mathematical form of the model is

$$y_i = L_0(d_i, h_i) + L_1(d_i, h_i) f(w_i) + e_i, \tag{7.1}$$

where $i = 1, 2, , N$, and N is the total number of observations in the training dataset;
y_i is the actual load;
d_i is the day of the week, $d_i = 1, 2, \ldots, 7$;
h_i is the hour of the day, $h_i = 0, 1, \ldots, 23$;
w_i are weather data, which are vectors whose components include the current temperature and relative humidity (observed or forecasted depending on the application), some of their powers, and some of the similar data from the previous 72 h;
e_i is the random error.

Holidays are treated as weekend days. If a holiday is followed by a business day, it is treated as Sunday and as Saturday otherwise. Similarly, the first business day after the holiday is treated as Monday, and the last business day before a holiday is treated as Friday.

Like for other statistical methods, we first estimate the model using the historical weather and load information. The hourly weather information, including ambient temperature and relative humidity, is provided by the NCDC (National Climatic Data Center). The hourly weather forecasts are purchased from some commercial weather forecast vendor contracted by the local utility. The hourly historical load data are obtained from the SCADA and EMS (Energy Management System).

Note that in $L_0(d, h)$ and $L_1(d, h)$ are discrete parameters, each taking 168 values that correspond to the 168 h during a week. The weather dependent factor, $f(w)$, takes the form of multiple linear regression. Given k weather parameters we need to estimate $k + 1$ coefficients (including the intercept). So, in total we need to estimate $337 + k$ parameters for model (7.1).

To estimate the parameters, we use the least square method and minimize the total squared residues, that is,

$$\min \sum_{i} [y_i - L_0(d_i, h_i) - L_1(d_i, h_i) f(w_i)]^2 , \tag{7.2}$$

where $i = 1, \ldots, N$, and N is the total number of observations in the training dataset.

This is an unconstrained nonlinear programming problem. Due to the excessive number of parameters and the mixture of discrete and continuous parameters in the model, traditional methods, such as trust region method, Newton–Raphson method, quasi-Newton method, double dogleg method, conjugate gradient method, and Levenberg–Marquardt (LM) method, are not very efficient. Instead we adopt a recursive algorithm that estimates the parameters sequentially.

The outline of the algorithm is as follows:

Step 1 Initialize L_0 and L_1. Set tolerance ε;
Step 2
- a. Estimate f from y, L_0 and L_1 using the least square method;
- b. Estimate L_1 from y, L_0 and f;
- c. Estimate L_0 from y, L_1 and f;

Step 3 Compute some measure of goodness-of-fit. If the difference between two successive iterations is less than ε, the model is finished. Otherwise go back to step 2a.

In step 3 the measure of goodness-of-fit typically includes Mean Absolute Deviation (MAD), Mean Absolute Percentage Error (MAPE), and Mean Square Error (MSE).

The above algorithm first updates f, then L_1 and finally L_0. We call it "$f - L_1 - L_0$." If we permute the order of L_0, L_1 and f in the above algorithm, we get six variants. Numerical results show that they do not differ too much but consistently some show better performance than others. According to our calculations. The order of "$L_1 - f - L_0$" gives the smallest model MAPE.

7.3 Computational Results

The proposed model has been evaluated on different level of loads. At the highest level we used it to model and forecast the aggregate load served by an electric utility. The peak load level is around 5,000 MW. We also tried it on several smaller levels including load pocket, substation and feeder levels. The actual performance changes with the scale of the application being the most accurate for the system level. For the system level, the hourly MAPE for a day-ahead forecast during the summer period is around \$2.35%. However, averaging the forecast with the less accurate forecast produced by the ANN (Artificial Neural Network), reduced the MAPE to around 2%. An important feature of the described model is that it is simple to apply it and

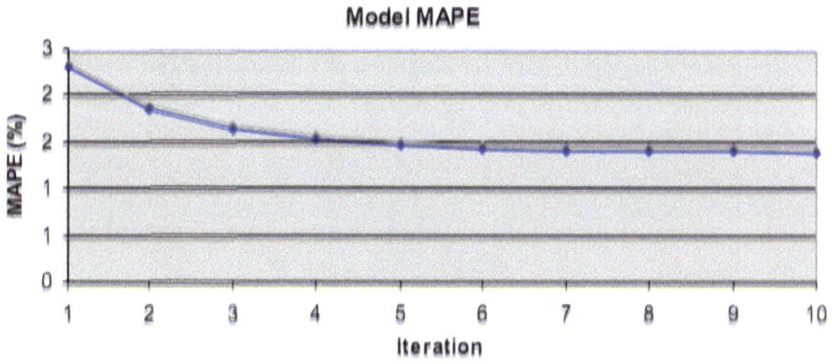

Fig. 7.1 Model MAPE at different iteration steps

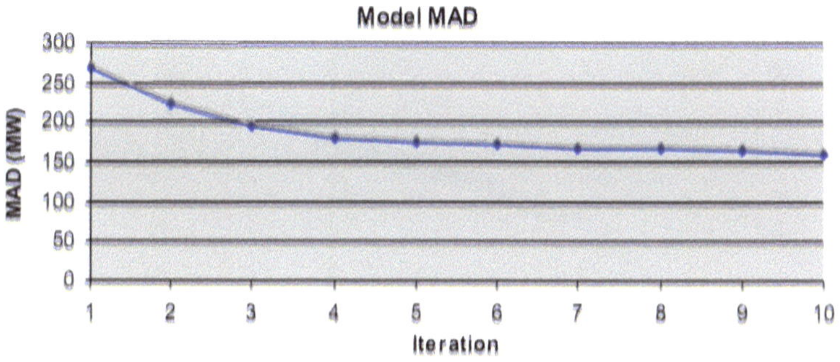

Fig. 7.2 Model MAD at different iteration steps

the algorithm efficiently estimates the model parameters. The algorithm is simple to implement, and it provides fast convergence to model parameter.

The data were hourly loads provided by an electric utility located in the Northeast of US for the summer seasons (May 1 through September 30). The hourly load values range from 1,500 to 5,000 MW. The weather data are described in the previous section. Efficiency is illustrated in Figs. 7.1, 7.2 and 7.3, where we plot the model MAPE, MAD and regression R-squared at different iteration steps. The algorithm provides a quick convergence in fewer than 10 iteration steps.

Figure 7.4 shows the scatter plot relating the model results and actual loads for the system level. The final model MAPE is 1.4%.

7.4 Applications in a Smart Grid Setting

The proposed model can be applied to load at different levels: at the system level for the entire utility, at the load pocket level (aggregation of several close substations or transformers), at the substation/transformer level, at the feeder level, and even at the customer level. At the customer levels, we were dealing with integrated customer

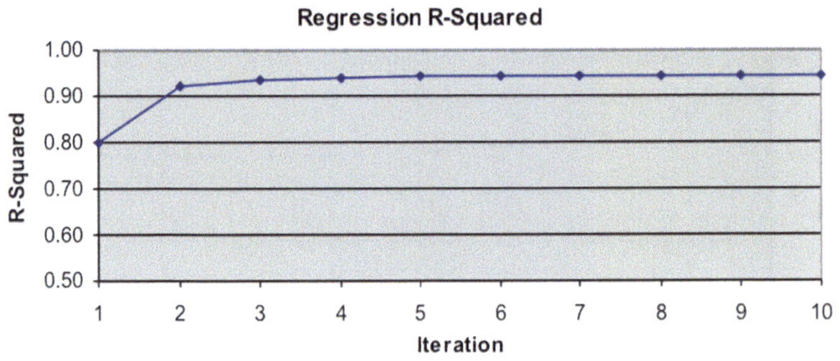

Fig. 7.3 Regression R2 at different iteration steps

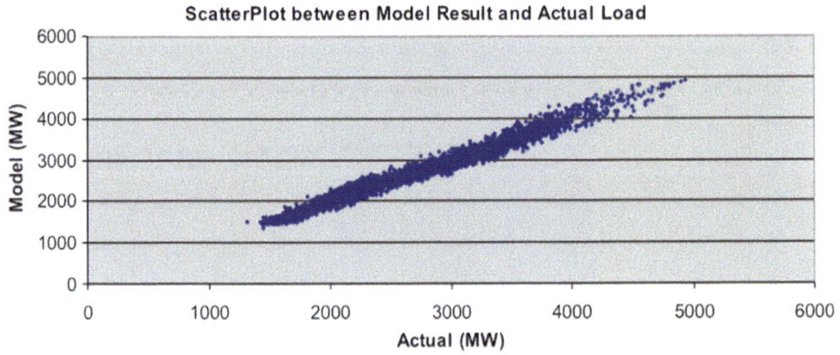

Fig. 7.4 Scatter plot between model result and actual load

loads recorded by AMIs. During the actual implementation we periodically update the model with the most recent data, usually every 1–2 weeks in addition to daily calculations. We simulated the real time run on the daily basis using the forecasted weather.

At the system level the forecasting of aggregate load helps a utility make long-term capital budgeting decisions and short-term forward purchase decisions. An accurate forecast may save a big amount of capital expenditure and cost of purchasing electricity in the spot market. The system level load forecast has been pretty accurate and the reported MAPE for 1–2 day-ahead forecast is between 2–3%. For example, during the real time run in the summer of 2008 our forecasting software gave an MAPE of 2.25% for the first-day forecast and 2.98% for the second-day forecast. Another competing forecaster gave an MAPE of 2.78 and 3.96%, for day 1 and day 2, respectively. If we focus on the summer peak time, 4–6 PM during which load forecasts are usually the least accurate and the most important, our accuracies were 2.23 and 2.79% for day 1 and day 2 while the competing forecaster reported 4.82 and 5.29%. Similar comparisons were demonstrated in other years too, which indicates that unlike many other forecasting methods the described method does not suffer from the accumulation and propagation of errors from prior hours.

Load pockets refer to the aggregate of several close geographic areas [4]. An area usually consists of a few distribution substations. Long-term load pocket load forecasting can be used for area load planning [23]. For example, if the predicted load for the next few years is close to or exceeds the current capacity of transformers, the utility should consider adding new transformers or upgrading the existing ones or diverting some load to a neighboring load pocket. The additional analysis should be conducted at the transformer and feeder levels. The developed load pocket forecasting software predicts the next year peak load. The software has been used in a local utility company for many years. It is used by area planners to compute Weather Normalized Factors and next year capital expenditure allocation.

Load forecasting at the substation/transformer level can be used to estimate the transformer rating and protect transformers from being overloaded and overheated [22]. Combined with the feeder level load forecasting, the load forecasting at the substation/transformer level can also be used for load switching, feeder reconfiguration, load reductions, and voltage control. Due to the large number of feeders in the system, some advanced computing techniques such as parallel computing should be adopted. In our application, it took about 20 min to complete the calculations for 140 feeders and the MAPE of the forecast was around 6%.

In the traditional modeling of an electric power grid, the load is modelled up to the feeder level. The power flow calculation shows the current in the feeder and voltage at the two ends of the feeder. Nothing is known about the details below the feeder level, for example, the current and voltage at the customer level. In the Smart Grid setting, with the installation of Advanced Meter Infrastructure (AMI), the load and voltage information are measured by AMIs, and it can be used to improve load forecasting for feeders. By using loads at the customer level, it is possible to model the electric system at the subsection/customer levels. This additional information can be used in multiple applications. For example, it is possible to analyze the voltage at customer levels and adjust settings at the transformers to change the voltage for some or all customers to make it closer to the nominal voltage such as 120/240V used in the USA. In dynamic settings, this information can be used for voltage and reactive power control [24].

We applied the model to AMI readings. This was done for cumulative customer loads for feeder segments and for large customer loads in order to simulate segment and customer loads under various weather conditions. The purpose was to provide recommendations to improve customer load profiles and load qualities (primarily load factor and voltage profiles). The accuracy of forecasts for feeder segments and for large customers was not as good as at the system and load pocket levels, but accurate enough for the purpose of providing recommendations to improve customer load profiles and load quality by adjusting settings of transformer taps and cap banks. The load models were used to simulate customer loads under historical weather conditions. This was done because actual AMI weather data were available just for a few seasons. The available AMI data were used to build models, and the models were run on the historical weather data for 30 years and more.

7.5 Discussions and Conclusions

This paper presents a mixed additive-multiplicative model for short-term load forecasting and proposes an efficient algorithm for estimating the model parameters. This model has been used to develop load forecasting software that had been evaluated on real data for system level, transformer level and feeder level in the Northeastern part of the USA and used by a utility [22, 23]. Unlike many other forecasting methods, this method does not suffer from the accumulation and propagation of errors from prior hours.

This paper then discusses applications of the proposed model in a Smart Grid setting. The model is flexible to be applied to load at different levels, from the system level to the feeder segments and even customer level. At different levels, the forecasts can be used to achieve different objectives such as capital budgeting, forward purchase, transformer rating, load switching, feeder reconfiguration, load reductions, voltage control, and customer power quality monitoring and improvement.

References

1. H.L. Willis, *Spatial Electric Load Forecasting* (Marcel Dekker, New York, 1996)
2. E.A. Feinberg, D. Genethliou, Load forecasting, in *Applied Mathematics for Restructured Electric Power Systems: Optimization, Control, and Computational Intelligence* ed. by J.H. Chow, F.F. Wu, J.J. Momoh (Springer, New York, 2005), pp. 269-285
3. A.D.P. Lotufo, C.R. Minussi, Electric power systems load forecasting: a survey, in *The IEEE Power Tech Conference*, Budapest, Hungary, 1999
4. E. Paparoditis, T. Sapatinas, Short-term load forecasting: the similar shape functional time-series predictor. IEEE Trans. Power syst. **28**(4), 3818–3825 (2013)
5. S. Ruzic, A. Vuckovic, N. Nikolic, Weather sensitive method for short-term load forecasting in electric power utility of Serbia. IEEE Trans. Power Syst. **18**, 1581–1586 (2003)
6. A. Goia, C. May, G. Fusai, Functional clustering and linear regression for peak load forecasting. Int. J. Forecast. **26**(4), 700–711 (2010)
7. N. Abu-Shikhah, F. Elkarmi, O.M. Aloquili, Medium-term electric load forecasting using multivariable linear and non-linear regression. Smart Grid Renew. Energy **2**(02), 126 (2011)
8. Y. Goude, R. Nedellec, N. Kong, Local short and middle term electricity load forecasting with semi-parametric additive models. IEEE Trans. Smart Grid **5**, 440–446 (2014)
9. C.M. Lee, C.N. Ko, Short-term load forecasting using lifting scheme and ARIMA models. Expert Syst. Appl. **38**(5), 5902–5911 (2011)
10. S.S. Pappas, L. Ekonomou, P. Karampelas, D.C. Karamousantas, S.K. Katsikas, G.E. Chatzarakis, P.D. Skafidas, Electricity demand load forecasting of the Hellenic power system using an ARMA model. Electr. Power Syst. Res. **80**(3), 256–264 (2010)
11. Y. Chakhchoukh, P. Panciatici, L. Mili, Electric load forecasting based on statistical robust methods. IEEE Trans. Power Syst. **26**(3), 982–991 (2011)
12. F. Yu, X. Xu, A short-term load forecasting model of natural gas based on optimized genetic algorithm and improved BP neural network. Appl. Energy **134**, 102–113 (2014)
13. H. Quan, D. Srinivasan, A. Khosravi, Short-term load and wind power forecasting using neural network-based prediction intervals. IEEE Trans. Neural Netw. Learn. Syst. **25**(2), 303–315 (2014)

14. C.N. Ko, C.M. Lee, Short-term load forecasting using SVR (support vector regression)-based radial basis function neural network with dual extended Kalman filter. Energy **49**, 413–422 (2013)
15. D. Niu, Y. Wang, D.D. Wu, Power load forecasting using support vector machine and ant colony optimization. Expert Syst. Appl. **37**(3), 2531–2539 (2010)
16. W.C. Hong, Application of chaotic ant swarm optimization in electric load forecasting. Energy Policy **38**(10), 5830–5839 (2010)
17. A. Khosravi, S. Nahavandi, D. Creighton, D. Srinivasan, Interval type-2 fuzzy logic systems for load forecasting: a comparative study. IEEE Trans. Power Syst. **27**(3), 1274–1282 (2012)
18. V.H. Hinojosa, A. Hoese, Short-term load forecasting using fuzzy inductive reasoning and evolutionary algorithms. IEEE Trans. Power Syst. **25**(1), 565–574 (2010)
19. H. Hagras, C. Wagner, Towards the wide spread use of type-2 fuzzy logic systems in real world applications. IEEE Comput. Intell. Mag. **7**(3), 14–24 (2012)
20. E.E. Elattar, J. Goulermas, Q.H. Wu, Electric load forecasting based on locally weighted support vector regression. IEEE Trans. Syst. Man Cybern. Part C (Applications and Reviews) **40**(4), 438–447 (2010)
21. E. Ceperic, V. Ceperic, A. Baric, A strategy for short-term load forecasting by support vector regression machines. IEEE Trans. Power Syst. **28**(4), 4356–4364 (2013)
22. E.A. Feinberg, J. Fei, J.T. Hajagos, R.R. Rossin, Smart grid software applications for distribution network load forecasting, in *Proceedings of the First International Conference on Smart Grids, Green Communications and IT Energy-Aware Technologies*, Venice, 22–27 May 2011
23. E.A. Feinberg, D. Genethliou, J.T. Hajagos, B.G. Irrgang, R.R. Rossin, Load pocket forecasting software, in *Proceedings of 2004 IEEE Power Systems Conference & Exposition*, New York, 10–13 October 2004
24. E.A. Feinberg, J. Hu, E. Yuan, A stochastic search algorithm for voltage and reactive power control with switching costs and ZIP load model. Electr. Power Syst. Res. **133**, 328–337 (2016)

Chapter 8
A Generic Method for Density Forecasts Recalibration

Jérôme Collet and Michael Richard

Abstract We address the calibration constraint of probability forecasting. We propose a generic method for recalibration, which allows us to enforce this constraint. It remains to be known the impact on forecast quality, measured by predictive distributions sharpness, or specific scores. We show that the impact on the Continuous Ranked Probability Score (CRPS) is weak under some hypotheses and that it is positive under more restrictive ones. We used this method on temperature ensemble forecasts and compared the quality of the recalibrated forecasts with that of the raw ensemble and of a more specific method, that is Ensemble Model Output Statistics (EMOS). Better results are shown with our recalibration rather than with EMOS in this case study.

Keywords Density forecasting · Rosenblatt transform · PIT series · Calibration · Bias correction

8.1 Introduction

Due to the increasing need for risk management, forecasting is shifting from point forecasts to density forecasts. Density forecast is an estimate of the conditional probability distribution. Thus, it provides a complete estimate of uncertainty, in contrast to point forecast, which is not concerned with uncertainty.

Two alternative ways to evaluate density forecast exist.

- The first one was proposed by T. Gneiting: *Probabilistic forecasting aims to maximize the sharpness of the predictive distributions, subject to calibration, on the*

J. Collet · M. Richard (✉)
EdF R&D, 7 Boulevard Gaspard Monge, 91120 Palaiseau, France
e-mail: richard.23.michael@gmail.com

J. Collet
e-mail: jerome.collet@edf.fr

M. Richard
University of Orléans, Château de la source, 45100 Orléans, France

P. Drobinski et al. (eds.), *Renewable Energy: Forecasting and Risk Management*, Springer Proceedings in Mathematics & Statistics 254,
https://doi.org/10.1007/978-3-319-99052-1_8

basis of the available information set. Calibration means predictive distributions are consistent with observations, it is more formally defined in [6]; sharpness refers to the concentration of the density forecast, and even in the survey paper of T. Gneiting, it is not formally defined. An important feature of this framework is that we face a multi-objective problem, which is difficult.

- The second way is the use of a scoring rule, which assesses simultaneously calibration and sharpness. Concerning the well-known CRPS scoring rule, Hersbach [10] showed that it can be decomposed into three parts: reliability (or calibration) part, resolution (or sharpness) part, and uncertainty, which measures the intrinsic difficulty of the forecast. Bröcker [2] generalized this result to any proper score, that is any score which is minimal if the forecasted probability distribution is the true one (w.r.t the available information). Recently, Wilks [14] proposed to add an extra miscalibration penalty, in order to enforce calibration in ensemble postprocessing. Nevertheless, even if the score we use mixes calibration and sharpness, the framework is essentially different from the first one.

Besides these two alternative ways of evaluation, probabilistic forecast is mainly used in two different contexts: finance and economics, and weather forecast. In finance and economics, calibration is the unique objective, so a recent survey on "Predictive density evaluation" [3] is in fact entirely devoted to the validation of the calibration, without any hint of sharpness. In weather forecast, both ways of evaluation are used. For a quick view on forecasting methods in atmospheric sciences, one can look at [13]. In the works of T. Gneiting [7, 8], and in the seminal work of Krzysztofowicz [11], the goal is to improve sharpness, while preserving calibration. Nevertheless, one can state that there is no formal test of calibration in these works. In [4], the only measure used is the CRPS, and [9] addresses exclusively the calibration issue.

Here, we are interested in the first method of evaluation: calibration constraint and sharpness objective. Indeed, risk management involves many stakeholders and thus, calibration is a key feature of trust between stakeholders since it impacts all of them. For example, EDF also faces a regulatory constraint: the French technical system operator imposes that the probability of employing exceptional means (e.g., load shedding) to meet the demand for electricity must be lower than 1% for each week (RTE, 2004), so EDF has to prove the calibration of its forecasts. Even inside EDF, many different business units may be involved in the management of a given risk, so calibration is compulsory to obtain confidence between risk management stakeholders.

The consequence is that we face a multi-criterion problem, the goal of our contribution is to allow us to enforce the calibration constraint, in a generic way. Furthermore, we show that, even if the evaluation framework is the proper score use, recalibrating leads in many cases to an improvement, and to a very limited loss in other cases.

The remainder of this chapter will be organized as follows. The next section explains the principle of the method. The third part provides some theoretical results while the fourth is devoted to a case study.

8.2 Principle of the Method

The Probability Integral Transform (PIT, Rosenblatt, 1952) is usually a measure of the calibration of density forecasts. Indeed, if $Y \sim F$ and is continuous, the random variable $F(Y)$ satisfies $F(Y) \sim U[0, 1]$. Thus, we can find in the literature many tests based on this transformation to evaluate the correct specification of a density forecast. In our case, it is used firstly to recalibrate the forecasts.

Let's look at the following case: let E be the set of all possible states of the world; for each forecasting time j the forecaster knows the current state of the world $e(j)$, and uses it to forecast. For example, in the case of a statistical regression model, E is the set of the possible values of the regressors, in the case of the post-processing of a weather forecasting model, E is the ensemble. The conditional **estimated** distribution function is G_e, whereas the **true** one is F_e. So the PIT series is:

$$\text{PIT} \equiv \left(G_{e(j)}(Y_j)\right)_j .$$

- **Assumption A.2.1**: G_e is invertible $\forall\, e \in E$.

If E is discrete, we assume that the frequency of appearance of each state of the world e is p_e. Then, under the Assumption **A.2.1**, the c.d.f of the PIT is:

$$C(y) \equiv \Pr(G(Y) \leq y) \equiv \sum_e p_e\, F_e \circ G_e^{-1}(y).$$

Note that all the results obtained under the hypothesis that E is discrete are still valid in continuous case, even if we only treat the discrete case in this article.

- **Assumption A.2.2**: F is invertible.

We propose to use C to recalibrate the forecasts. For each quantile $\tau \in [0, 1]$, we use the original model to forecast the quantile τ_C, such that $\Pr(G(Y) \leq \tau_c) = \tau$. We remark that this implies $\tau_c = C^{-1}(\tau)$.

This correction makes sense since under the Assumptions **A.2.1** and **A.2.2**:

$$\begin{aligned}\Pr(C \circ G(Y) \leq y) &= \Pr(G(Y) \leq C^{-1}(y)) \\ &= C \circ C^{-1}(y) \\ &= y,\end{aligned}$$

which means that the recalibrated forecasts are uniformly distributed on the interval [0,1].

Note that this method is close to the quantile-quantile correction as in [12] but here, we are concerned by PIT recalibration, which allows us to consider the conditional case.

8.3 Impact on Global Score

If we evaluate our method on the basis of calibration, it ensures this constraint is enforced. But it is important to know if our method is still useful even if one of the probability forecasting users prefers to use scores, for example the Continuous Ranked Probability Score (CRPS).

The CRPS:

$$CRPS(G, x) = \int_{-\infty}^{+\infty} (G(y) - \mathbf{1}_{\{x \le y\}})^2 \, dy ,$$

with G a function and x the observation, is used to evaluate the whole distribution, since it is minimized by the true c.d.f of X.

However, since we have:

$$CRPS(G, x) = 2 \int_0^1 L_\tau(x, G^{-1}(\tau)) \mathrm{d}\tau , \tag{8.1}$$

as shown in [1], with L_τ the Pinball-Loss function:

$$L_\tau(x, y) = \tau(x - y)\mathbf{1}_{\{x \ge y\}} + (y - x)(1 - \tau)\mathbf{1}_{\{x < y\}} ,$$

with y the forecast, x the observation and $\tau \in [0, 1]$ a quantile level, and that L_τ is easier to work with, we use this scoring rule to obtain results on CRPS.

L_τ is used to evaluate quantile forecasts. Indeed, it is a proper scoring for the quantile of level τ, since its expectation is minimized by the true quantile of the distribution of X.

To begin with, we will prove that under some hypotheses, our correction improves systematically the quality of the forecasts in an infinite sample. Then we will show that under less restrictive hypotheses, our correction deteriorates only slightly—in the worst case—the quality of the forecasts in a more realistic case, e.g. finite sample.

8.3.1 Impact on Score: Conditions for Improvement

To assess conditions for improvements, we need to consider:

$$\mathrm{E}_Y[\, L_\tau - L_{\tau_c} \,] \equiv \mathrm{E}_{Y,e}[\, L_\tau(\boldsymbol{Y}, G_e^{-1}(\tau)) \,] - \mathrm{E}_{Y,e}[\, L_\tau(\boldsymbol{Y}, G_e^{-1}(\tau_c)) \,].$$

Here, under the Assumption **A.2.1**, $G_e^{-1}(\tau)$ corresponds to the estimated conditional quantile of level $\tau \in [0, 1]$ and $G_e^{-1}(\tau_c)$ to the corrected conditional quantile. Denote: $\eta_e \equiv G_e - F_e$. We make the following assumptions which ensure small errors of specification and regularity conditions on the estimated c.d.f G_e, the true one F_e and their derivatives g_e and f_e:

- **Assumption A.3.1.1**: G_e are $C^3 \ \forall\, e \in E$.
- **Assumption A.3.1.2**: F_e are C^3 and invertible $\forall\, e \in E$.
- **Assumption A.3.1.3**: η_e, f_e and their derivatives are bounded $\forall\, e \in E$ by a constant which doesn't depend on e.
- **Assumption A.3.1.4**: $\forall\, \tau \in [0, 1]$, $\forall\, e \in E$, η_e, its first, second and third derivatives are finite in $F_e^{-1}(\tau)$,

Using functional derivatives, directional derivatives and the implicit function theorem (proof in Appendix) we can rewrite (adding the Assumption **A.2.1**):

$$\begin{aligned} \mathrm{E}_Y[\, L_\tau - L_{\tau_c}\,] \sim & \left(\sum_e \frac{p_e \eta_e(F_e^{-1}(\tau))}{f_e(F_e^{-1}(\tau))}\right)\left(\sum_e p_e \eta_e(F_e^{-1}(\tau))\right) \\ & - \left(\sum_e \frac{p_e}{2 f_e(F_e^{-1}(\tau))}\right)\left(\sum_e p_e \eta_e(F_e^{-1}(\tau))\right)^2 \\ & \text{as } max\ \eta_e \to 0, \end{aligned} \tag{8.2}$$

with p_e the frequency of appearance of the state e.

This result allows us to find conditions for improvement of the expectation of the Pinball-Loss score, under one of the following alternative additional conditions.

- **Assumption A.3.1.5**: η or f^{-1} is a constant, or $max_e(\bullet)/min_e(\bullet) < 3 + 2\sqrt{2}$ for both η_e and f_e^{-1}, $\forall\, e \in E$,
- **Assumption A.3.1.6**: the correlation between η and f^{-1}, $\sigma_{f^{-1}}$ or σ_η is null. Here the correlation is used as a descriptive statistics notation, even if the series η and f^{-1} are deterministic. The null correlation means that the difference between the true probability distribution function and the model have the same magnitude in low and in high density regions.

Under the Assumption **A.3.1.5** or **A.3.1.6**, if $\exists\ \nu \geq 0$ (sufficiently small) $\forall\, e \in E$ $\forall\, y \in \mathbf{R}$; $|\eta_e(y)| \leq \nu$, we show that (proof in Appendix):

$$0 \leq \mathrm{E}_Y[\, L_\tau - L_{\tau_c}\,] \text{ and} \tag{8.3}$$

$$0 \leq \mathrm{E}_Y[\, CRPS_{G,C\circ G}\,], \tag{8.4}$$

with $\mathrm{E}_Y[\, CRPS_{G,C\circ G}\,] \equiv \mathrm{E}_Y[\, CRPS(G, \mathbf{Y}) - CRPS(C \circ G, \mathbf{Y})\,]$. In other words, with those restrictions, our recalibration systematically improves the quality of the forecasts. Indeed, remember that the expectation of the Pinball-Loss score is minimized by the true quantile of the distribution of $\mathbf{Y}$ and negatively oriented. Thus, the lower the expectation of the Pinball-Loss score, the better.

8.3.2 Impact on Score: Bounds on Degradation

In reality, we cannot obtain the corrected probability level $\tau_c \in [0, 1]$, and we need to estimate it. If we want to upper bound the score change, we can study the more realistic case of

$$\mathrm{E}_Y[L_{\widehat{\tau_c}} - L_\tau] \equiv \mathrm{E}\left[\frac{1}{n}\sum_{j=1}^{n} L_\tau(y_j, G_j^{-1}(\widehat{\tau_c})) - L_\tau(y_j, G_j^{-1}(\tau))\right], \quad (8.5)$$

with $\tau, \widehat{\tau_c} \in [0, 1]$. In our case study, $\widehat{\tau_c}$ is obtained empirically, on the basis of the available PIT values. Thus, we have a consistent estimator of τ_c and one can rewrite (8.5) such as $\mathrm{E}_Y[\,L_\tau(Y, G^{-1}(Q_\tau))\,] - \mathrm{E}_Y[\,L_\tau(Y, G^{-1}(\tau))\,]$, with Q_τ a random variable converging in distribution to a Normal distribution with mean τ_c and a variance decreasing at the rate $\frac{1}{n}$.

In such a case, it is still possible to obtain bounds concerning the error induced by our correction. We make the following assumptions.

- **Assumption A.3.2.1**: F_e and G_e are C^2 $\forall\, e \in E$.
- **Assumption A.3.2.2**: $\forall\, y \in \mathbf{R}, \forall\, e \in E, |F_e(y) - G_e(y)| \leq \varepsilon$, with $\varepsilon \in [0, 1]$.
- **Assumption A.3.2.3**: the derivatives of G_e are lower bounded $\forall\, e \in E$, $\forall\, \tau \in [0, 1]$ by $1/\xi$, on the intervals $[G_e^{-1}(0 \vee (\tau - \varepsilon)), G_e^{-1}(1 \wedge (\tau + \varepsilon))]$, with $\xi \in]0, +\infty[$.
- **Assumption A.3.2.4**: $\forall\, e \in E, \forall\, \tau \in [0, 1], f_e(G_e^{-1}(\tau_c)) \leq \beta$, with $\beta \in]0, +\infty[$ and f_e the derivatives of F_e.
- **Assumption A.3.2.5**: f_e are continuous over the interval $[-\infty, G_e^{-1}(\tau_c)]$ $\forall\, e \in E$ and their derivatives are bounded, i.e $\forall y \in \mathbf{R}, \forall\, e \in E, |f_e^{'}(y)| \leq M$, with $M \in]0, +\infty[$ and f_e the derivative of F_e.
- **Assumption A.3.2.6**: the derivatives of g_e are bounded, i.e $\forall y \in \mathbf{R}, \forall\, e \in E$, $|g_e^{'}(y)| \leq \alpha$, with $\alpha \in]0, +\infty[$ and g_e the derivative of G_e.

Under the Assumptions **A.2.1, A.3.2.1–A.3.2.6**, we prove (proof in Appendix):

$$\left|\mathrm{E}_Y[L_{\widehat{\tau_c}} - L_\tau]\right| \leq 2\,\varepsilon^2\,\xi + \frac{C\,\lambda}{n} \quad \text{and} \quad (8.6)$$

$$\left|\mathrm{E}_Y[\,CRPS_{G,C\circ G}\,]\right| \leq 2\left(2\,\varepsilon^2\,\xi + \frac{C\,\lambda}{n}\right) \quad (8.7)$$

with $C = \frac{(1-\tau)\alpha\xi^3}{2} + C_{Int} + C_{abs}$, $C_{Int} = \frac{\xi^2\beta}{2}\left[1 + \alpha\xi^2 + \frac{\alpha^2\xi^4}{4}\right]$,

$C_{abs} = \frac{M\xi^3}{6}\left[1 + \frac{3\xi^3\alpha}{2} + \frac{3\xi^3\alpha^2}{4} + \frac{\xi^3\alpha^3}{8}\right]$ and $\frac{\lambda}{n}$, the variance of Q_τ.

This inequality shows that our recalibration deteriorates only slightly the quality of the forecasts in the worst case. Obviously, it also shows that our method improves only slightly the quality, but remember that our goal is to enforce the calibration constraint, which is achieved.

8.4 Case Study

We use our method on ensemble forecasts data set from the European Centre for Medium-Range Weather Forecasts (ECMWF). One can see in [5] that the statistical post-processing of the medium range ECMWF ensemble forecast has been addressed many times. The extended range (32 days instead of 10 days) has been addressed in some studies, but with the same methods and tools. We will show here that our recalibration method, despite its generic nature, is competitive with a standard post-processing method. We dispose of temperature forecasts in a 3-dimensional array. The first one represents the date of forecasts delivery. The forecasts were made every Monday and Thursday from 11/02/13 to 02/02/17. Since 3 observations are missing, we have 413 dates of forecasts delivery. The second dimension is the number of the scenario in the ensemble member, and we have 51 scenarios. The third dimension is the forecast horizon. Since we have 32 days sampled with a forecast every 3 h, it produces 256 horizons.

We study the calibration and compare the CRPS expectation using directly the ensemble forecast, the so-called Ensemble Model Output Statistics (EMOS) method and our recalibration method with a Cauchy Kernel dressing for the ensembles.

We choose a Cauchy Kernel in order to address problems with the bounds of the ensembles. Indeed, a lot of observations were out of the bounds of the ensemble, which produces a lot of PIT with value 0 or 1. Thus, to avoid this problem, we need to use a Kernel with heavy tail.

During the last 12 years, the ECMWF has changed its models 27 times, which means a change every 162 days on average. Thus, it is important to use a train sample significantly smaller than 162 days. However, it is also important to dispose of enough observations to obtain a consistant estimator of τ_c. Our method obtains good results with 30 days used for the recalibration but the algorithm to minimize in order to find the parameters of the EMOS in the *R* package *EnsembleMOS* doesn't converge if we use less than 60 days (at least with our data set). Thus, we chose to use 60 days for the recalibration.

To recalibrate the forecasts for a particular forecasting day and a particular horizon (remember that we have 256 horizons), we use the forecasts made for the same horizon, over the 60 previous dates of forecast delivery for the two methods. However, with our method, we use a linear interpolation based on the PIT series formed by these 60 previous days to recalibrate the forecasts. The linear interpolation is also used to calculate the different quantile levels when we are not working with EMOS (in that case, for the recalibration or to calculate the quantile, we use the Normale distribution with the fitted parameters). Note that the hypotheses concerning only

Table 8.1 Success rate to 5% K–S test

	Raw ensemble	EMOS	Our method
Success rate in %	14	0.39	96

We have calculated the PIT series for each horizon (256), and use 5% K–S test for each of them. The success rate is the percentage of horizons passing the test

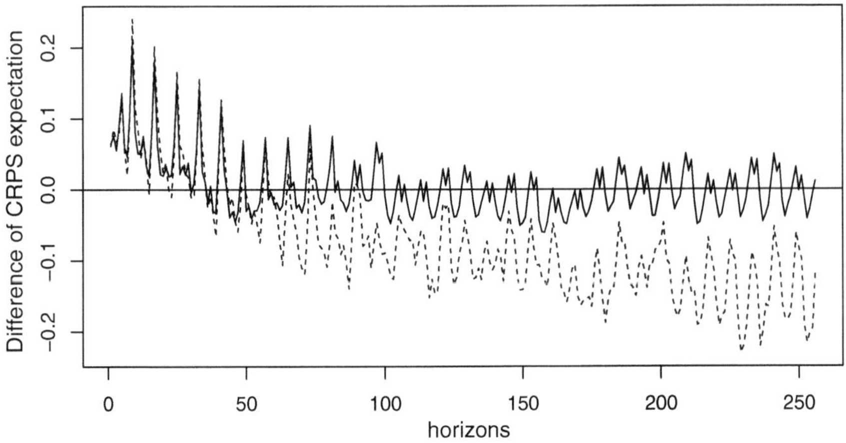

Fig. 8.1 Comparison of EMOS and our method CRPS expectation with that of raw ensemble. The empty line corresponds to our method and the dashed one to the EMOS

G_e are verified $\forall\ e \in E$. Besides, even if we cannot verify the other hypotheses, we show expected results.

Let's start with the calibration property:

As expected, we can see in Table 8.1 that our method allows us the test of validity to be passed while the use of the raw ensemble fails. The EMOS also failed to pass the test. Clearly, our method is useful to ensure the calibration property. But how about the quality of the density forecast? In order to evaluate the impact of our correction on the forecast quality, we are interested in the CRPS expectation.

We can see in Fig. 8.1 that EMOS as well as our method are more efficient than the raw ensemble for little horizons. However, the EMOS deteriorates clearly the quality of the forecasts when the horizon grows, contrarily to our method which deteriorates only slightly the quality of the forecasts, when it is the case.

Thus, this study highlights perfectly the usefulness of our method, which is very simple to use. Indeed, it shows that it allows us to ensure the validity constraint, with a limited negative impact on the quality.

Acknowledgements This research was supported by the ANR project FOREWER (ANR-14-CE05-0028).

8.5 Appendix

Here are gathered all the proofs concerning the results presented in the chapter. The first section is concerned by proofs of results in an infinite sample and the second by result in a finite sample.

Lemma 8.1

$$\mathrm{E}_Y[\,L_\tau - L_{\tau_c}\,] = \sum_e p_e \int_{G_e^{-1}(\tau_c)}^{G_e^{-1}(\tau)} (F_e(y) - \tau)dy\,,$$

with $\tau,\tau_c \in [0,1]$ and p_e the frequency of appearance of the state e. Under the Assumption **A.2.1**, we prove Lemma 8.1.

Proof We have:

$$\mathrm{E}_Y[\,L_\tau - L_{\tau_c}\,] = \sum_e p_e \left(\mathrm{E}_Y[\,L_\tau(\boldsymbol{Y}, G_e^{-1}(\tau))\,] - \mathrm{E}_Y[\,L_\tau(\boldsymbol{Y}, G_e^{-1}(\tau_c)\,]\right). \quad (8.8)$$

First, we only focus on a particular e. Thus, we are interested in:

$$\mathrm{E}_Y[\,L_\tau(\boldsymbol{Y}, G_e^{-1}(\tau))\,] - \mathrm{E}_Y[\,L_\tau(\boldsymbol{Y}, G_e^{-1}(\tau_c)\,] \equiv \mathrm{E}_{Y,e}[\,L_{\tau,\tau_c}\,].$$

For ease of notation and comprehension, we suppress e in the notation since there is no confusion. Moreover, we suppose, for ease of notation again (and since we obtain the same result if we inverse the inequality) that $G^{-1}(\tau) \leq G^{-1}(\tau_c)$. So, we have:

$$\begin{aligned}
\mathrm{E}_Y[\,L_{\tau,\tau_c}\,] &= \int_{-\infty}^{+\infty} \left([y - G^{-1}(\tau)]\,\tau + [G^{-1}(\tau) - y]\,\mathbf{1}_{\{y \leq G^{-1}(\tau)\}}\right) f_Y(y)\,\mathrm{d}y \\
&\quad - \int_{-\infty}^{+\infty} \left([y - G^{-1}(\tau_c)]\,\tau + [G^{-1}(\tau_c) - y]\,\mathbf{1}_{\{y \leq G^{-1}(\tau_c)\}}\right) f_Y(y)\,\mathrm{d}y \\
&= [G^{-1}(\tau_c) - G^{-1}(\tau)]\,\tau + [G^{-1}(\tau) - G^{-1}(\tau_c)]\,F \circ G^{-1}(\tau) \\
&\quad - G^{-1}(\tau_c)\,[\,F \circ G^{-1}(\tau_c) - F \circ G^{-1}(\tau)\,] + \int_{y=G^{-1}(\tau)}^{G^{-1}(\tau_c)} \underbrace{y}_{v}\ \underbrace{f_Y(y)}_{u'}\,\mathrm{d}y\,.
\end{aligned}$$

Using integral by parts, we have:

$$\begin{aligned}
\mathrm{E}_Y[\,L_{\tau,\tau_c}\,] &= [G^{-1}(\tau_c) - G^{-1}(\tau)]\,\tau + \int_{y=G^{-1}(\tau_c)}^{G^{-1}(\tau)} F(y)\,\mathrm{d}y \\
&= \int_{y=G^{-1}(\tau_c)}^{G^{-1}(\tau)} [\,F(y) - \tau\,]\,\mathrm{d}y\,.
\end{aligned}$$

Replacing it in (8.8) finishes the demonstration. □

8.5.1 Impact on Score: Conditions for Improvement

In this section, the reader can find the proofs of results mentioned in Sect. 8.3.1 of the chapter. We first demonstrate how to approximate the difference of L_τ expectation before showing that under some hypotheses, our correction improves systematically the quality of the forecasts.

8.5.1.1 Rewriting the Difference of L_τ Expectation

Under the Assumptions **A.2.1**, **A.3.1.1–A.3.1.4** and using functional derivatives and the implicit function theorem, we prove (8.2).

Proof Remember: Let H be a functional, h a function, α a scalar and δ an arbitrary function.

We can write the expression of the functional evaluated at $f + \delta\alpha$ as follow:

$$H[h+\delta\,\alpha] = H[h] + \frac{\mathrm{d}H[h+\delta\,\alpha]}{\mathrm{d}\alpha}|_{\alpha=0}\,\alpha + \frac{1}{2}\frac{\mathrm{d}^2 H[h+\delta\,\alpha]}{\mathrm{d}\alpha^2}|_{\alpha=0}\,\alpha^2 + \cdots + \mathrm{Rem}(\alpha),$$

with $\mathrm{Rem}(\alpha)$ the remainder. Denote:

$$\begin{aligned}\Delta PL[h] &= \sum_e p_e \int_{h_e^{-1}(\tau_e)}^{h_e^{-1}(\tau)} (F_e(y)-\tau)\mathrm{d}y \\ &= \sum_e p_e \Delta PL_e[h_e]\,.\end{aligned}$$

For ease of notation, denote $\Delta PL_e[F_e + \delta_e\alpha] \equiv \Delta PL_{F,\delta,e}$. Choosing $H = \Delta PL_e$, $h = F_e$ and $\eta_e = \alpha\delta_e$ (even if we use $\alpha\delta_e$ in the development in order to use functional derivatives, directional derivatives and the implicit function theorem), we have:

$$\begin{aligned}\Delta PL_{F,\delta,e} &\sim \Delta PL_e[F_e] + \frac{\mathrm{d}\Delta PL_{F,\delta,e}}{\mathrm{d}\alpha}|_{\alpha=0}\,\alpha + \frac{1}{2}\frac{\mathrm{d}^2\Delta PL_{F,\delta,e}}{\mathrm{d}\alpha^2}|_{\alpha=0}\,\alpha^2 + \mathrm{Rem}_e(\alpha) \\ &= \left[\frac{\partial\Delta PL_{F,\delta,e}}{\partial\alpha}|_{\alpha=0\,,\,\tau_c=\tau} + \frac{\partial\Delta PL_{F,\delta,e}}{\partial\tau_c}|_{\alpha=0\,,\,\tau_c=\tau}\frac{\mathrm{d}\tau_c}{\mathrm{d}\alpha}\right]\alpha \\ &+ \left[\frac{\partial^2\Delta PL_{F,\delta,e}}{\partial\alpha^2}|_{\alpha=0\,,\,\tau_c=\tau} + 2\frac{\partial^2\Delta PL_{F,\delta,e}}{\partial\alpha\partial\tau_c}|_{\alpha=0\,,\,\tau_c=\tau}\frac{\mathrm{d}\tau_c}{\mathrm{d}\alpha}\right]\frac{\alpha^2}{2} \\ &+ \left[\frac{\partial^2\Delta PL_{F,\delta,e}}{\partial\tau_c^2}|_{\alpha=0\,,\,\tau_c=\tau}\left(\frac{\mathrm{d}\tau_c}{\mathrm{d}\alpha}\right)^2 + \frac{\partial\Delta PL_{F,\delta,e}}{\partial\tau_c}|_{\alpha=0\,,\,\tau_c=\tau}\frac{\mathrm{d}^2\tau_c}{\mathrm{d}\alpha^2}\right]\frac{\alpha^2}{2} \\ &+ \mathrm{Rem}_e(\alpha)\,.\end{aligned}$$

To calculate $\frac{\mathrm{d}\tau_c}{\mathrm{d}\alpha}$, we will use the equation which link τ_c and α:

$$\sum_e p_e F_e \circ (F_e + \delta_e \alpha)^{-1}(\tau_c) = \tau.$$

Using the implicit function theorem, we find:

$$\frac{\mathrm{d}\tau_c}{\mathrm{d}\alpha} = \sum_e p_e \delta_e \circ F_e^{-1}(\tau)$$

Now, we need to calculate partial derivatives:

$$\frac{\partial \Delta PL_{F,\delta,e}}{\partial \alpha}|_{\alpha=0,\,\tau_c=\tau} = \frac{\partial \left(\int_{(F_e+\delta_e\alpha)^{-1}(\tau_c)}^{(F_e+\delta_e\alpha)^{-1}(\tau)} (F_e(y) - \tau)\mathrm{d}y \right)}{\partial \alpha}|_{\alpha=0,\,\tau_c=\tau} = 0;$$

$$\frac{\partial \Delta PL_{F,\delta,e}}{\partial \tau_c}|_{\alpha=0,\,\tau_c=\tau} = 0 \; ; \quad \frac{\partial^2 \Delta PL_{F,\delta,e}}{\partial \tau_c^2}|_{\alpha=0,\,\tau_c=\tau} = -\frac{1}{f_e \circ F_e^{-1}(\tau)};$$

$$\frac{\partial^2 \Delta PL_{F,\delta,e}}{\partial \alpha^2}|_{\alpha=0,\,\tau_c=\tau} = 0 \; ; \quad \frac{\partial^2 \Delta PL_{F,\delta,e}}{\partial \alpha \partial \tau_c}|_{\alpha=0,\,\tau_c=\tau} = \frac{\delta_e \circ F_e^{-1}(\tau)}{f_e \circ F_e^{-1}(\tau)}.$$

Thus, we have:

$$\Delta PL_e[F_e + \delta_e \alpha] \sim \left[\left(\frac{\delta_e \circ F_e^{-1}(\tau)}{f_e \circ F_e^{-1}(\tau)} \right) \sum_e p_e \delta_e \circ F_e^{-1}(\tau) \right] \alpha^2 - \left[\frac{(\sum_e p_e \delta_e \circ F_e^{-1}(\tau))^2}{2 f_e \circ F_e^{-1}(\tau)} \right] \alpha^2 + \mathrm{Rem}_e(\alpha),$$

and hence:

$$\Delta PL[F + \delta\alpha] \sim \left(\sum_e \frac{p_e \delta_e (F_e^{-1}(\tau))}{f_e(F_e^{-1}(\tau))} \right) \left(\sum_e p_e \delta_e (F_e^{-1}(\tau)) \right) \times \alpha^2 - \left(\sum_e \frac{p_e}{2 f_e(F_e^{-1}(\tau))} \right) \left(\sum_e p_e \delta_e (F_e^{-1}(\tau)) \right)^2 \times \alpha^2 + \sum_e p_e \mathrm{Rem}_e(\alpha).$$

Now, let's focus on the remainders. Following the Taylor–Lagrange inequality, if M such that $\left| \frac{\mathrm{d}^3 \Delta PL_{F,\delta,e}}{\mathrm{d}\alpha^3} \right| \leq M$ exists, we have $|\mathrm{Rem}_e(\alpha)| \leq \frac{M|\alpha^3|}{3!}$. Let' s find conditions for the existence of M. The third derivative is:

$$\frac{\mathrm{d}^3 \Delta PL_{F,\delta,e}}{\mathrm{d}\alpha^3} = \frac{\partial \Delta PL_{F,\delta,e}}{\partial \tau_c}\frac{\mathrm{d}^3\tau_c}{\mathrm{d}\alpha^3} + 3\frac{\partial^2 \Delta PL_{F,\delta,e}}{\partial \tau_c \partial \alpha}\frac{\mathrm{d}^2\tau_c}{\mathrm{d}\alpha^2} + 3\frac{\partial^2 \Delta PL_{F,\delta,e}}{\partial \tau_c^2}\frac{\mathrm{d}^2\tau_c}{\mathrm{d}\alpha^2}\frac{\mathrm{d}\tau_c}{\mathrm{d}\alpha}$$
$$+ \frac{\partial^3 \Delta PL_{F,\delta,e}}{\partial \alpha^3} + 3\frac{\partial^3 \Delta PL_{F,\delta,e}}{\partial \tau_c \partial \alpha^2}\frac{\mathrm{d}\tau_c}{\mathrm{d}\alpha} + 3\frac{\partial^3 \Delta PL_{F,\delta,e}}{\partial \tau_c^2 \partial \alpha}\left(\frac{\mathrm{d}\tau_c}{\mathrm{d}\alpha}\right)^2$$
$$+ \frac{\partial^3 \Delta PL_{F,\delta,e}}{\partial \tau_c^3}\left(\frac{\mathrm{d}\tau_c}{\mathrm{d}\alpha}\right)^3 .$$

Let's calculate the partial derivatives of order 3:

$$\frac{\partial^3 \Delta PL_{F,\delta,e}}{\partial \alpha^3}|_{\alpha=0,\,\tau_c=\tau} = 0 \ ; \quad \frac{\partial^3 \Delta PL_{F,\delta,e}}{\partial \tau_c^3}|_{\alpha=0,\,\tau_c=\tau} = 2\frac{f_e^{'} \circ F_e^{-1}(\tau)}{f_e \circ F_e^{-1}(\tau)} ;$$
$$\frac{\partial^3 \Delta PL_{F,\delta,e}}{\partial \tau_c^2 \partial \alpha}|_{\alpha=0,\,\tau_c=\tau} = -2\frac{f_e^{'} \circ F_e^{-1}(\tau)}{f_e \circ F_e^{-1}(\tau)^3}(\delta_e \circ F_e^{-1}(\tau)) - 2\frac{\delta_e^{'} \circ F_e^{-1}(\tau)}{f_e \circ F_e^{-1}(\tau)^2} ;$$
$$\frac{\partial^3 \Delta PL_{F,\delta,e}}{\partial \tau_c \partial \alpha^2}|_{\alpha=0,\,\tau_c=\tau} = \frac{f_e^{'} \circ F_e^{-1}(\tau)}{f_e \circ F_e^{-1}(\tau)^3}\left(\delta_e \circ F_e^{-1}(\tau)\right)^2$$
$$- 2\frac{\delta_e^{'} \circ F_e^{-1}(\tau)}{f_e \circ F_e^{-1}(\tau)^2}\left(\delta_e \circ F_e^{-1}(\tau)\right).$$

Moreover, we have:

$$\frac{d^2\tau_c}{d\alpha^2} = \sum_e p_e \left(\frac{2\delta_e^{'} \circ F_e^{-1}(\tau) - f_e^{'} \circ F_e^{-1}(\tau)}{f_e \circ F_e^{-1}(\tau)}\right)\delta_e \circ F_e^{-1}(\tau).$$

Since η_e, its first, second and third derivatives are finite in $F_e^{-1}(\tau)$, it is also the case for δ_e and the partial derivatives are finite. Furthermore, f_e, δ_e and their derivatives are bounded (since η_e and their derivatives are bounded), which implies that the second derivatives of $\Delta PL_e[F_e + \delta_e\alpha]$ are also bounded. Thus, under these conditions, M exists. Then, we can write $\frac{\mathrm{d}^3 \Delta PL_{F,\delta,e}}{\mathrm{d}\alpha^3} = M_1\delta_e^3$ and hence $|\mathrm{Rem}_e(\alpha)| \leq \frac{|M_1||\alpha\delta_e|^3}{3!}$ which implies that $\lim \frac{\mathrm{Rem}_e(\alpha)}{(\alpha\delta_e)^2} = 0$, $\alpha\delta_e \to 0$, which shows that $\mathrm{Rem}_e(\alpha)$ is negligible compared to $\frac{\mathrm{d}^2 \Delta PL_{F,\delta,e}}{\mathrm{d}\alpha^2}$.

Moreover, since $\forall\ e \in E$ the functions F_e are C^3 and the functions f_e and their derivatives are bounded by a constant which doesn't depend on e, $\forall\ e \in E$, the development is valid for all directions and thus, since $\eta_e = G_e - F_e$, we have:

$$\mathrm{E}_Y[\, L_\tau - L_{\tau_c}\,] \sim \left(\sum_e \frac{p_e\eta_e(F_e^{-1}(\tau))}{f_e(F_e^{-1}(\tau))}\right)\left(\sum_e p_e\eta_e(F_e^{-1}(\tau))\right)$$
$$-\left(\sum_e \frac{p_e}{2f_e(F_e^{-1}(\tau))}\right)\left(\sum_e p_e\eta_e(F_e^{-1}(\tau))\right)^2$$
$$\text{as } max\ \eta_e \to 0.$$

To finish the demonstration, remark that Lemma 8.1 proves that:

$$\Delta PL[G] = \mathrm{E}_Y[\, L_\tau - L_{\tau_c}\,]. \qquad \square$$

8.5.1.2 Systematic Improvement of the Quality

Under the Assumption **A.3.1.5** or **A.3.1.6**, if $\exists\ \nu \geq 0$ (sufficiently small) $\forall\ e \in E$ $\forall\ y \in \mathbf{R}$; $|\eta_e(y)| \leq \nu$, we show (8.3) and (8.4):

Proof Prove (8.3) is equivalent to show that $\Delta PL[G]$ is positive, and if we rewrite:

$$\Delta PL[G] \sim (2\mathrm{E}[f^{-1}\eta] - \mathrm{E}[f^{-1}]\mathrm{E}[\eta])\mathrm{E}[\eta]\,,$$

it is clear that the Assumption **A.3.1.6** ensures the positivity of $\Delta PL[G]$.

However, we need more argumentation to understand the complete utility of the Assumption **A.3.1.5**. Let's look at one of the two worst cases: only two states of the world, the correlation coefficient $\rho = -1$, $\eta > 0$ (the other case is when $\rho = 1$ and $\eta < 0$) and at each bound of the support of δ and f^{-1}, there is half of the probability mass. We also consider that the ratios between max and min of the supports are equal. If we define $max_e = M$ and $min_e = \frac{M}{r}$, one has the following equation:

$$\frac{1}{2} = \frac{2(r^2+1)}{(r+1)^2} - 1.$$

Solving this equation in r produces the expected result concerning the ratio between max and min values of η and f^{-1}.

Now, let's prove (8.4). According to (8.1), we have:

$$\mathrm{E}_Y[\, CRPS_{G,C\circ G}\,] = 2\int_{-\infty}^{+\infty}\left(\int_0^1 L_\tau(y, G^{-1}(\tau)) - L_\tau(y, G^{-1}\circ C^{-1}(\tau))\mathrm{d}\tau\right) f_Y(y)\mathrm{d}y.$$

We can rewrite:

$$\begin{aligned}\mathrm{E}_Y[\, CRPS_{G,C\circ G}\,] = {} & 2\int_{-\infty}^{+\infty}\int_0^1 L_\tau(y, G^{-1}(\tau))\, f_Y(y)\,\mathrm{d}\tau\,\mathrm{d}y \\ & - 2\int_{-\infty}^{+\infty}\int_0^1 L_\tau(y, G^{-1}\circ C^{-1}(\tau))\, f_Y(y)\,\mathrm{d}\tau\mathrm{d}y\,,\end{aligned}$$

and using the Fubini–Tonelli theorem, one obtains:

$$\begin{aligned} \mathrm{E}_Y[\,CRPS_{G,C\circ G}\,] &= 2\int_0^1 \mathrm{E}_Y[\,L_\tau - L_{\tau_c}\,]\mathrm{d}\tau \\ &\geq 0. \end{aligned} \tag{8.9}$$

□

8.5.2 *Impact on Score: Bounds on Degradation*

Under the Assumptions **A.2.1, A.3.2.1–A.3.2.6** we prove (8.6) and (8.7).

Proof adding and substracting $\mathrm{E}_Y[\,L_\tau(Y, G^{-1}(\tau_c))\,]$ to $\mathrm{E}_Y[L_{\widehat{\tau}_c} - L_\tau\,]$, we obtain:

$$\begin{aligned} \mathrm{E}_Y[L_{\widehat{\tau}_c} - L_\tau\,] &= \mathrm{E}_Y[\,L_\tau(Y, G^{-1}(Q_\tau))\,] - \mathrm{E}_Y[\,L_\tau(Y, G^{-1}(\tau_c))\,] \\ &\quad + \mathrm{E}_Y[\,L_\tau(Y, G^{-1}(\tau_c))\,] - \mathrm{E}_Y[\,L_\tau(Y, G^{-1}(\tau))\,]\,, \end{aligned}$$

and finally:

$$\mathrm{E}_Y[L_{\widehat{\tau}_c} - L_\tau\,] = \mathrm{E}_{Y,e}[\,L_\tau(Y, G_e^{-1}(Q_\tau))\,] - \mathrm{E}_{Y,e}[\,L_\tau(Y, G_e^{-1}(\tau_c))\,] - \mathrm{E}_Y[\,L_\tau - L_{\tau_c}\,].$$

To begin with, we treat the third term on the right side. We have:

$$\mathrm{E}_{Y,e}[\,L_{\tau,\tau_c}\,] = \int_{y=G_e^{-1}(\tau_c)}^{G_e^{-1}(\tau)} [\,F_e(y) - \tau\,]\,\mathrm{d}y\,.$$

Using the change of variable $y = G_e^{-1}(z)$ and taking the absolute value, we find:

$$\left|\mathrm{E}_{Y,e}[\,L_{\tau,\tau_c}\,]\right| = \left|\int_{z=\tau_c}^{\tau} (F_e \circ G_e^{-1}(z) - \tau)\,\frac{1}{g_e(G_e^{-1}(z))}\,\mathrm{d}z\right|.$$

Now, one needs to distinguish two cases.

If $\tau > \tau_c$, one has:

$$\begin{aligned} \left|\,\mathrm{E}_{Y,e}[\,L_{\tau,\tau_c}\,]\right| &= \int_{z=\tau_c}^{\tau} \left|(F_e \circ G_e^{-1}(z) - \tau)\,\frac{1}{g_e(G_e^{-1}(z))}\right| \mathrm{d}z \\ &\leq \int_{z=\tau_c}^{\tau} \left|(F_e \circ G_e^{-1}(z) - \tau)\right|\,\xi\;\mathrm{d}z\,. \end{aligned}$$

Since $|\,F_e(z) - G_e(z)\,| \leq \varepsilon$, $\forall z \in \mathbf{R}$, $\forall e \in E$, one obtains $|\,F_e \circ G_e^{-1}(z) - z\,| \leq \varepsilon$, $\forall z \in [0, 1]$, $\forall e \in E$ and then:

- if $z = \tau$, one has $\left| F_e \circ G_e^{-1}(\tau) - \tau \right| \leq \varepsilon$,
- if $z = \tau_c$, $\left| F_e \circ G_e^{-1}(\tau_c) - \tau) \right| = \left| F_e \circ G_e^{-1}(\tau_c) - \tau_c + \tau_c - \tau \right|$.

Moreover, one has:

$$\begin{aligned} |\tau_c - \tau| &= \left| \sum_e p_e \left(\tau_c - F_e \circ G_e^{-1}(\tau_c) \right) \right| \\ &\leq \sum_e p_e \left| F_e \circ G_e^{-1}(\tau_c) - \tau_c \right| \\ &\leq \varepsilon\,, \end{aligned}$$

and finally:

$$\begin{aligned} \left| F_e \circ G_e^{-1}(\tau_c - \tau) \right| &\leq \left| F_e \circ G_e^{-1}(\tau_c) - \tau_c \right| + |\tau_c - \tau| \\ &\leq 2\,\varepsilon\,. \end{aligned}$$

One deduces, when $\tau > \tau_c$:

$$\left| \mathrm{E}_{Y,e}[\, L_{\tau,\tau_c}\,] \right| \leq 2\,(\,\tau - \tau_c\,)\,\varepsilon\,\xi.$$

When $\tau < \tau_c$, one obtains:

$$\left| \mathrm{E}_{Y,e}[\, L_{\tau,\tau_c}\,] \right| \leq \int_{z=\tau}^{\tau_c} \left| (F_e \circ G_e^{-1}(z) - \tau) \right| \,\xi\, \mathrm{d}z\,,$$

and using the same arguments as previously:

$$\left| \mathrm{E}_{Y,e}[\, L_{\tau,\tau_c}\,] \right| \leq 2\,(\,\tau_c - \tau\,)\,\varepsilon\,\xi.$$

Hence, one concludes that:

$$\left| \mathrm{E}_{Y,e}[\, L_{\tau,\tau_c}\,] \right| \leq 2\,|\,\tau - \tau_c\,|\,\varepsilon\,\xi.$$

To finish, replacing $\mathrm{E}_{Y,e}[\, L_{\tau,\tau_c}\,]$ in (8.8), we have:

$$\left| \mathrm{E}_{Y}[\, L_{\tau} - L_{\tau_c}\,] \right| \leq 2\,\varepsilon^2\,\xi.$$

Now let's focus on the remainder on the right side. First, we only focus on a particular e. Thus, we are interested in:

$$\mathrm{E}_Y[\, L_\tau(Y, G_e^{-1}(Q_\tau))\,] - \mathrm{E}_Y[\, L_\tau(Y, G_e^{-1}(\tau_c))\,] \equiv \mathrm{E}_{Y,e}[L_{\widehat{\tau_c}} - L_{\tau_c}\,].$$

For ease of notation and comprehension, we suppress e in the notation since there is no confusion. So, we have:

$$\mathrm{E}_Y[L_{\widehat{\tau_c}} - L_{\tau_c}] = \left(\frac{1}{2} - \tau\right) \mathrm{E}_Y\left[G^{-1}(Q_\tau) - G^{-1}(\tau_c)\right] + \frac{1}{2}\mathrm{E}_Y\left[\,|Y - G^{-1}(Q_\tau)| - |Y - G^{-1}(\tau_c)|\right].$$

We find:

$$\left|\mathrm{E}_Y[L_{\widehat{\tau_c}} - L_{\tau_c}]\right| \le \left|\frac{1}{2}\mathrm{E}_Y\left[|Y - G^{-1}(Q_\tau)| - |Y - G^{-1}(\tau_c)| - G^{-1}(Q_\tau) + G^{-1}(\tau_c)\right]\right| + (1-\tau)\left|\mathrm{E}_Y\left[G^{-1}(Q_\tau) - G^{-1}(\tau_c)\right]\right|.$$

Let's focus on the second term on the right side. Using a Taylor series approximation around $\tau_c \in [0, 1]$ and the Taylor–Lagrange formula for the remainder, one has:

$$G^{-1}(Q_\tau) = G^{-1}(\tau_c) + \frac{1}{g(G^{-1}(\tau_c))}(Q_\tau - \tau_c) + \frac{g^{'}(\gamma)}{g(\gamma)^3}\frac{(Q_\tau - \tau_c)^2}{2},$$

with $\gamma = \tau_c + (Q_\tau - \tau_c)\,\theta$, and $0 < \theta < 1$.

And so

$$(1-\tau)\left|\mathrm{E}_Y\left[G^{-1}(Q_\tau) - G^{-1}(\tau_c)\right]\right| \le \frac{(1-\tau)\,\alpha\,\xi^3}{2}\frac{\lambda}{n}.$$

Now, one can study the first term on the right side. Some useful remarks before the next: one can easily see that the study of such a function can be restricted to a study on the interval $I_y :=]-\infty, G^{-1}(\tau_c)]$, since we can find results on the interval $[G^{-1}(\tau_c), \infty[$ using the same arguments.

Let's define $G^{-1}(Q_\tau) \equiv Z_\tau$, $G_{\tau_c}^{-1} \equiv G^{-1}(\tau_c)$ and $f_Y^{G_{\tau_c}^{-1}} \equiv f_Y(G^{-1}(\tau_c))$, for ease of notation.

Thus, we are interested in calculating:

$$\frac{1}{2}\int_{y=-\infty}^{G_{\tau_c}^{-1}} f_Y(y)\,\underbrace{\left(\mathrm{E}_{Z_\tau}[\,|G_{\tau_c}^{-1} - Z_\tau| + |Z_\tau - y|\,] - G_{\tau_c}^{-1} + y\right)}_{=\mathrm{E}_{Z_\tau}[\,|Z_\tau - y| - Z_\tau\,] + y}\,\mathrm{d}y\,. \tag{8.10}$$

However, the function studied in the integral is complicated to work with. So, one will prefer to use its integral version, that is,

$$\mathrm{E}_{Z_\tau}[\,|Z_\tau - y| - Z_\tau\,] + y = \int_{u=-\infty}^{y}\frac{\mathrm{d}}{\mathrm{d}u}\left(\mathrm{E}_{Z_\tau}[\,|Z_\tau - u| - Z_\tau\,] + u\right)\mathrm{d}u\,.$$

For the bounds of the integral, the upper one is obvious. To justify the lower one, it is important to note that lim $\mathrm{E}_{Z_\tau}[\,|Z_\tau - y| - Z_\tau\,] + y = 0,\ y \to -\infty$.

Indeed, one has:

$$\begin{aligned}
\mathrm{E}_{Z_\tau}[\,|Z_\tau - y| - Z_\tau\,] + y &= \int_{z=-\infty}^{y} (y-z)\,h(z)\,dz + \int_{z=y}^{\infty} (z-y)\,h(z)\,\mathrm{d}z \\
&\quad + \int_{z=-\infty}^{\infty} (y-z)\,h(z)\,\mathrm{d}z \\
&= \int_{z=-\infty}^{y} 2(y-z)\,h(z)\,\mathrm{d}z \\
&= 2y\,H(y) - \int_{z=-\infty}^{y} 2z\,h(z)\,\mathrm{d}z\,,
\end{aligned}$$

with h and H the p.d.f and the c.d.f of the variable Z_t. If the variable Z_τ has a finite mean, lim, $h(y) = 0,\ y \to -\infty$, and thus it is clear that the choice of $-\infty$ for the lower bound of the integral is the good one.

At this stage, it is not easy to see the usefulness of the transformation, but it will be after the following calculus:

$$\begin{aligned}
\frac{d}{du}(\mathrm{E}_{Z_\tau}[\,|Z_\tau - u| - Z_\tau\,] + u) &= 1 + \frac{d}{du}\left(\int_{z=-\infty}^{u} (u-z)\,h(z)\,\mathrm{d}z\right) \\
&\quad + \frac{d}{du}\left(\int_{z=u}^{\infty} (z-u)\,h(z)\,\mathrm{d}z\right).
\end{aligned}$$

Finally, we have:

$$\begin{aligned}
\frac{d}{du}(\mathrm{E}_{Z_\tau}[\,|Z_\tau - u| - Z_\tau\,] + u) &= \int_{z=-\infty}^{u} h(z)\,dz - \int_{z=u}^{\infty} h(z)\,\mathrm{d}z + 1 \\
&= H(u) - (1 - H(u)) + 1 \\
&= 2\,H(u).
\end{aligned}$$

Now, it is clear that this transformation could help us for the calculus of (8.10) since it is equivalent to study:

$$\int_{y=-\infty}^{G_{\tau_c}^{-1}} f_Y(y) \left(\int_{u=-\infty}^{y} H(u)\,du\right) \mathrm{d}y \equiv \text{Half Int.}$$

A difficulty remains, though. Indeed, f_Y in unknown, and in consequence, not easy to work with. That's why, at first, one will use $f_Y^{G_{\tau_c}^{-1}}$ for our calculus, and then we will study the impact of such a manipulation.

Let's start with the first task. Using an integral by part on Half Int:

$$\int_{y=-\infty}^{G_{\tau_c}^{-1}} \underbrace{f_Y^{G_{\tau_c}^{-1}}}_{u'} \underbrace{\left(\int_{u=-\infty}^{y} H(u)\,du\right)}_{v} \mathrm{d}y\,.$$

One obtains:

$$\begin{aligned}
\text{Half Int} &= \left[y f_Y^{G_{\tau_c}^{-1}} \left(\int_{u=-\infty}^{y} H(u)\,du\right)\right]_{y=-\infty}^{G_{\tau_c}^{-1}} - \int_{y=-\infty}^{G_{\tau_c}^{-1}} y f_Y^{G_{\tau_c}^{-1}} H(y)\,\mathrm{d}y \\
&= \int_{u=-\infty}^{G_{\tau_c}^{-1}} f_Y^{G_{\tau_c}^{-1}} \underbrace{[G_{\tau_c}^{-1} - u]}_{u'} \underbrace{H(u)}_{v} \,\mathrm{d}u \\
&= \left[f_Y^{G_{\tau_c}^{-1}} \left(u\, G_{\tau_c}^{-1} - \frac{u^2}{2}\right) H(u)\right]_{u=-\infty}^{G_{\tau_c}^{-1}} \\
&\quad - \int_{u=-\infty}^{G_{\tau_c}^{-1}} f_Y^{G_{\tau_c}^{-1}} \left(u\, G_{\tau_c}^{-1} - \frac{u^2}{2}\right) h(u)\,\mathrm{d}u\,.
\end{aligned}$$

Since $\left(u\, G_{\tau_c}^{-1} - \frac{u^2}{2}\right) = \left(\frac{(u-G_{\tau_c}^{-1})^2}{2} - \frac{(G_{\tau_c}^{-1})^2}{2}\right)$, we have:

$$\text{Half Int} = f_Y^{G_{\tau_c}^{-1}} \left(\int_{u=-\infty}^{G_{\tau_c}^{-1}} \frac{(u - G_{\tau_c}^{-1})^2}{2} h(u)\,\mathrm{d}u\right).$$

Now, using the change of variable $G(u) = z$, a Taylor series approximation around τ_c and the Taylor–Lagrange formula, one has the following approximation for Half Int:

$$\frac{f_Y^{G_{\tau_c}^{-1}}}{2} \int_{z=0}^{\tau_c} \left[\frac{1}{g(G_{\tau_c}^{-1})^2}(z-\tau_c)^2 + \frac{g'(\gamma)}{g(G_{\tau_c}^{-1})\, g(\gamma)^3}(z-\tau_c)^3 + \frac{g'(\gamma)^2}{4g(\gamma)^6}(z-\tau_c)^4\right]\phi(y)\mathrm{d}y\,,$$

with ϕ the p.d.f of the random variable Q_τ. Using the Jensen inequality and since $0 \leq z \leq \tau_c$, we find:

$$\begin{aligned}
|\text{Half Int}| &\leq \frac{f_Y^{G_{\tau_c}^{-1}}}{2}\left[\frac{\xi^2}{2}\frac{\lambda}{n} + \frac{\alpha\xi^4}{2}\frac{\lambda}{n} + \frac{\alpha^2\xi^6}{8}\frac{\lambda}{n}\right] \\
&\leq \frac{1}{2}\frac{C_{int}\,\lambda}{n}.
\end{aligned}$$

Since $\frac{\lambda}{n}$, which is the variance of the random variable Q_τ, is decreasing with n, let's study:

$$\Delta_{\int f} \equiv \left| \int_{u=-\infty}^{G_{\tau_c}^{-1}} (f_Y(y) - f_Y^{G_{\tau_c}^{-1}}) \left(\int_{u=-\infty}^{y} H(u)\,du \right) dy \right| .$$

Since one supports the hypothesis that $f_Y^{'}$ is bounded, using the mean value theorem, one has:

$$\begin{aligned} \Delta_{\int f} &\leq \int_{y=-\infty}^{G_{\tau_c}^{-1}} | f_Y(y) - f_Y^{G_{\tau_c}^{-1}} | \left(\int_{u=-\infty}^{y} H(u)\,du \right) dy \\ &\leq \int_{y=-\infty}^{G_{\tau_c}^{-1}} M \underbrace{(G_{\tau_c}^{-1} - y)}_{u'} \underbrace{\left(\int_{u=-\infty}^{y} H(u)\,du \right)}_{v} dy , \end{aligned}$$

and thus,

$$\begin{aligned} \Delta_{\int f} &\leq M \left(\left[\left(y\, G_{\tau_c}^{-1} - \frac{y^2}{2} \right) \int_{u=-\infty}^{y} H(u)\,du \right]_{y=-\infty}^{G_{\tau_c}^{-1}} - \int_{y=-\infty}^{G_{\tau_c}^{-1}} \left(y\, G_{\tau_c}^{-1} - \frac{y^2}{2} \right) H(y) dy \right) \\ &= M \left(\frac{(G_{\tau_c}^{-1})^2}{2} \int_{u=-\infty}^{G_{\tau_c}^{-1}} H(u)\,du + \int_{u=-\infty}^{G_{\tau_c}^{-1}} \left(\frac{(u - G_{\tau_c}^{-1})^2}{2} - \frac{(G_{\tau_c}^{-1})^2}{2} \right) H(u)\,du \right) \\ &= M \int_{u=-\infty}^{G_{\tau_c}^{-1}} \underbrace{H(u)}_{v} \underbrace{\frac{(u - G_{\tau_c}^{-1})^2}{2}}_{u'} du \\ &= M \left(\left[\frac{(u - G_{\tau_c}^{-1})^3}{6} H(u) \right]_{u=-\infty}^{G_{\tau_c}^{-1}} - \int_{u=-\infty}^{G_{\tau_c}^{-1}} \frac{(u - G_{\tau_c}^{-1})^3}{6} h(u)\,du \right) . \end{aligned}$$

Finally, we obtain with the same change of variable and Taylor approximation as previously:

$$\begin{aligned} \Delta_{\int f} &\leq \frac{M}{6} \int_{z=0}^{\tau_c} \left[\frac{1}{g(G_{\tau_c}^{-1}))} (\tau_c - z) + \frac{g^{'}(\gamma)}{2g(\gamma)^3} (\tau_c - z)^2 \right]^3 \phi(z)\,dz \\ &\leq \frac{M}{6} \left[\frac{\xi^3}{2} \frac{\lambda}{n} + \frac{3\xi^3 \alpha}{4} \frac{\lambda}{n} + \frac{3\xi^3 \alpha^2}{8} \frac{\lambda}{n} + \frac{\xi^3 \alpha^3}{16} \frac{\lambda}{n} \right] \\ &\leq \frac{1}{2} \frac{C_s\, \lambda}{n} . \end{aligned}$$

Thus, one has $\left| \mathrm{E}_{Y,e}[L_{\hat{\tau}_c} - L_{\tau_c}] \right| \leq \frac{(C_{int} + C_s)\lambda}{n}$. Since C_{int} and C_s do not depend on e, this result remains meaningful when we are interested in the conditional expectation with respect to the random variable E and so $\left| \mathrm{E}_Y[L_{\hat{\tau}_c} - L_\tau] \right| \leq 2\,\varepsilon^2\,\xi + \frac{C\lambda}{n}$.

Moreover, using (8.9), we prove (8.7). □

References

1. S. Ben Taieb, R. Huser, R.J. Hyndman, M.G. Genton, Forecasting uncertainty in electricity smart meter data by boosting additive quantile regression. IEEE Trans. Smart Grid **7**, 2448–2455 (2016)
2. J. Bröcker, Reliability, sufficiency, and the decomposition of proper scores. Q. J. R. Meteorol. Soc. (2009). https://doi.org/10.1002/qj.456
3. V. Corradi, N.R. Swanson, Predictive density evaluation, in *Handbook of Economic Forecasting* ed. by G. Elliott, C.W.J. Granger, A. Timmermann (Elsevier, Amsterdam, 2006), pp. 197–284
4. V. Fortin, A.-C. Favre, M. Said, Probabilistic forecasting from ensemble prediction systems: improving upon the best-member method by using a different weight and dressing kernel for each member. Q. J. R. Meteorol. Soc. (2006). https://doi.org/10.1256/qj.05.167
5. T. Gneiting, Calibration of medium-range weather forecasts, Technical Memorandum, European Centre for Medium-Range Weather Forecasts. https://www.ecmwf.int/en/elibrary/9607-calibration-medium-range-weather-forecasts (2014)
6. T. Gneiting, M. Katzfuss, Probabilistic forecasting. Annu. Rev. Stat. Appl. (2014). https://doi.org/10.1146/annurev-statistics-062713-085831
7. T. Gneiting, A.E. Raftery, A.H. Westveld III, T. Goldman, Calibrated probabilistic forecasting using ensemble model output statistics and minimum CRPS estimation. Mon. Weather Rev. (2005). https://doi.org/10.1175/MWR2904.1
8. T. Gneiting, F. Balabdaoui, A.E. Raftery, Probabilistic forecasts, calibration and sharpness. J. R. Stat. Soc. Ser. B (Stat. Methodol.) (2007). https://doi.org/10.1111/j.1467-9868.2007.00587.x
9. A. Gogonel, J. Collet, A. Bar-Hen, Improving the calibration of the best member method using quantile regression to forecast extreme temperatures. Nat. Hazards Earth Syst. Sci. (2013). https://doi.org/10.5194/nhess-13-1161-2013
10. H. Hersbach, Decomposition of the continuous ranked probability score for ensemble prediction systems. Weather Forecast. (2000). https://doi.org/10.1175/1520-0434(2000)015<0559:DOTCRP>2.0.CO;2
11. R. Krzysztofowicz, Bayesian processor of output: a new technique for probabilistic weather forecasting, in *17th Conference on Probability and Statistics in the Atmospheric Sciences* (American Meteorological Society, 2004). https://ams.confex.com/ams/pdfpapers/69608.pdf
12. P.-A. Michelangeli, M. Vrac, H. Loukos, Probabilistic downscaling approaches: application to wind cumulative distribution functions. Geophys. Res. Lett. (2009). https://doi.org/10.1029/2009GL038401
13. D.S. Wilks, *Statistical Methods in the Atmospheric Sciences*. International Geophysics Series ed. by D.S. Wilks, vol. 100 (Academic Press, Cambridge, 2011)
14. D.S. Wilks, Enforcing calibration in ensemble postprocessing. Q. J. R. Meteorol. Soc. (2017). https://doi.org/10.1002/qj.3185

Part II
Renewable Energy: Risk Management

Chapter 9
Anticipating Some of the Challenges and Solutions for 60% Renewable Energy Sources in the European Electricity System

Vera Silva, Miguel López-Botet Zulueta, Ye Wang, Paul Fourment, Timothee Hinchliffe, Alain Burtin and Caroline Gatti-Bono

Abstract In this study, EDF R&D used the EU "high RES" (RenewableEnergy Sources) scenario of the 2011 European Energy Roadmap, reaching 60% of renewables generation by 2030 including 40% from variable RES (such as wind and solar), and analysed its implications on system development and operation. The analysis was based on an in-house chain of power system planning, dispatch and simulation tools. The study indicates that a strong development of variable RES generation would imply significant changes to the thermal generation mix required to balance supply and demand, with the need for less base load power plants and for more flexible units. The study shows that conventional plants are still required to ensure security of supply and, in order to reach a high level of decarbonation, low carbon base plants are essential. Furthermore, the results also underline the strong interest of deploying a certain level of interconnections, especially around the North Sea and France: it is a very efficient way to optimize the systems costs since these ensure that electricity generated by RES can reach demand and curtailment can be avoided, while also enabling the sharing of backup plants and of RES and demand diversity. Storage and flexible demand play a complementary role as flexibility providers, as a complement to thermal plants and RES curtailment. The potential for cost effective additional storage will however depend on the zone and on the possibility to deploy the other existing levers. Storage is particularly interesting in island systems with limited flexibility such as the UK. Load generation balancing will be highly dependent on weather conditions and its associated uncertainty that will increase the need for operation margins at different lead times and reserves. In order to limit the

V. Silva · M. López-Botet Zulueta · Y. Wang · P. Fourment · T. Hinchliffe · A. Burtin
C. Gatti-Bono (✉)
EDF R&D, 7 Boulevard Gaspard Monge, 91120 Palaiseau, France
e-mail: caroline.bono@edf.fr

P. Drobinski et al. (eds.), *Renewable Energy: Forecasting and Risk Management*, Springer Proceedings in Mathematics & Statistics 254,
https://doi.org/10.1007/978-3-319-99052-1_9

impact of this uncertainty, forecasting tools and the operational practices will play an important role. An increase of variable RES in the mix leads to challenges in terms of dynamic stability, with frequency excursion potentially reaching security limit. These challenges are linked to the fact that variable RES are interfaced with the system by power electronics and do not naturally contribute to system inertia, which is a key factor in maintaining system security. In order to maintain system security, some curtailment or the deployment of innovative solutions such as fast frequency response from battery storage and RES are required. Lastly, the economics of such a system would be a significant challenge, as the cost of the infrastructure is high while the market profitability of RES decreases with RES penetration since it is exposed to a "cannibalisation effect".

Keywords Renewable energy · Low-carbon Europe · Roadmap 2050 · RES-E variability · Power system reliability · Dynamic stability · Thermal backup Storage · Active demand response · Interconnections · Synthetic inertia

9.1 Introduction and Hypotheses

European countries have committed to an important change in their energy system to reduce carbon emissions and foster greater energy efficiency. The power sector will be a key contributor namely with an increase of renewables for many decades to follow. The European Union has issued a series of climate and energy packages that define milestones until 2050.

The HiRes scenario of the EC Energy Roadmap 2011 [1] constitutes the base of the hypotheses used in this paper. The share of renewables in the mix reaches 60% of the European Union gross electricity consumption by 2030.[1] Unlike today where the largest share of renewable energy is produced by hydraulic plants, in 2030, the highest share of renewable production will come from wind and solar plants. This comes from the fact that there are very few new sites where hydraulic plants could be built and the costs of wind and solar are plummeting, allowing for a mass development of these technologies for the next decades. In this scenario, 40% of the European Union gross electricity consumption would come from wind and solar technologies. This quantitative scenario is used to illustrate the issues of the large deployment of variable renewable generation in the European system.

The original HighRES scenario is the result of a global energy modeling exercise commissioned by the EU. The EU roadmap also provides the electricity generation from low carbon sources (energy generation for wind, PV, biomass, hydro, other RES (Renewable Energy Sources) and installed capacity of nuclear) as well as the commodity and CO2 prices. The TIMES model [2], used to develop the original scenarios, is a bottom-up model that produces least-cost energy systems under a given

[1] This is to be compared to a share of 29.6% of renewable energy in the EU28 power system in 2016 (Eurostat, SHARES 2016).

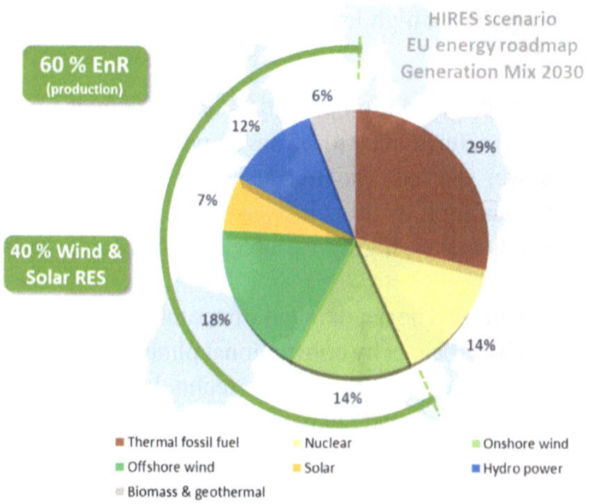

High RES 2030	GW	Load factor (h/yr)
Solar (PV)	220	1100
Onshore wind	280	1900
Offshore wind	205	3200
Hydro	120	3800

Fuel	Price
Coal	86 €/t
Gas	10 €/MMBtu
Oil	107 €/baril
CO_2	35 €/t

Fig. 9.1 European union energy roadmap

set of constraints at different time horizons, using linear programming. Therefore, it encompasses the whole European energy sectors and demands, and it cannot rely on the same level of details that is used in state-of-the-art studies of the power system [3, 4]. In particular, it provides an average view for the contribution of variable RES to demand supply using few time-slices, while in reality, the supply and demand must be balanced for every hour (and less) whatever the weather conditions and resulting RES generations happen to be.

EDF R&D constructed a detailed dataset for the European energy mix respecting the global energy volumes per energy source described in EU energy roadmap 2050 [1] including the geographical distribution of the development of the different technologies. A significant body of work was conducted to build a realistic representation of the future European interconnected system. The model used covers 17 countries: the European continental area, the UK, Ireland, and the Nordic system. For each country, we represent hydro-generation (run of the river and lake), pumped storage, thermal generation, demand, variable RES (wind, PV), other RES (biomass, geothermal,) and the interconnection capacity between countries. The geographical distribution and installed capacity of variable RES (onshore wind, offshore wind and PV) are optimized given the resource potential, land usage and social acceptance using a TIMES based model. We find that the level of wind and PV reaches 705 GW as detailed in the table of Fig. 9.1 with 220 GW of Solar, 280 GW of onshore wind and 205 GW of offshore wind. This brings new challenges to the operation of the power system through the variability of the wind and solar generation and the interface with power electronics. It will lead to new ways of operating the power system as well as new market designs, and will require important infrastructure developments and a transformation of the conventional mix. A complete report with the details of the modelling is available in [5] and also in [12–14] and [15].

The four main challenges of integrating a high level of renewables in a mix, that will be developed in this paper, are:

- Connecting RES and load: Renewable generation potentials do not coincide with demand location. Therefore, infrastructures need to be developed to bring renewable generation to where the power is used. Having these additional networks will help smooth the renewable generation by allowing for a bigger geographical diversity. The generation remains, however, highly variable which requires more flexibility.
- Bringing flexibility to handle variability: The net demand (the total demand minus the variable RES production) that has to be met by conventional plants is exhibiting new features and becoming increasingly variable. Storage technologies can help but the conventional mix will also see profound changes.
- Keeping the lights on: To keep the same level of reliability that consumers have come to expect, RES-E will need to provide new services since conventional plants that have historically ensured the stability of the network will not always be online at the most delicate times unless RES-E are curtailed.
- Balancing the economics: Finally, the integration of massive RES-E also changes the economics of the system with the marginal prices exhibiting the shape of a duck curve or Nessie curve, as already seen today in California and Hawaii.

9.2 Connecting RES and Load

9.2.1 Integrating a Large Share of Variable RES Requires a Coordinated Development of RES and Networks

Renewable generation potentials do not coincide with power needs. Solar potential are highest in the south of Europe and wind potential are highest around the North Sea. EDF R&D constructed a detailed dataset where the wind and solar capacities were distributed by country. Interconnections are necessary to connect production and load, pool thermal backup and share the variability of intermittent RES-E. EDF R&D performed a cost-benefit analysis to assess the reinforcements required by the new generation mix (see Fig. 9.2). This analysis finds similar reinforcements to the ones planned in the TYNDP 2014 [6]. In particular, as exhibited in other studies such as e-Highways 2050 [7], we see a North-South corridor[2] going through France to link electric peninsulas and share wind in the north and solar in the south, as well as a triangle around the North Sea to bring back the offshore wind to the continent.

[2]From the UK and the Netherlands all the way to Spain and Italy.

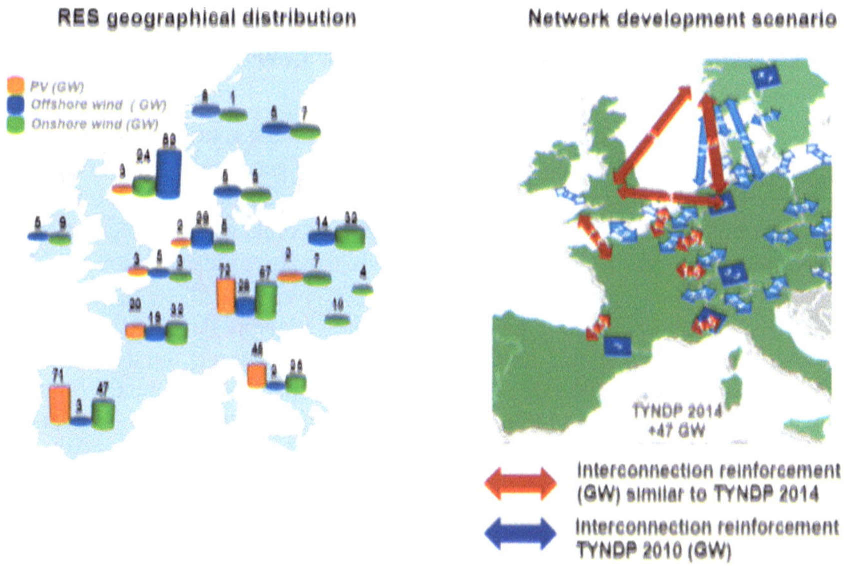

Fig. 9.2 Geographic distribution of RES and network development

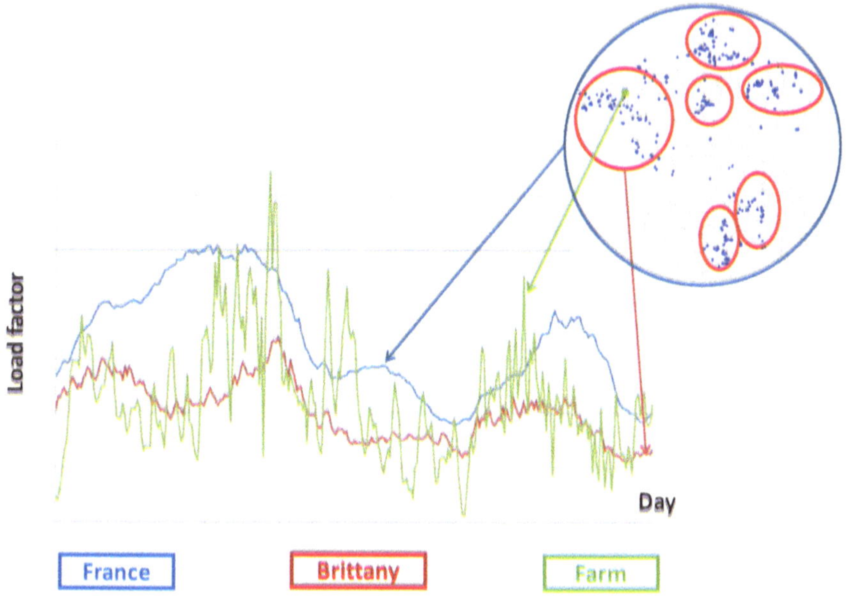

Fig. 9.3 Onshore wind generation for different geographical scales. (source RTE)

9.2.2 Geographical Diversity Does Help, but There Is Still Significant Variability at European Level

The development of infrastructures to bring the renewable generation to where the power is used will also help smooth out the renewable generation by allowing for a bigger geographical diversity as shown on Fig. 9.3. The generation is very uneven on the scale of a wind farm (green curve). It becomes smoother when aggregated over a region (red curve) and even smoother when aggregated over the entire country (blue curve).

Despite the generation becoming smoother, the variability remains significant at the European level as shown on Fig. 9.4. The figure represents the daily onshore wind generation for 30 different climatic years. A climatic year is created by using climate characteristics of a year in the past, such as wind speeds and solar radiation and temperature, to project the generation of wind and solar plants into the future. It therefore preserves the correlation with demand.[3] In the figure, there are 30 dots for each day of the year. We can observe a very large dispersion from one year to the next, as well as from one day to the next. It is largest over the winter where wind production is highest with 90 GW on average and still reaches a significant level over the summer. The same can be observed for solar plants with the highest dispersion occurring over the summer where the production is largest.

9.3 Flexibility to Handle the Variability

9.3.1 Not only Conventional Generation, but also Variable RES, Will Contribute to Balancing and Ancillary Services

Wind and PV generation increase the variability that needs to be managed by conventional generation. The net demand[4] profile, supplied by conventional generation, is more variable than demand alone, increasing the solicitation of the flexibility of conventional plants (see Fig. 9.5). This impact on flexibility needs is expressed mainly in terms of a higher frequency of large variations in net demand. At European level, upward hourly variations larger than 20 GW and downward variations larger than 10 GW increase by 50% and extreme hourly variations (>70 GW), which do not happen for demand, are present in net demand.

[3]For the rest of the dataset, the climatic year will take into account the rain and snow patterns of the year for hydraulics, as well as load pattern.

[4]Net demand = Demand − Variable RES.

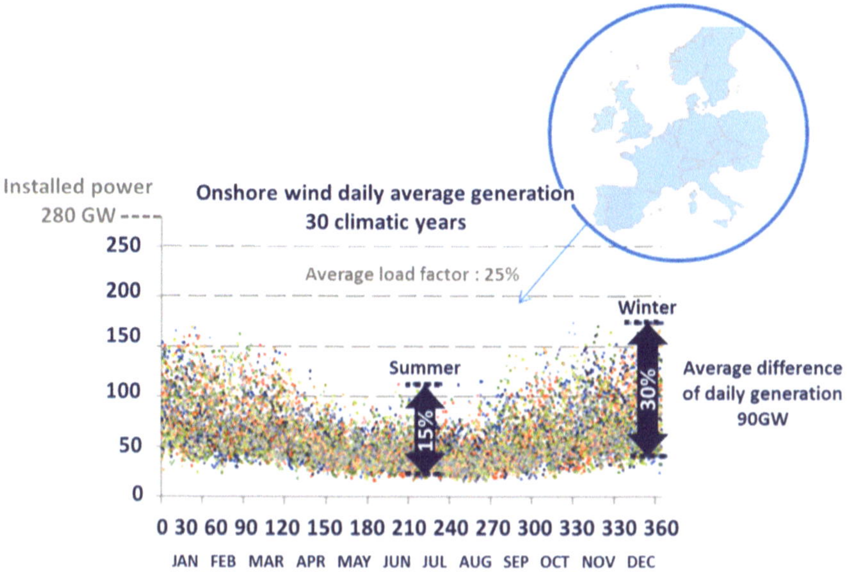

Fig. 9.4 Dispersion of daily onshore wind generation for 30 climatic years for the European power system. The average load factor (ratio between the generation output and the total installed capacity) varies from 15% in summer to 30% in winter

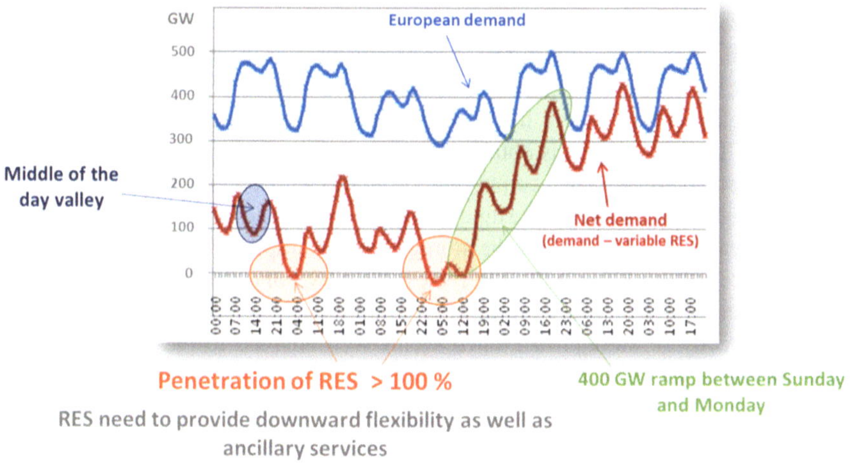

Fig. 9.5 Load-generation balancing becomes quite complex for periods with high net demand variability

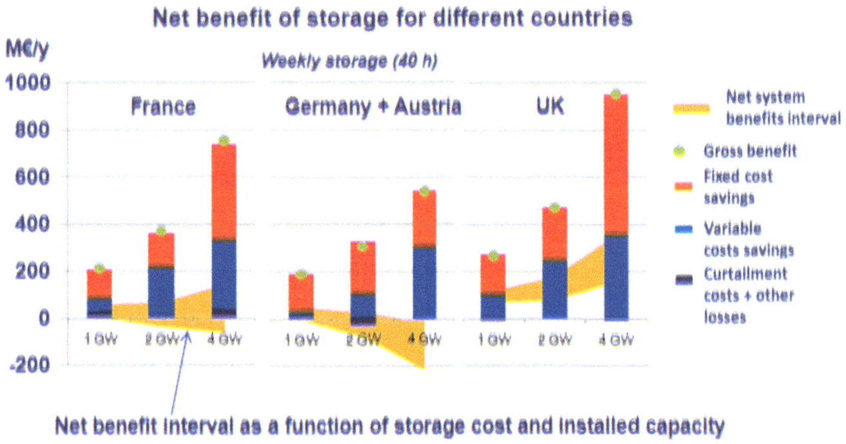

Fig. 9.6 Evolution of net and gross storage benefits in md/year for with a 40 h reservoir

9.3.2 *Storage and Active Demand May to a Certain Extent Supplement Generation to Balance Supply and Demand*

The results of the performance of the European system integrating a high penetration of variable RES do not provide clear economic justifications for further wide-scale development of centralized storage for managing the generation-demand balance, given the volume of storage that already exists. The potential for development of storage will vary across the different zones in the European systems. Figure 9.6 presents the net value of storage for a 40 h reservoir. This net value is obtained using a cost-benefit analysis where the cost of storage is subtracted to the gross value of storage. The gross value is obtained in terms of system cost savings (fixed plus variable costs) obtained when comparing a scenario with a scenario without storage. The net value is represented in the figure using a yellow band that presents the interval of the net value as a function of the storage cost assumptions. We have considered different scenarios of this cost (see for example [8]) and the band represents the potential net value depending on the cost assumptions. The results indicate that the current capacity of storage in France seems well adapted to the optimization of the generation-load balance. The region of Germany/Austria does not seem to hold great promise for the development of storage. This is in contrast to the UK in which the strong potential for offshore wind generation could make storage an interesting proposition. The interest in such deployment can only increase if storage contributes to ancillary services and reserves. The conclusions are similar if we consider a smaller storage with a 2 h reservoir, with slightly better perspectives for intra-day storage in Germany.

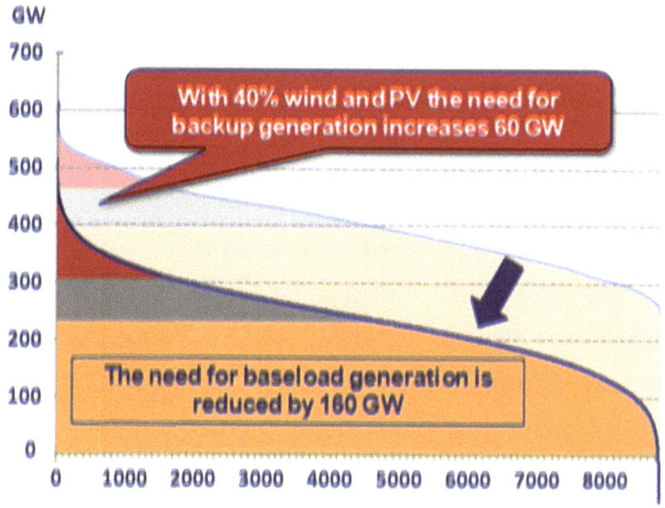

Fig. 9.7 European load duration curve of demand and net demand with 60% RES. Baseload generation = orange, semi-baseload generation = grey, peaking generation = red

9.3.3 Fuel Plant Are Needed for Backup Capacity for Security of Supply with High Level of Decarbonisation

Generation from wind and PV contributes mainly to the supply of energy. The stochastic nature of this production means that its output does not always coincide with periods of high demand and as a consequence it makes a minor contribution to capacity. A simple statistical calculation, based on a load duration curve at the European system level, can illustrate this issue. Figure 9.7 (transparent colors) illustrates demand by a stack of conventional generation technologies (thermal and hydro) without variable generation in the mix. Figure 9.7 (solid colors) illustrates the same but with the presence of 40% variable RES. In this case the conventional technologies stack aims at covering net demand (demand minus variable RES). The conventional generation technology stack is represented in the area below the duration curve of demand and net demand.

From the Fig. 9.7, one can observe the following:

- The energy produced by wind and PV displaces base load generation: the 700 GW of wind and PV displace 160 GW of base load generation, equivalent to 40% of the annual demand in energy.
- The development of variable RES entails a need for backup capacity, required during periods when wind and PV are not available: in the 60% RES scenario, 60 GW of additional backup capacity (called on for very short durations) are required to respect the capacity adequacy criteria of an expected loss of load of 3h/year.
- Overall, the development of 700 GW of wind and PV would lead to a reduction in conventional generation capacity in the order of 100 GW (160−60 = 100 GW).

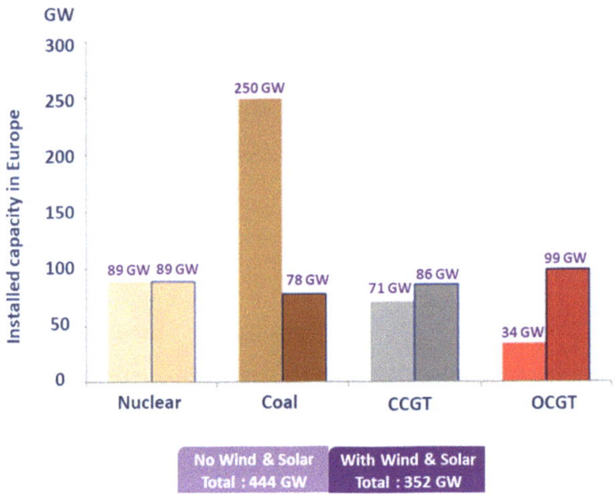

Fig. 9.8 Structure of the generation mix with and without wind and pv generation. CCGT = Combined Cycle Gas Turbine, OCCGT = Open Cycle Gas Turbine

This capacity credit comes solely from wind generation, since in Europe, PV generation is not present during winter peak.

- Saying this, periods with an offer of variable RES higher than 100% of demand are observed at the European level. During these periods, when all demand can be covered by must-run RES, curtailment may be required to maintain demand-generation balance as well as to allow the provision of reserves and ancillary services, required to ensure the security of the system.

The 60% RES scenario, represents an annual CO2 savings in the order of 1 Gt when compared to a scenario without variable RES. These savings come from wind and PV generation and from the reduction of carbon-emitting base load plants in the conventional generation mix (See Fig. 9.8). The decarbonisation of European base load generation is achieved with a mix of RES and nuclear plants. The average CO2 content per kWh produced with 60% RES is 125 gCO2/kWh, a value significantly lower than todays 350 gCO2/kWh. The additional replacement of coal with gas plants would allow CO2 output to be reduced to as low as 73 gCO2/kWh. Above a certain share of RES, however, the marginal efficiency of CO2 reductions drop and the marginal cost of this reduction increases (as a result of an increase in curtailment of wind and PV and the reduction of capacity credits and of fossil fuel savings).

9.4 Keeping the Lights on: Variable RES Production Should Potentially Provide New Services Like Fast Frequency Response (Inertia)

Wind and PV farms differ from conventional generation and other RES because of their power electronics interface with the system, often designated as asynchronous (Fig. 9.9). The connection of wind farms and PV via power electronic interfaces will lead to a reduction in the inertia of the system.

This reduction of inertia impacts the dynamic robustness of the system, namely the frequency[5] following an incident. For low to moderate penetration of variable RES, the synchronously interconnected European grid today has high inertia, which ensures that it has the capacity to accept a significant number of sources of production connected through power electronics interfaces. In order to quantify the impact of close to 40% variable RES in the European synchronous system, we have performed a large number of dynamic simulations. With 40% variable RES, for the majority of cases, the overall European network appears to be sufficiently robust, as illustrated in Fig. 9.10. The figure presents the frequency nadir, following a reference incident of 3.5 GW, for all hours of the studied years (close to 100 resulting from combining 30 weather years with generation availability scenarios).

However, critical situations with a frequency nadir lower than 49 Hz, triggering under frequency load shedding, and with a frequency lower than the security level of 49.2 Hz could happen. These are observed for periods with 25% instantaneous penetration of RES, when the overall system demand is low (<250 GW). A similar incident occurring during periods of high demand would not seem to pose a problem even for instantaneous penetration of RES as high as 70%, given that the load self-regulating effect will contribute naturally to the re-establishment of the system frequency.

9.5 Balancing the Economics: The Pace of Deployment of Variable RES Should Be Optimised in Order to Limit Costs of Storage or Excessive Curtailment

The analysis of the revenues touched by variable RES, considering that they are paid at the system marginal cost, shows their revenue decreases with the scale of their deployment. This effect has been designated in literature as the "cannibalization effect". This is translated by a difference between the system yearly base load price and the average revenue of variable RES, designated here as "market revenue gap".

[5]Keeping the frequency within a prescribed range is essential for the safety of the electric grid. The power that is received by a consumer stems from generators several hundreds of kilometers away sending electricity through a maze of lines at a given frequency. If the frequency shifts, there can be serious consequences for the network and consumer equipments, as well as for the electric grid, that can lead in the most extreme case to a blackout.

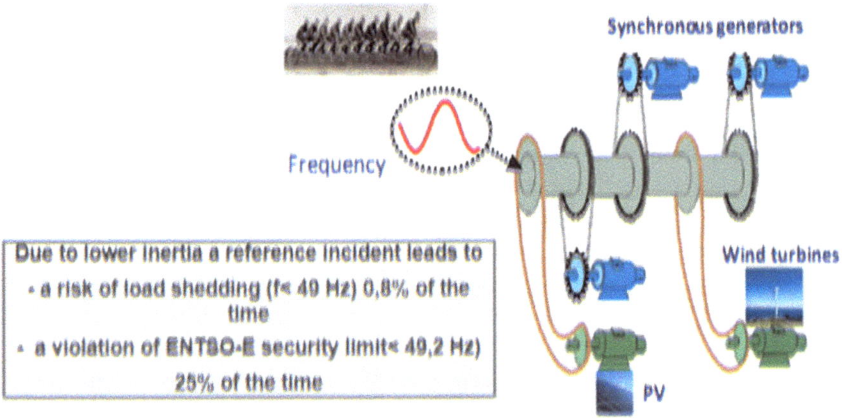

Fig. 9.9 Inertia and system frequency stability. Wind turbines and PV differ from synchronous generators in part because their power electronics interface is decoupled from the grid

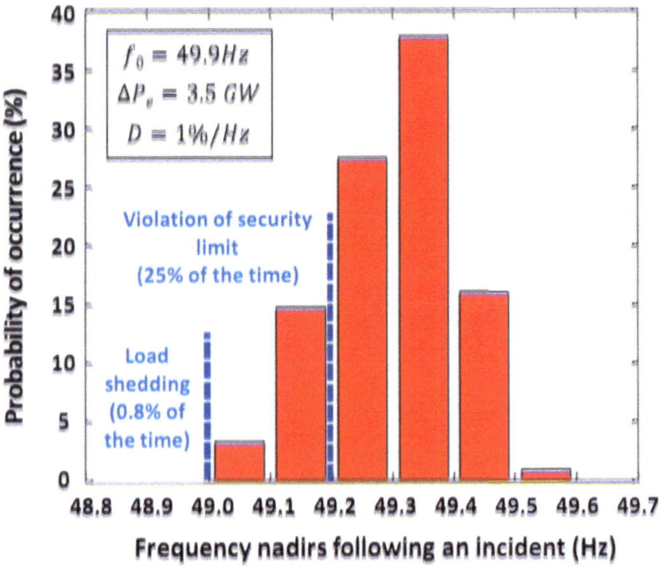

Fig. 9.10 Analysis of frequency stability in the European continental synchronous area with 35–38% share of variable RES

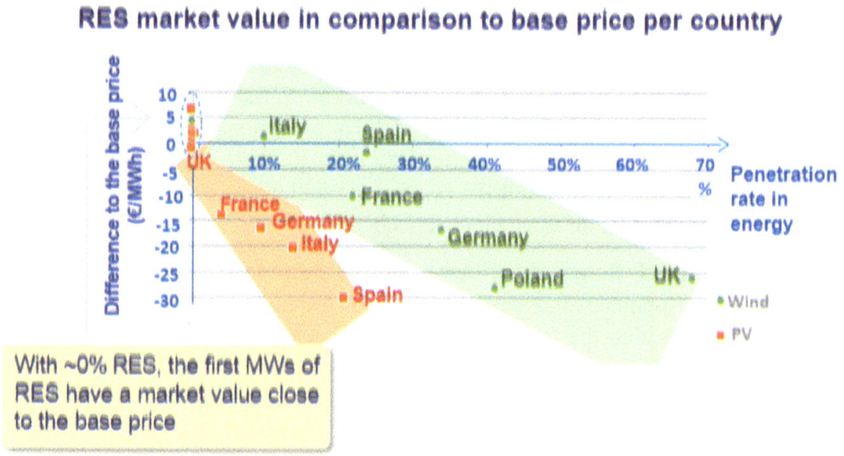

Fig. 9.11 Market revenue gap of Wind and PV in the 60% RES scenario (y-axis shows the gap between the yearly base load price and the market revenues of wind and PV and the x-axis presents the relative penetration of wind and PV in the zone in question)

Similar findings have been published in literature for the German, the British system and some parts of Continental Europe [9–11].

The analysis of the "market revenue gap" for wind and PV, for different countries, for the "60% RES" scenario, is presented in Fig. 9.11. The figure presents the evaluation of the incremental value of the service provided by variable RES to the system by comparing the marginal value of the first kW with the value of "40% variable RES". We can see that the value gap is very low or positive for the first MW of wind or PV (while their presence is marginal to the formation of the system marginal cost). Instead, for the higher penetrations of wind and PV for the "60% RES" scenario the gap becomes significant.

This result, which may seem counterintuitive, is easily explained. A technology is usually said to be mature when its levelised cost of production appears competitive compared with traditional thermal technologies or with a benchmark price for electricity. Joskow [9] notes however that for variable RES this comparison is misleading because the variable generation of a RES unit may be statistically biased towards periods when wholesale spot prices are higher or lower than the benchmark (see also [10]). In our approach, we capture this effect since the system marginal costs are outputs of the model and depend on the amount of RES capacity and on their generation patterns. A noticeable contribution of our approach is to reveal a telling pattern for how market value for RES decreases with their deployment.

9.6 Summary and Conclusion

Europe is leading an energy transition to lower carbon emissions. The HiRes scenario of the EU Energy roadmap 2050 is based on the introduction of massive volumes of renewable energies with a share reaching 60% in 2030, with 40% wind and solar in the mix. Such levels of wind and PV bring new challenges to the operation of the power system with the four main challenges being:

- **Connecting RES (Renewable Energy Sources) and load**
  Network developments at a local level within the distribution network and at a national level within the transmission networks along with new interconnectors may be needed if it is wished to capitalize on the natural diversity in demand and the production from the different RES sites. Nevertheless, climatic phenomena, which can have a simultaneous impact across the European continent, can result in marked changes in wind production as seen across the entire system. In addition, network development costs may be too high if variable RES is developed too far away for the load centers.
- **Bringing flexibility to handle the variability**
  If RES penetration reaches 60%, out of which 40% is variable RES, close to 500 GW of conventional generation (thermal, hydro and biomass) will still be required. The European electrical system will be required to cope with the variations in variable RES production. For instance, an installed capacity of 705 GW of wind and PV could see its daily production vary by a volume equivalent to 50% of total European demand within a 24 hour period. For an installed on-shore wind capacity of 280 GW, the average hourly generation on a winter's day could vary from one year to the next between 40 and 170 GW depending on specific weather conditions.
  Near-term flexibility needs will be important, and extreme hourly variations (>70 GW) that do not occur in demand can be found in net demand.
  There does not appear to be a business case in the next 15-year for a wide-scale development of storage as a means to manage intermittency, given the existing volume of storage in the European electrical system. In addition to backup capacity, demand response mechanisms should also be developed to contribute to generation/load balancing. Nonetheless, while load shifting could play a role in extreme situations as means to limit peak demand, it will not be capable of dealing solely with the variability introduced by wind and PV production.
- **Keeping the lights on: variable RES production should potentially provide new services like fast frequency response (inertia)**
  The most critical periods for frequency stability are those when the demand is low. During these periods, it will be necessary to limit the instantaneous penetration of RES in order to maintain the security of the system. Innovative solutions such as the creation of synthetic inertia from wind farms or the contribution of wind generation to frequency regulation are expected to reduce the severity of some of these limits.
  Smaller systems such as Ireland limit already the instantaneous penetration of RES

in order to preserve the security of their system and are looking to require new wind generation capacity to provide synthetic inertia and frequency regulation services. It is essential that the variable RES production which is displacing conventional generation is also able to contribute to the provision of ancillary services and also potentially provide new services (e.g. inertia).

- **Balancing the economics: the pace of deployment of variable RES should be optimised in order to limit costs of storage or excessive curtailment**
  We showed earlier in this document that variable RES displace base generation and increase the need for flexible backup. This difference in the service provided to the system is translated by a market value loss when compared to other technologies. This effect is quantified in terms of the gap between the average system marginal price and the average market revenue of wind and PV.
  Our results show that for the "60% RES" scenario this value gap for wind and PV ranges from 10 to 30% depending on the country. The gap presents a degree of correlation with the penetration rate of variable RES. Moreover, this energy value gap increases with the variable RES penetration ("cannibalisation" effect). In Europe, this "cannibalisation" effect is more pronounced for PV.

The study shows that variable and conventional generation should be viewed as complementary. Wind and PV are an important component in the EUs decarbonisation strategy, thermal generation is necessary to maintain system reliability and security of supply. Furthermore, low carbon base load generation is needed in order to deliver the reduction in the average carbon factor of European electricity.

Acknowledgements A team of dedicated EDF R&D engineers with a wide area of expertise worked on this study:

Vera Silva (Technical and Project Lead) from the Economic and Technical Analysis of Energy systems department (EFESE)

Nicolas Chamolet, Frédéric Dufourd, Marianne Entem, Laurent Gilotte, Grégory Fayet, Timothée Hinchliffe, Marie Perrot, Yann Rebours, François Rehulka, Jean-Matthieu Schertzer, Ye Wang, from the Economic and Technical Analysis of Energy systems department (EFESE),

Clémence Alasseur, Jérôme Boujac, Dominique Daniel, Paul Fourment, Jérémy Louyrette, and Miguel López-Botet Zulueta, from the Optimisation Simulation Risks and Statistics Department (OSIRIS),

Marie Berthelot and Vincent Maupu from the Applied Meteorology and Atmospheric Environment department (MFEE).

References

1. European Commission, Energy Road Map 2050, December 2011, http://ec.europa.eu/energy/en/topics/energy-strategy-and-energy-union/2050-energy-strategy
2. http://iea-etsap.org/index.php/etsap-tools/model-generators/times
3. Y. Rebours, M. Trotignon, V. Lavier, T. Derbanne, F. Meslier, How much electric interconnection capacities are needed within Western Europe? in *7th International Conference on the European Energy Market (EEM)*, 23–25 June 2010, pp. 1–6

4. r2b energy consulting GmbH, Ermittlung des Marktwertes der deutschlandweiten Stromerzeugung aus regenerativen Kraftwerken (Los 1). Gutachten fr die TransnetBW GmbH in Vertretung der deutschen Übertragungnetzbetreiber, 11 October 2013
5. V. Silva, A. Burtin, Technical and economic analysis of the European system with 60% RES, EDF Technical report, June 2015 (in English and in French). EDF report, https://www.edf.fr/sites/default/files/Lot%203/CHERCHEURS/Portrait%20de%20chercheurs/summarystudyres.pdf
6. TYNDP 2014, https://www.entsoe.eu/Documents/TYNDP
7. e_Highways 2050, http://www.e-highway2050.eu/e-highway2050/
8. A. Zucker, T. Hinchliffe, Optimum sizing of PV-attached electricity storage according to power market signals a case study for Germany and Italy. Appl. Energy **127**, 141–155 (2014)
9. Joskow, Comparing the costs of intermittent and dispatchable electricity generating technologies. Am. Econ. Rev. Pap. Proc. (2011)
10. L. Hirth, The market value of variable renewables: the effect of solar wind power variability on their relative price. Energy Econ. **38** (2013)
11. R.J. Green, T.-O. Léautier, Do costs fall faster than revenues? Dynamics of renewables entry into electricity markets, Working paper, July 2015
12. M. Lopez-Botet et al., Methodology for the economic and technical analysis of the European power system with a large share of variable renewable generation, in *IEEE PES General Meeting*, Washington, USA, 27–31 July 2014
13. Y. Wang, V. Silva, A. Winkels, Impact of high penetration of wind and PV generation on frequency dynamics in the continental Europe interconnected system, in *13th International Workshop on Large-scale Integration of Wind Power into Power Systems*, Berlin, October 2014 (Best paper award)
14. M. Perrot, V. Silva, T. Hinchliffe, P. Fourment, M. Lopez-Botet Zulueta, Economic and technical analysis of the European system with high RES scenarios, in *10th Conference on The Economics of Energy and Climate Change*, Toulouse, September 2015
15. Y. Wang, V. Silva, M. Lopez-Botet Zulueta, Impact of high penetration of variable renewable generation on frequency dynamics in the continental Europe interconnected system. IET Renew. Power Gener. **10**(1), 10–16 (2016)

Chapter 10
A Joint Model for Electricity Spot Prices and Wind Penetration with Dependence in the Extremes

Thomas Deschatre and Almut E. D. Veraart

Abstract This article analyses the dependence between electricity spot prices and the wind penetration index in the European energy market. The wind penetration index is given by the ratio of the wind energy production divided by the total electricity production. We find that the wind penetration has an impact on the intensity of the spike occurrences in the electricity prices, and we formulate a joint model for electricity prices and wind penetration and calibrate it to recent data. We then use the new joint model in an application where we assess the impact of the modelling assumptions on the potential income of an electricity distributor who buys electricity from a wind farm operator.

Keywords Dependence modelling · Spikes · Doubly stochastic Poisson process CAR processes · Electricity prices · Wind penetration index

10.1 Introduction

Motivation

The trend to increase renewable energy production all over the world with the possibility of phasing out conventional energy sources while at the same time ensuring reliability of energy networks constitutes one of the key challenges of modern societies. Due to their high variability renewable sources of energy tend to be more difficult to predict and for instance their precise impact on electricity prices is far

T. Deschatre
Université Paris-Dauphine, PSL Research University, CNRS, Ceremade, 75016 Paris, France
e-mail: thomas.deschatre@gmail.com

T. Deschatre
EDF Lab, OSIRIS, 91120 Palaiseau, France

A. E. D. Veraart (✉)
Department of Mathematics, Imperial College London, 180 Queen's Gate,
London SW7 2AZ, UK
e-mail: a.veraart@imperial.ac.uk

P. Drobinski et al. (eds.), *Renewable Energy: Forecasting and Risk Management*,
Springer Proceedings in Mathematics & Statistics 254,
https://doi.org/10.1007/978-3-319-99052-1_10

from being understood. The recent literature presents various attempts to model and characterise the impact of wind energy production on electricity prices, see for instance [7, 11–13] for some recent accounts. In these studies, wind is often considered as an exogenous variable for the electricity price. However, for many applications, one of which will be presented in this article, we in fact need a joint model for electricity prices and wind energy generation (and possibly other sources of electricity as well). Hence the goal of this article is to formulate and estimate a joint model for electricity spot prices and wind energy production, more precisely wind penetration, which is the ratio between wind energy production and total energy production.

Our modelling idea which is guided by our statistical analysis rests on the hypothesis that increasing wind penetration typically results in lower electricity spot prices and that high wind penetration might increase the frequency at which negative price spikes occur. In order to formalise these ideas we draw on new statistical methodology for estimating jump intensities and mean reversion rates in mean-reverting jump processes, which has very recently been developed in [6] and Chap. 4 of [5].

The goal of this paper is twofold:

(i) First, we carry out a statistical study to assess the impact of the wind penetration on electricity spot spikes occurrences.
(ii) Second, we quantify the impact of this dependence on the distribution of a distributor's income when buying wind production.

Main Results

Statistical studies show that the probability to have an extreme return in electricity spot time series seems to increase with the wind penetration index, see Sect. 10.2. From this statistical evidence, the following joint model is introduced for the German and Austrian electricity spot price S and the wind penetration index WP:

$$\begin{cases} S_t &= \Gamma_{1,t} + X_{1,t} + Y_t, \\ Y_t &= \int_0^t e^{-\beta(t-s)} dM_s, \\ WP_t &= expit\left(\Gamma_{2,t} + X_{2,t}\right) \end{cases}$$

for $t \in [0, T]$ (for some finite $T > 0$) where Γ_1 and Γ_2 are seasonality functions, X_1 and X_2 are continuous autoregressive processes, $expit\,(x) = \frac{1}{1+e^{-x}}$, $x \in \mathbb{R}$, $\beta > 0$ and M is a compound Poisson process with stochastic intensity $(\lambda_t)_{0 \le t \le T}$. The process Y is a mean reverting compound Poisson process which models the spike part of the electricity spot price. The linear dependence between S and WP is modelled by a correlation between the two Brownian motions driving X_1 and X_2. However, it does not capture any dependence between extreme values of the electricity spot price and the wind penetration.

While the marginal modelling of electricity spot price and the linear dependence are rather traditional, the novelty of our model lies in the fact that it takes the dependence between the intensity of the spikes in the spot price and the wind penetration level into account. This dependence is introduced by modelling the intensity of the

Poisson process, representing the frequency of spike occurrence, as a function of the wind penetration level:

$$\lambda_t = q\left(WP_t\right), t \in [0, T]$$

with $q : \mathbb{R} \to (0, \infty)$.

After using the method of [6] for jump detection, the function q is inferred non-parametrically from the data using the estimation procedure proposed in Chap. 4 of [5]. The statistical results confirm our intuition: the intensity of negative spikes is influenced by the wind penetration level and increases with it. However, the wind penetration has no impact on the frequency of the positive spikes. Statistical tests are used to support these hypotheses. In addition, we describe the calibration procedure used for estimating the entire model.

For our second contribution, let us consider the point of view of an electricity distributor buying electricity from a wind farm and selling it to the market at price S. The distribution of their income depends jointly on the electricity spot price and the wind penetration index. In order to quantify the impact of our statistical results concerning the dependence between the occurrence of negative spikes and the wind penetration, we compare two different models. The first model consists of modelling the intensity of negative spikes as a constant, that is assuming there is no dependence between it and the wind penetration level. The second model consists of modelling the intensity of negative spikes as a simple parametric function of the wind penetration index. Our quantitative results show that this dependence can not be neglected: the risk measures of the distributor's income such as the value at risk and the expected shortfall diminish significantly from the first model to the second one.

Structure of the Paper

The outline of the remainder of this article is as follows. Section 10.2 gives a detailed description of the data and data sources used in our statistical analysis. The key contributions are then collected in Sect. 10.3, where we introduce and estimate the joint model for hourly electricity spot prices from Germany and Austria and the corresponding wind penetration index. Finally, in Sect. 10.4, the impact of our joint model on an electricity distributor's income buying wind production is studied.

10.2 Data Description and Exploratory Study

10.2.1 Data Description

In this article we analyse German and Austrian wind energy production data, electricity price data and load data covering a period from the 1st of January 2012 to the 31st of December 2016. The data have been downloaded from the following website https://data.open-power-system-data.org/. The precise data sets considered are the following ones:

- The German and Austrian hourly electricity spot prices (from the day-ahead auction) from the 1st of January 2012 to the 31st December 2016,
- the German and Austrian hourly load data from the 1st of January 2012 to the 31st December 2016,
- the German and Austrian hourly wind energy production data from the 1st of January 2012 to the 31st December 2016. Note that the data has been aggregated over the four German transmission system operators 50 Hertz Transmission, Amprion, Tennet TSO and EnBW Transportnetze and the Austrian transmission system operator APG.

Throughout this article, we will be analysing and modelling the hourly data from the day-ahead market rather than the daily data, which is often done in the literature. The hourly data are revealed in the daily auction at the same time, but it has been found in e.g. [9] that when taking seasonality into account, they can be successfully modelled by classical time series models at an hourly frequency. An alternative approach, which we leave for future research, would be to model the hourly data as a 24-dimensional vector of daily data, as it has for instance been done by e.g. [10] in the discrete-time setting and by [21] in the continuous-time setting.

Motivated by the work by [11] we use our data to compute the German/Austrian wind penetration index, defined as the ratio of the wind energy production and the electricity load. In [11] the authors found a strong association between the wind penetration index and the corresponding electricity prices. Figure 10.1a, b depict the time series of the German/Austrian spot price and the wind penetration, respectively. We observe that the wind penetration index takes values between zero and one, except for eleven values in 2016 when it exceeds one. This can happen because of the interconnection between the transmission networks of the various European countries. Since the total electricity production can also be used in other countries in Europe, a wind penetration index exceeding one tells us that in these particular hours all of the German/Austrian electricity came from wind energy.

10.2.2 *Exploratory Data Analysis*

Next, we aim to study the relation between the prices and the wind penetration index. To this end, let us look at Fig. 10.1c. Here we observe that high negative values of spot prices happen when the wind penetration crosses a certain threshold. More precisely, Fig. 10.1d reveals that high negative price returns appear when the wind penetration is over 0.4. Next, Fig. 10.1e, f depict the returns of the spot price against the returns of the wind penetration index, where extreme values have been removed in the latter picture. (The procedure of how the extremes were removed is described below.) The corresponding correlation coefficients for the data in Fig. 10.1e, f are given by -0.125 and -0.148, respectively. We want to investigate now whether or not there is any association between wind penetration and extreme spot returns. We recall from Fig. 10.1a that both positive and negative spikes appear in the data. Hence when we

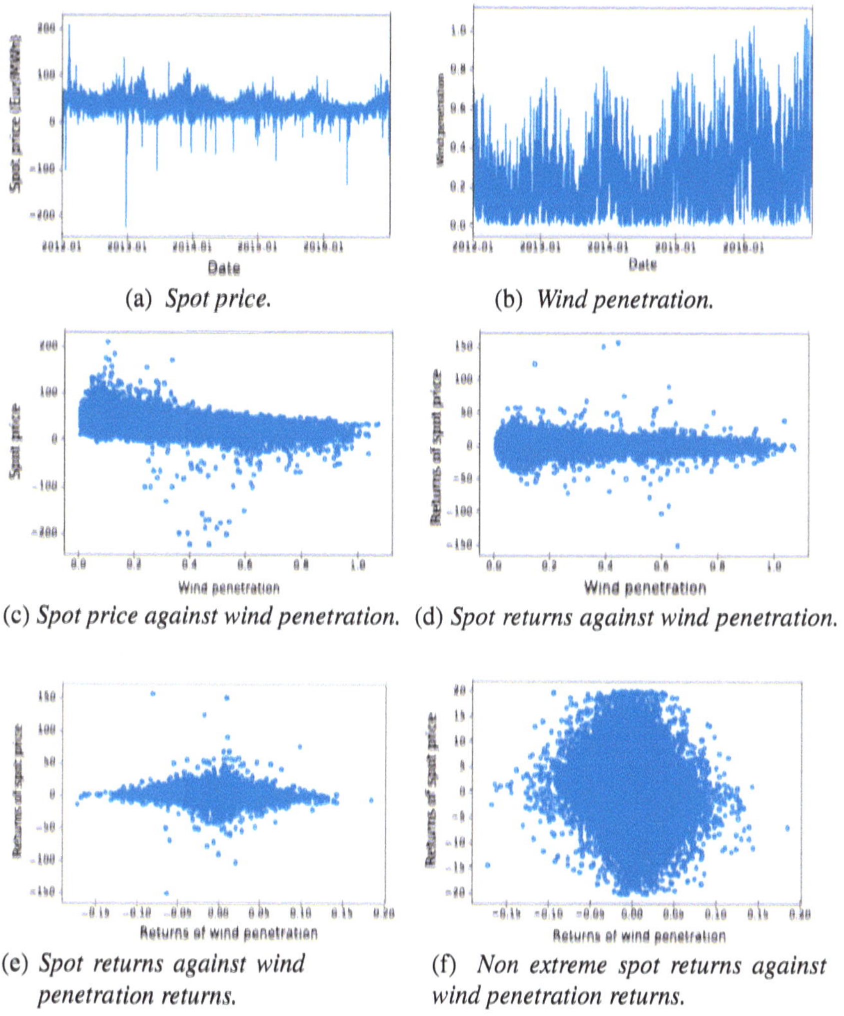

(a) *Spot price.* (b) *Wind penetration.*

(c) *Spot price against wind penetration.* (d) *Spot returns against wind penetration.*

(e) *Spot returns against wind penetration returns.* (f) *Non extreme spot returns against wind penetration returns.*

Fig. 10.1 Hourly German spot price and wind penetration index

are talking about extreme values in the following we mean the absolute value of the corresponding spot returns. Note that as a positive spike (resp. negative) leads to a negative (resp. positive) extreme return due to the high mean reversion, we do not consider to study negative and positive extreme values separately: in this case, the negative (resp. positive) extreme value is caused by the positive (resp. negative) one and not by the wind.

In order to identify the extreme values in the spot returns, we establish the QQ-Plot of the spot returns against a normal law: we obtain Fig. 10.2a. Removing the values having an absolute value greater than 20, we obtain the QQ-Plot in Fig. 10.2b. This new plot shows a better correspondence with the normal distribution and we consider

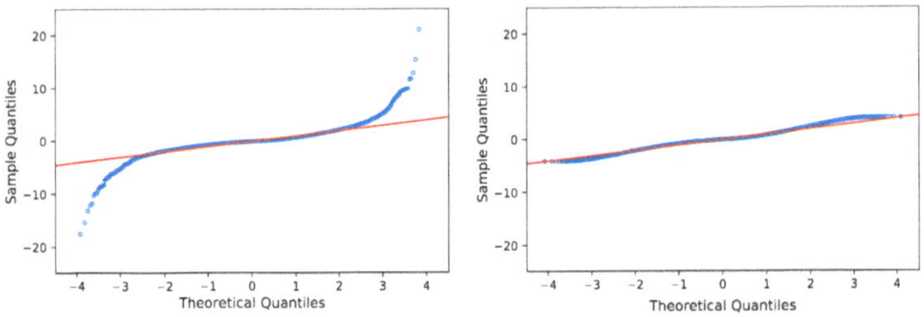

(a) *QQ-Plot with all the values.* (b) *QQ-Plot without extremal returns.*

Fig. 10.2 QQ-Plot of spot returns against a normal law, with and without certain extreme values

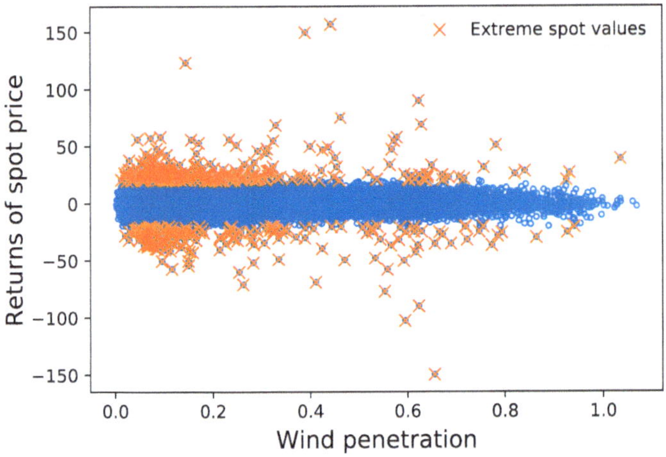

Fig. 10.3 Returns of the spot price against the wind penetration with extreme spot returns

the returns of the spot having an absolute value greater than 20, corresponding to the quantile of order 99.1%, as extreme values. Figure 10.3 represents the extreme values of the spot return against the wind penetration.

We remark that our procedure of filtering out extreme values can be regarded as a hard thresholding approach, whereas one could also use more sophisticated techniques based on methods from extreme value theory as in [14, 21].

In order to study the impact of wind penetration on spot returns extreme values, one indicator, omitting the time dependency aspect, is the quantity $\mathbb{P}(|R| > 20|WP > x)$ for a given x where R corresponds to the return of the spot and WP is the wind penetration. Figure 10.4 corresponds to the function $x \to \mathbb{P}(|R| > 20|WP > x)$. For $x = 0$, the function takes value 0.82% where it reaches level of order 2% when x increases: the probability to have an extreme value in the time series of the spot price is higher with higher values of wind penetration; this dependence has to be taken into account if we want to model the joint distribution of the spot price and the wind penetration.

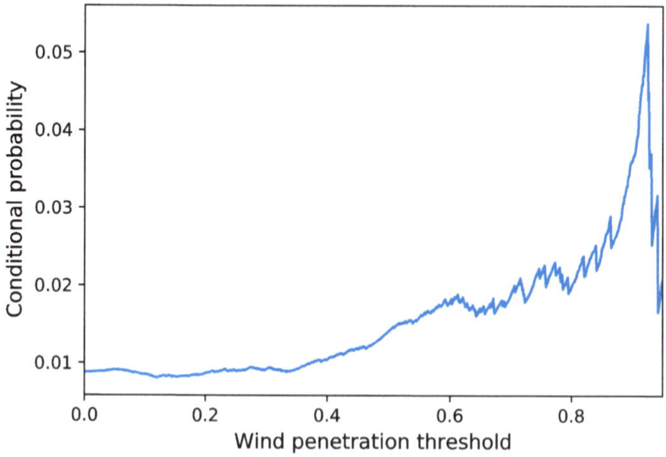

Fig. 10.4 Probability to have an extreme returns conditionally on wind production greater than a certain threshold

10.3 A Joint Model for the Electricity Spot Price and the Wind Penetration Index

The main contribution of this article is that we present and estimate a joint stochastic model for the hourly electricity spot price (denoted by S) and the corresponding wind penetration index (denoted by WP). We will now proceed by explaining step-by-step the stochastic model and the corresponding estimation procedure.

10.3.1 Model for the Electricity Spot Price

10.3.1.1 Model

Let $(S_t)_{0 \leq t \leq T}$ denote the electricity spot price. Motivated by our exploratory study, we decompose S in the following way:

$$S_t = \Gamma_{1,t} + X_{1,t} + Y_t$$

where Γ_1 is a seasonality function, X_1 represents a continuous stochastic part of the spot price and Y represents a spike part.

Modelling of the Seasonality Function Γ_1

The seasonality Γ_1 is assumed to be of the form

$$\Gamma_{1,t} = c_{0,1} + c_{1,1}t + c_{2,1}t^2 + c_{3,1}\cos\left(\frac{\tau_{0,1} + 2\pi t}{365 \times 24}\right) + c_{4,1}\cos\left(\frac{\tau_{1,1} + 2\pi t}{7 \times 24}\right) + c_{5,1}\cos\left(\frac{\tau_{2,1} + 2\pi t}{24}\right). \quad (10.1)$$

Modelling of the Continuous Stochastic Part X_1

As working in a continuous-time framework, a continuous autoregressive process (CAR) of order 24, see for instance Chap. 4 in [4], is chosen to model X_1. The dynamics of X_1 is modelled by

$$X_1 = \mathbf{b}^T \mathbf{X}_1,$$

$$d\mathbf{X}_{1,t} = \mathbf{A}_1 \mathbf{X}_{1,t} dt + \mathbf{e}\sigma_1 dW_{1,t},$$

where

$$\mathbf{A}_1 = \begin{pmatrix} 0 & 1 & 0 \cdots & 0 \\ 0 & 0 & 1 \ddots & \vdots \\ \vdots & & \ddots \ddots & 0 \\ 0 & \cdots & \cdots 0 & 1 \\ -\alpha_{24,1} & -\alpha_{23,1} & \cdots\cdots & -\alpha_{1,1} \end{pmatrix}, \quad \mathbf{e} = \begin{pmatrix} 0 \\ 0 \\ \vdots \\ 0 \\ 1 \end{pmatrix}, \quad \mathbf{b} = \begin{pmatrix} 1 \\ 0 \\ \vdots \\ 0 \end{pmatrix},$$

and $(W_{1,t})_{0 \le t \le T}$ is a standard Brownian motion. The choice of this modelling is motivated by the statistical study in a discrete time framework done below.

Modelling the Spike Part Y

While Sect. 10.2.2 gives statistical evidence of a dependence in the extreme values of electricity spot returns and wind penetration, in this section, we add the temporal aspect to our analysis. We note that spikes are characteristics of electricity prices and correspond to extreme values in the time series. We are interested in studying the relation between the intensity of the spike occurrence and the wind penetration index. Furthermore, a spike can be modelled as a jump going back to 0 with a strong mean reversion and with a typically stochastic size. To this end, we start by modelling the spikes by a mean-reverting Poisson-type process with stochastic intensity function depending on the wind penetration. Thus, the spike part of the spot Y is modelled as

$$Y_t = \int_0^t e^{-\beta(t-s)} dM_s,$$

where $M_t = \sum_{i=1}^{N_t} K_i$ denotes a compound Poisson process with $(K_i)_{i \ge 1}$ being a sequence of i.i.d. real random variables corresponding to the jump sizes and $(N_t)_{0 \le t \le T}$ a Poisson process with stochastic intensity $(\lambda_t)_{0 \le t \le T}$ corresponding to the number of jumps in the interval $[0, t]$. Since we are interested in studying the dependence between the intensity λ and the wind penetration, we assume a functional relationship of the form:

$$\lambda_t = q\left(WP_t\right), \; t \in [0, T],$$

where $q : \mathbb{R} \mapsto (0, \infty)$ is a deterministic function. Conditional on the wind penetration index, N is an inhomogeneous Poisson process. Also, $\beta > 0$ is the speed of the mean reversion.

10.3.1.2 Calibration

Estimation of the Times of the Spikes

First of all we need to identify the times of the spikes before we can estimate the sizes of the spikes, the mean reverting parameter β and the intensity function q. Reference [6] establishes a method for estimating the arrival times of the jumps in the above setting as soon as the process $\Gamma_1 + X_1$ is a continuous semi-martingale, which is the case here. We remark that in [6], the intensity of the Poisson process is not stochastic. However, adding the assumption that λ is bounded below and above, the result can easily be extended to the stochastic case using the same arguments as in the original proof.

Let us briefly recall the key ideas: Suppose the price S is observed over the time interval $[0, T]$ for some $T > 0$ at discrete times $t_i = i\Delta_n$, for $i = 0, 1, \ldots, \lfloor T/\Delta_n \rfloor$. Here $\Delta_n > 0$ and $\Delta_n \to 0$ as $n \to \infty$. Note that the asymptotic framework, contrarily to classical high-frequency one, is not limited to $\Delta_n \to 0$; in [6], λ and β can either grow with n or be bounded.

The method for identifying the jumps is based on the use of a threshold of the form $C\Delta_n^{\varpi}$ for the returns, where $C > 0$ and $\varpi \in (0, 0.5)$ are constants which we specify below. This method is classic for the estimation of jumps in continuous jump diffusion models with discrete observations, see [2, Section 10.4] for instance, and has been adapted to the spike case when β is large in [6]. Indeed, as the mean reversion is large and can be of the size of a jump, we need to add the condition that $\Delta_i^n S \Delta_{i+1}^n S < 0$, where $\Delta_i^n S = S_{t_i} - S_{t_{i-1}}$. It indicates that after a jump (in the interval $((i-1)\Delta_n, i\Delta_n])$, the subsequent increment will be of the opposite sign of the jump. We choose a threshold equals to $5\bar{\sigma}\Delta_n^{0.49}$ with $\bar{\sigma}^2$ the multipower variation estimator of the volatility with order 20; the reader can refer to [3, 19] for more details about multipower variation. The particular choice of our threshold is motivated by the following findings in the literature: First, both [1], see Sect. 5.3, and [2], see Sect. 6.2.2, p. 187, recommend using a constant of the form $C = \tilde{C}\bar{\sigma}$, where $\tilde{C}$ is a constant and $\bar{\sigma}$ is an 'average' of the volatility. Moreover they advise choosing $\varpi \in (0, 0.5)$ close to 0.5. In addition, [1] also suggests choosing $\tilde{C}$ between 3 and 5, see also [16] ($\tilde{C} = 3$) and [17] ($\tilde{C} = 4$).

In order to avoid too much change in volatility, the data are segmented in time series of one year in order to identify the jumps. That means that $\bar{\sigma}$ does not stay constant over the entire sample, but just over each of the five years. For the considered data, we find an estimated number of 114 jumps, 30 of which are negative and 84 positive. Figure 10.5a corresponds to the spot price time series with the times of the

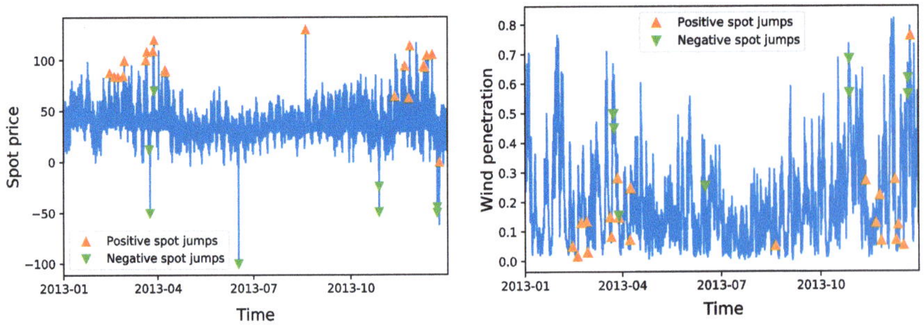

(a) *Spot.* (b) *Wind penetration.*

Fig. 10.5 Jumps in German spot price and wind penetration time series

jumps marked by upward triangles for positive jumps and by downward triangles for negative jumps.

Remark 10.1 The multipower variation estimator of the volatility is chosen with order 20 which is high compared to the orders typically chosen in the literature. Contrarily to classical jumps, spikes have a strong mean reversion. Using for instance bipower variation in such a scenario is not suitable because the impact of the jump is not diminished by the increments that follows. One also needs to compensate the effect of the mean reversion, that can be present in two or three increments following the jumps. Moreover, some spikes have a large amplitude, and then have a strong impact on the value of the volatility estimator if the order is low. Having 8760 observations per year allows us to choose an order 20 with a good estimation quality. Some simulations have been performed on simple models such as an arithmetic or geometric Brownian motions plus spikes and order 20 appears to be a good choice. The choice of 20 corresponds to a threshold of around 20 Euros per MWh per year, corresponding to the one used to select the extreme values in the quantile-quantile plot. Note that we found that the choice of the order of the multipower does not influence the results significantly in any case. Decreasing the order of the estimator leads to higher values of the estimated volatility and then also for the threshold. A threshold of 10 leads to a detection of 61 positive spikes and 27 negative spikes. The number of negative spikes, which is our key object of interest, does not change much. Also, the results concerning the dependence with the wind penetration are not affected.

Estimation of the Sizes of the Spikes

An estimator of the size of the jth jump K_j is given by $\Delta_i^n S$ if the jth jump is identified in the interval $((i-1)\Delta_n, i\Delta_n]$. However, as β is large, this estimator is biased: indeed, as noticed in [6], if the jumps are well identified, $\Delta_i^n S$ is equivalent to $e^{-\beta\left(\lfloor \frac{T_j}{\Delta_n} \rfloor \Delta_n+\Delta_n-T_j\right)} K_j$ with T_j the time of the jth jump. In the case $\beta\Delta_n$ is not small, the term before the jump size is not closed to one. As T_j cannot be estimated because of the discrete observations, we cannot estimate this bias term to deduce the exact jump size K_j. Nevertheless, as the number of jumps is low, we do not consider a model for the jump sizes and keep the empirical one, knowing that there can be a bias.

Estimation of the Speed of Mean Reversion of the Spikes β

Reference [6] presents a method for estimating the speed of mean reversion of the spikes given by β, assuming that β is large enough. The parameter β is estimated using the slope of the process after a jump, which is of order $K_j\left(1-e^{-\beta_n \Delta_n}\right)$ for the jth jump. We find $\beta = 7718.84\ year^{-1}$. As for the jump detection, we note that while the results of [6] consider a constant λ, the proofs of the consistency of the estimator (and possibly also of the central limit theorem) can be generalised to the case of a stochastic intensity, provided the stochastic intensity is bounded below and above.

Estimation of the Spike Intensity as a Function of the Wind Penetration Index

Figure 10.5b depicts the positive and negative jump times of the spot price superimposed on the time series plot of the wind penetration index. We observe that negative jumps appear more often when the wind penetration is high, whereas positive jumps also appear for small values of wind penetration. In the following, we will separate the positive and the negative jumps in order to study the impact of the wind penetration on each type of jumps independently.

To this end, let us consider two point processes corresponding to the positive and the negative jumps, respectively. We want to study the dependence between the point processes and the wind penetration. In order to simplify the exposition, we do not use superscripts for the two different Poisson processes we are considering, but just the generic N for the Poisson process governing either the positive or the negative spikes. Our goal is to estimate the function q on a given time interval I where wind penetration data is available.

In the following, let us assume that every jump has been identified meaning that we have an estimated sample path of N. Chapter 4 in [5] proposes a local polynomial estimator of the function q, which is a generalisation of a kernel estimator. This estimator depends on a bandwidth parameter h belonging to $(0, \infty)$. Chapter 4 in [5] gives a method for choosing this bandwidth parameter h in an optimal way relatively to the L_2 error on I between q and its estimator. The optimal bandwidth is chosen among a discrete set $\mathcal{H} \subset (0, \infty)$ given by the statistician. The method leads to a trade-off between bias and variance, and minimises the sum of an estimator of the bias and an estimator of the variance which is weighted by an hyperparameter κ. Choosing this bandwidth leads to an oracle inequality, justifying this particular choice. This method is similar to the one used by [15] for the kernel estimation of a density, which is an extension of the Goldenshluger–Lepski method [8]. For the estimation procedure to work, a few assumptions are required on the exogenous process, which is here WP. The most important one is that WP must be a semimartingale, which is the case in the model proposed below. We remark that Chap. 4 in [5] is casted in a continuous-time framework, whereas our estimation is done in discrete time. However, since $\Delta_n \to 0$ and T is large, the resulting approximation error which is

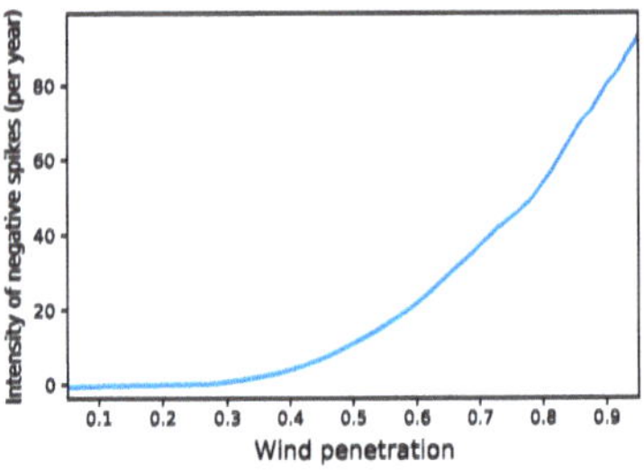

(a) *For negative jumps.*

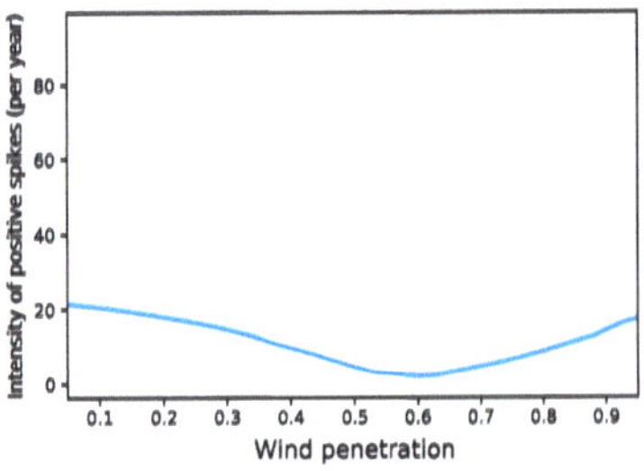

(b) *For positive jumps.*

Fig. 10.6 Kernel estimators of the intensity of the spot spikes as a function the wind penetration (in $year^{-1}$)

of order Δ_n is asymptotically negligible and the processes can be assumed to be continuously observed.

For the considered data, we use the Epanechnikov kernel $K(u) = \frac{3}{4}\left(1 - u^2\right) \mathbf{1}_{|u|\leq 1}$ and work with a kernel estimator, that is with a local polynomial estimator of degree 0. The hyperparameter κ is set to 1, leading to an asymptotic optimal oracle inequality. This choice of parameter cannot be optimal with respect to the data but no alternative method is proposed to choose it. Choosing $I = [0.05, 0.95]$, $\mathscr{H} = \{h = h_{\min} + 0.01 \times i,\ i \in \mathbb{N},\ h \leq 0.4\}$, $h_{\min} = 0.0225$ for negative jumps and $h_{\min} = 0.0089$ for positive ones, the optimal bandwidths given by this method are $\hat{h} = 0.25$ for negative jumps and $\hat{h} = 0.30$ for positive ones. The optimal estimator $\hat{q}_{\hat{h}}$ for negative jumps is given in Fig. 10.6a and the one for positive jumps in Fig. 10.6b.

Chapter 4 in [5] proposes a method to test if the intensity function belongs to some parametric function. Using this method, we test if the intensity function q is constant with respect to the wind penetration. The test is rejected for negative jumps at level 95% (with p-value equal to 0) but is not for positive jumps (with p-value equal to 0.47). These results confirm our intuition: the wind penetration index has an impact on the probability to have a negative spike but not on the probability to have a negative one. High wind penetration implies that other means of production are not used to satisfy the demand, and renewable energies constitute the cheapest means of energy production. Because of the non-storability of electricity, a producer owning a plant not used to satisfy the demand needs to pause it. Sometimes, it is cheaper for him to pay someone to consume the produced electricity rather than stop and restart his plant. This leads to negative jumps and negative prices for the electricity spot price.

Estimation of the Seasonality Function Γ_1

In order to estimate the seasonality function, we start by removing spot values corresponding to spikes and jumps. Values such that $|\Delta_i^n S| \geq 5\bar{\sigma}\Delta_n^{0.49}$, which are not always spikes, are removed; they correspond to extreme value returns. We also remove ten values following an extreme value, that insures for a spike to have reverted

Table 10.1 Estimated parameters of Γ_1

	$c_{0,1}$	$c_{1,1}$	$c_{2,1}$	$c_{3,1}$	$c_{4,1}$	$c_{5,1}$
Estimate	45.94	-8.2×10^{-4}	1.0×10^{-8}	−3.85	−6.00	−6.10
Standard error	0.16	1.7×10^{-5}	3.8×10^{-10}	7.6×10^{-2}	7.5×10^{-2}	7.6×10^{-2}

	$\tau_{0,1}$	$\tau_{1,1}$	$\tau_{2,1}$
Estimate	36744.54	1032.51	132.97
Standard error	172.85	2.12	0.30

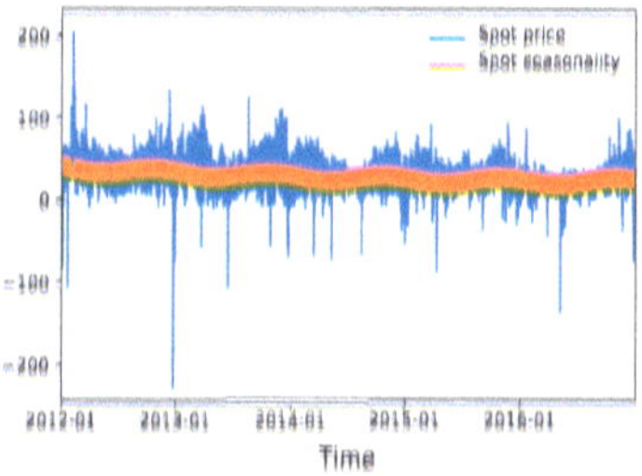

(a) *Seasonality function of the spot price.*

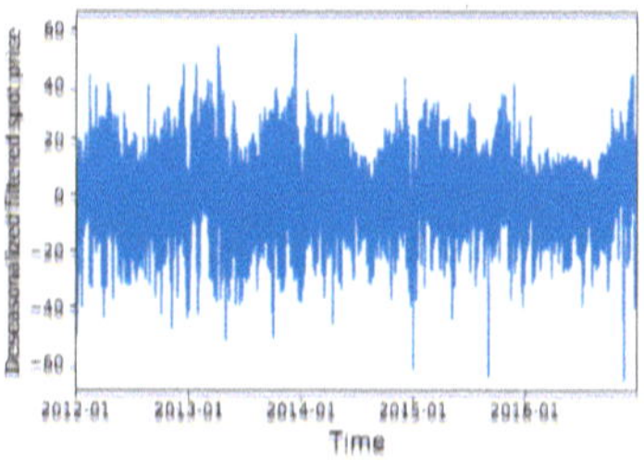

(b) *Deseasonalised and filtered spot price.*

Fig. 10.7 Seasonality function and deseasonalised spot price

to low value. After this filtering, a least square minimisation is done. Parameters and standard errors are provided in Table 10.1. The function Γ_1 and the deseasonalised and filtered spot are depicted in Fig. 10.7a and b, respectively.

Estimation of the Continuous Stochastic Part X_1

In a discrete time framework, X_1 is often modelled by an ARMA process. As for the deseasonalising, we remove the extreme values and ten values thereafter. The autocorrelation and the partial autocorrelation of the deseasonalised spot process without the extreme values are given in Fig. 10.8a, b. In a first instance, we choose to model X_1 in a discrete-time framework by an autoregressive process of order 24:

$$X_{1,t} = Q_1(D)\, X_{1,t} + \varepsilon_{1,t},$$

where D is the lag operator $DX_{1,t} = X_{1,t-1}$ and $Q_1(D) = \sum_{i=1}^{24} a_{i,1} D^i$ is a polynomial of degree 24 and $\varepsilon_{1,t}$ is a normal random variable with mean 0 and variance σ_1^2.

The parameters estimated by exact likelihood maximisation are given in Table 10.2 together with their standard errors. The autocorrelation function and the partial autocorrelation function of the residuals are given in Fig. 10.9a, b, respectively.

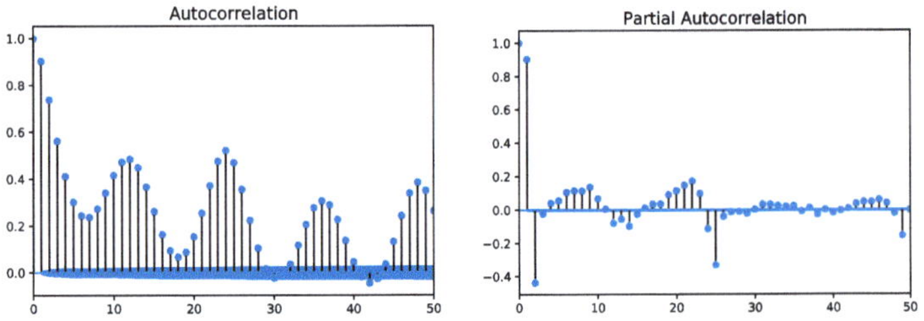

(a) *Autocorrelation function of the deseasonalized and filtered spot price.*

(b) *Partial autocorrelation function of the deseasonalised and filtered spot price.*

Fig. 10.8 Autocorrelation and partial autocorrelation of the deseasonalised and filtered spot price

Table 10.2 Estimated parameters of autoregressive process with order 24 and equivalent CAR parameters on the filtered deseasonalised spot price time series considering a step time of one hour

Parameter	Value	Standard error	CAR parameter ($\alpha_{i,1}$)
$a_{1,1}$	1.19	0.004	1.95
$a_{2,1}$	−0.35	0.006	38.77
$a_{3,1}$	−0.0098	0.007	67.71
$a_{4,1}$	0.027	0.007	631.48
$a_{5,1}$	−0.013	0.008	983.30
$a_{6,1}$	0.0074	0.008	5649.48
$a_{7,1}$	0.035	0.008	7804.14
$a_{8,1}$	−0.040	0.007	30499.38
$a_{9,1}$	0.048	0.007	37143.97
$a_{10,1}$	−0.0013	0.007	102868.84
$a_{11,1}$	0.081	0.007	109599.10
$a_{12,1}$	−0.080	0.007	217258.32
$a_{13,1}$	0.078	0.007	200565.84
$a_{14,1}$	−0.080	0.007	280634.70
$a_{15,1}$	−0.0084	0.007	221774.15
$a_{16,1}$	−0.030	0.008	210406.08
$a_{17,1}$	0.034	0.008	140047.42
$a_{18,1}$	−0.038	0.008	83291.14
$a_{19,1}$	0.0035	0.008	45551.82
$a_{20,1}$	0.0019	0.008	14618.89
$a_{21,1}$	−0.030	0.008	6246.01
$a_{22,1}$	0.0082	0.007	789.80
$a_{23,1}$	0.27	0.007	200.64
$a_{24,1}$	−0.14	0.004	2.71
σ_1^2	13.35	0.057	13.35

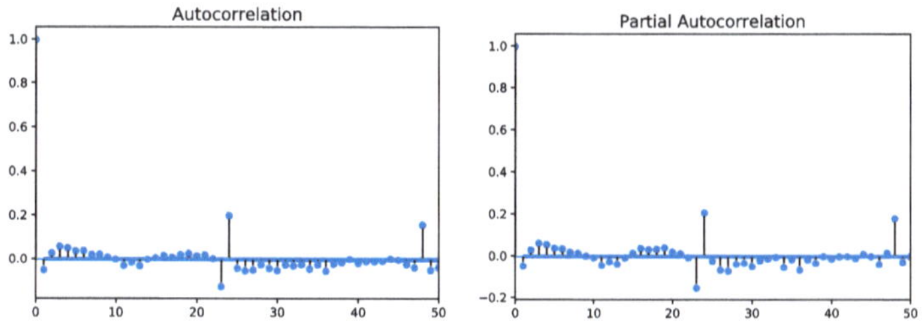

(a) *Autocorrelation function of the spot price residuals.*

(b) *Partial autocorrelation function of the spot price residuals.*

Fig. 10.9 Autocorrelation and partial autocorrelation of the spot price residuals

The parameters corresponding to the lags 3, 5, 6, 10, 15, 19, 20 and 22 are not significant at level 95% but fixing these parameters to 0 leads to non-convergence of the maximisation of the likelihood. Hence all the parameters are then considered. One observes that there remains a significant autocorrelation at lags 24 and 48, due to a seasonal (diurnal) effect. This seasonal effect is considered in [9], where a seasonal autoregressive process is used to model the spot price and fits the data better. However, considering a seasonal autoregressive process does not allow to have an equivalent stationary model in continuous time. Furthermore, contrarily to [9], we choose not to consider the fractional integrated part corresponding to a long term memory; indeed, it would consist in considering a fractional Brownian motion in an equivalent continuous-time setting and then to leave the semi-martingale framework.

Note that CAR processes are a generalisation of AR processes to a continuous-time setting. Indeed, the exact discretisation of a CAR process is a discrete-time process with AR dynamics. Thus, in order to obtain the parameters of the CAR process, one can infer the parameters of the AR process corresponding to the discretised data and map them to obtain the parameters of the CAR process using its exact discretisation. More details about the equivalence between those parameters are given in Chap. 4 of [4]. The vector $\left(\alpha_{i,1}\right)_{1\leq i\leq 24}$ corresponding to the CAR parameters is given in Table 10.2.

10.3.2 *Model for the Wind Penetration Index*

10.3.2.1 Model

As we have seen in Sect. 10.2.1, the wind penetration index takes values between 0 and 1, except for a small number of values exceeding 1. To avoid technical difficulties stemming from these values, they are set to a number close to 1 (here 0.999) below. We then write the wind penetration in the form

$$WP_t = expit\left(\Gamma_{2,t} + X_{2,t}\right),$$

with $expit(x) = \frac{1}{1+e^{-x}}$, for $x \in \mathbb{R}$. The function Γ_2 is a seasonality function and X_2 a stochastic process with mean 0. The process $X_2 + \Gamma_2$ is equal to $logit(WP_t)$ with $logit(x) = \log\left(\frac{x}{1-x}\right)$, for $x \in (0, 1)$ and is observable.

Modelling the Seasonality Function Γ_2

As for the spot price, the seasonality function is assumed to be of the form

$$\Gamma_{2,t} = c_{0,2} + c_{1,2}t + c_{2,2}t^2 + c_{3,2}\cos\left(\frac{\tau_{0,2} + 2\pi t}{365 \times 24}\right)$$
$$+ c_{4,2}\cos\left(\frac{\tau_{1,2} + 2\pi t}{7 \times 24}\right) + c_{5,2}\cos\left(\frac{\tau_{2,2} + 2\pi t}{24}\right). \quad (10.2)$$

Modelling of the Continuous Stochastic Part of the Logit Wind Penetration X_2

As for the spot modelling, the deseasonalised logit wind penetration X_2 is modelled by a CAR process of order 24. The dynamics of X_2 is modelled by

$$X_2 = \mathbf{b}^T \mathbf{X_2},$$

$$d\mathbf{X}_{2,t} = \mathbf{A_2}\mathbf{X}_{2,t}dt + \mathbf{e}\sigma_2 dW_{2,t},$$

where

$$\mathbf{A_2} = \begin{pmatrix} 0 & 1 & 0 & \cdots & 0 \\ 0 & 0 & 1 & \ddots & \vdots \\ \vdots & & \ddots & \ddots & 0 \\ 0 & \cdots & \cdots & 0 & 1 \\ -\alpha_{24,2} & -\alpha_{23,2} & \cdots & \cdots & -\alpha_{1,2} \end{pmatrix}$$

and $\left(W_{2,t}\right)_{0 \le t \le T}$ is a standard Brownian motion. This modelling is motivated by the statistical study done below.

10.3.2.2 Calibration

Estimation of the Seasonality Function Γ_2

The parameters are estimated by a least square minimisation. The parameter estimates are given in Table 10.3. The seasonality function and the deseasonalised time series X_2 are shown in Fig. 10.10a, b, respectively.

Estimation of the Continuous Stochastic Part of the Logit Wind Penetration X_2

As for the spot price, let us first consider a discrete-time modelling setting. The autocorrelation function and the partial autocorrelation function, see Fig. 10.11a, b,

Table 10.3 Estimated parameters of f_2

	$c_{0,2}$	$c_{1,2}$	$c_{2,2}$	$c_{3,2}$	$c_{4,2}$	$c_{5,2}$
Estimate	−1.73	-1.06×10^{-5}	8.82×10^{-10}	0.52	0.10	0.31
Standard error	0.015	1.5×10^{-6}	3.4×10^{-11}	6.9×10^{-3}	6.9×10^{-3}	6.9×10^{-3}

	$\tau_{0,2}$	$\tau_{1,2}$	$\tau_{2,2}$
Estimate	−2203.66	0.18	−94.62
Standard error	118.37	11.18	0.54

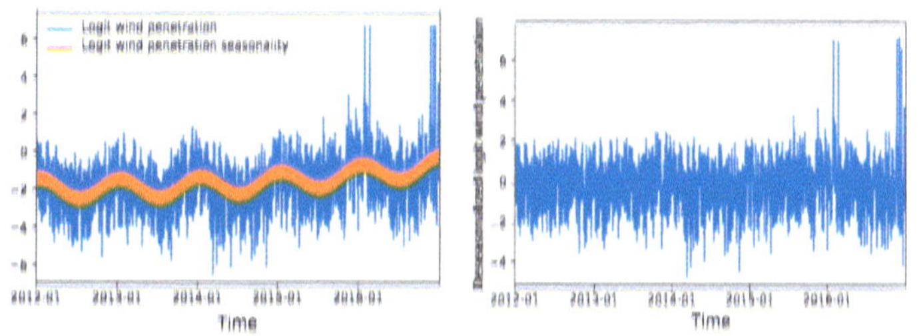

(a) *Seasonality function of the logit wind penetration.* (b) *Deseasonalised logit wind penetration.*

Fig. 10.10 Seasonality function and deseasonalised logit wind penetration

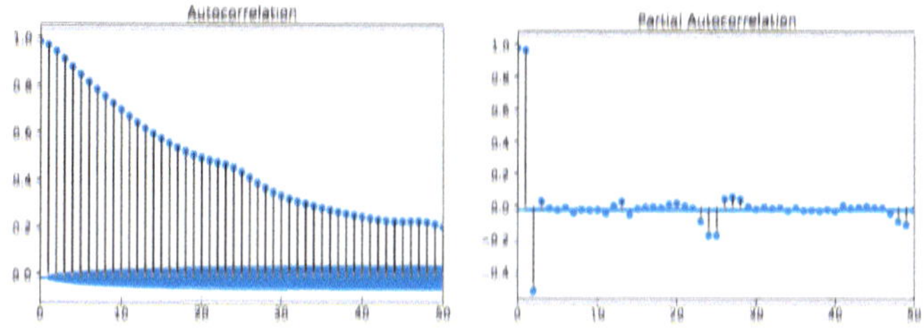

(a) *Autocorrelation function of the deseasonalised logit wind penetration.* (b) *Partial autocorrelation function of the deseasonalised logit wind penetration.*

Fig. 10.11 Autocorrelation and partial autocorrelation of the deseasonalised logit wind penetration

respectively, suggest that one suitable model in a discrete-time framework is an autoregressive process of the form:

$$X_{2,t} = Q_2(D) X_{2,t} + \varepsilon_{2,t},$$

where $Q_2(D) = \sum_{i=1}^{24} a_{i,2} D^i$ is a polynomial of degree 24 and $\varepsilon_{2,t}$ a normal random variable with mean 0 and variance σ_2^2. The parameters estimated with exact likelihood maximisation are given in Table 10.4. The parameters corresponding to lags 8, 15, 16, 17 and 18 are not significant, but as for the spot modelling, fixing them to 0 leads

Table 10.4 Estimated parameters of autoregressive process with order 24 and equivalent CAR parameters on the deseasonalised logit wind penetration time series considering a step time of one hour

Parameter	Value	Standard error	CAR parameter ($\alpha_{i,1}$)
$a_{1,2}$	1.49	0.001	1.86
$a_{2,2}$	−0.54	0.002	38.97
$a_{3,2}$	−0.025	0.004	63.16
$a_{4,2}$	0.023	0.004	636.17
$a_{5,2}$	−0.031	0.003	894.05
$a_{6,2}$	0.036	0.003	5687.46
$a_{7,2}$	−0.017	0.004	6869.22
$a_{8,2}$	0.0028	0.006	30590.17
$a_{9,2}$	−0.017	0.005	31372.93
$a_{10,2}$	0.044	0.002	102444.15
$a_{11,2}$	−0.0162	0.002	87798.34
$a_{12,2}$	−0.084	0.004	213952.50
$a_{13,2}$	0.097	0.005	149955.10
$a_{14,2}$	−0.025	0.006	271808.63
$a_{15,2}$	−0.0090	0.007	151212.31
$a_{16,2}$	−0.0063	0.008	198823.20
$a_{17,2}$	0.034	0.008	84071.24
$a_{18,2}$	−0.010	0.008	75735.70
$a_{19,2}$	−0.018	0.008	22726.15
$a_{20,2}$	0.018	0.008	12434.60
$a_{21,2}$	−0.031	0.007	2319.68
$a_{22,2}$	0.031	0.006	570.60
$a_{23,2}$	0.16	0.004	45.37
$a_{24,2}$	−0.16	0.002	1.33
σ_2^2	0.020	0.000032	0.020

to non-convergence of the likelihood maximisation. The autocorrelation function and the partial autocorrelation function of the residuals are given in Fig. 10.12a, b, respectively. We note that a significant autocorrelation at lag 24 remains due to a seasonal component, but we do not take this issue further for the same reasons as in the spot price model.

The parameters $\left(\alpha_{i,2}\right)_{1\leq i\leq 24}$ are given in Table 10.4 and are mapped from the parameters of the AR process using the exact discretisation of the CAR one.

10.3.3 Dependence Modelling

In order to make the joint model between the spot electricity price and the wind penetration index complete, we need to specify the dependence between the two components.

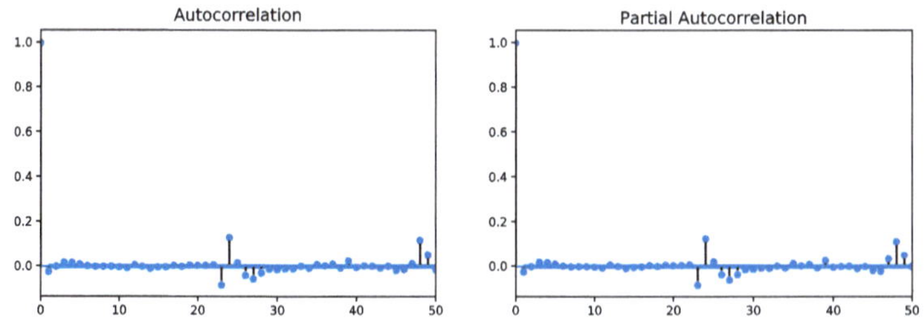

(a) *Autocorrelation function of the logit wind penetration residuals.* (b) *Partial autocorrelation function of the logit wind penetration residuals.*

Fig. 10.12 Autocorrelation and partial autocorrelation of deseasonalised logit wind penetration residuals

10.3.3.1 Model

Here we proceed by modelling the dependence between the spot price and the wind penetration by four parameters. The first parameter ρ models a linear dependence between the continuous part of the spot and the wind penetration which is (mildly) observed in Sect. 10.2.2. The three other parameters are used to model the intensity of the spike process as a two-valued function. This modelling is motivated by the kernel estimator computed in Sect. 10.3.1.1, where the test investigating whether or not the intensity of negative spikes is a constant function of the wind penetration was rejected. In the following, we will now assume that the spike process Y consists in fact of the sum of two (possibly doubly stochastic) independent Poisson-type processes: one for the negative spikes and one for the positive ones, both having the same mean reversion parameter β. Concerning the negative spikes, we observe that the intensity of negative spikes increases with the wind penetration. One simple way to take this dependence into account is to consider a regime-switching intensity function taking two different values depending on the value of the wind penetration, that is by a function of the form $\lambda : t \mapsto \lambda_{-,min}\mathbf{1}_{WP_t \leq WP_{thre}} + \lambda_{-,max}\mathbf{1}_{WP_t > WP_{thre}}$. There are two states: one state with a low intensity if the wind penetration is under a certain threshold WP_{thre}, a second state where the intensity is higher. This modelling is similar in spirit to the one proposed by [20], where two states were considered for the driving Lévy process: one for low wind penetration values and one for high wind penetration, where the distribution of the driving Lévy process in the regime with high wind penetration featured larger skewness and fatter tails than the one in the low wind penetration setting. Concerning the positive jumps, we choose to model the intensity by a constant function $\lambda : t \mapsto \lambda_+$, corresponding to a simple Poisson process.

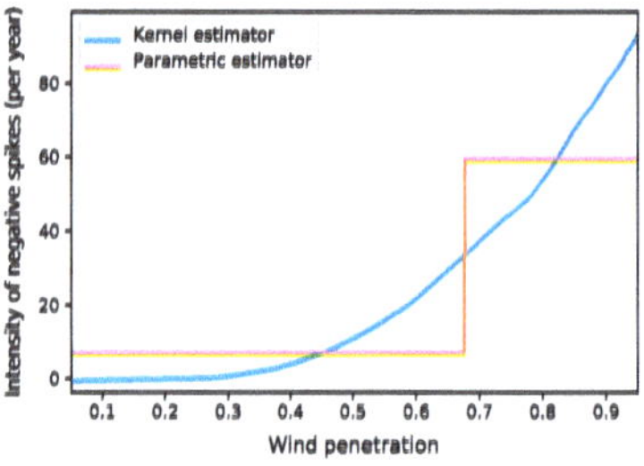

(a) *For negative jumps.*

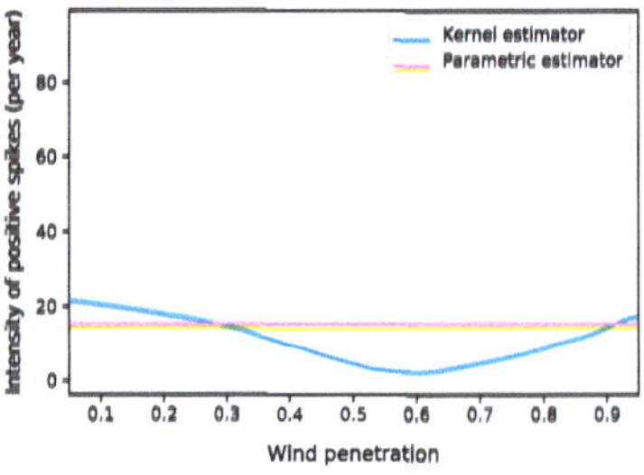

(b) *For positive jumps.*

Fig. 10.13 Parametric estimators of the intensity of the spot spikes as a function the wind penetration (in $year^{-1}$)

Table 10.5 Parameters for the dependence between the electricity spot price and the wind penetration index

Parameters	ρ	$\lambda_{-,min}$ ($year^{-1}$)	$\lambda_{-,max}$ ($year^{-1}$)	WP_{thre}	λ_+ ($year^{-1}$)
Estimate	−0.082	8.82	61.02	0.6773	16.80

10.3.3.2 Calibration

The parameters ρ are estimated by the empirical correlation between the spot residuals and the wind residuals. To estimate the parameters $\lambda_{-,min}$, $\lambda_{-,max}$, and WP_{thre}, we minimise the L_2 distance on I between the function $x \mapsto \lambda_{-,min}\mathbf{1}_{x \leq WP_{thre}} + \lambda_{-,max}\mathbf{1}_{x > WP_{thre}}$ and the kernel estimator given in Fig. 10.6a. An estimator of the intensity of the positive spikes is given by the number of positive jumps, which is equal to 84, divided by the number of years, 5 (unit in $year^{-1}$). The parametric intensity function of negative and positive jumps as a function of the wind penetration are respectively given in Fig. 10.13a, b. The parameters for the dependence are given in Table 10.5.

10.4 Application: Impact of the Dependence Between Electricity Spot Prices and Wind Penetration on the Income of an Electricity Distributor

In order to round off this article we consider an application where a joint model for electricity spot prices and wind penetration is needed. To this end, let us take the point of view of an electricity distributor. This distributor settles a contract with a wind farm that produces $Q\%$ of the German and Austrian wind production, assuming that the wind in the location considered resembles the one described by the aggregated wind data which is available to us. The distributor buys the electricity from the wind farm at a fixed price K, say. Its income on this contract over a time period $[0, T]$ is then equal to

$$P = Q \int_0^T (S_t - K)\, WP_t C_t dt,$$

where C_t is the German and Austrian load at each time t. As we want to study the impact of the wind penetration on the spot prices, let us assume that the load C_t is deterministic, knowing this assumption is rather strong. Indeed, the spot price and the load are in fact dependent, but only the dependence arising through the seasonality is considered in the following.

We are interested in the extreme values in the prices and their impact on the distributor's income. We shall consider two risk measures which are widely used in practice: the value at risk (VaR) and the expected shortfall (ES). The value at risk at level $\alpha \in (0, 1)$, denoted by $VaR_\alpha(P)$ corresponds to the maximal loss given the confidence level $1 - \alpha$, that is the quantile of order α of the distribution of P. The expected shortfall at level $\alpha \in (0, 1)$ corresponds to the expected loss in the tail distribution and is defined by $\mathbb{E}(P | P \leq VaR_\alpha(P))$. We choose to work with the levels $\alpha = 95\%$ and $\alpha = 99\%$ that are often considered in finance.

In order to study the impact of the dependencies between electricity spot prices and wind penetration, we consider two models. The first one is the one considered in Sect. 10.3 where the intensity of the negative electricity spot spikes can take two values depending on the wind penetration level. It is referred to as the two-state model in the following. The second one differs by considering a constant intensity function for the negative electricity spot prices and is referred to as the one-state model. The value of the estimated intensity in the second model is equal to 30 divided by 5.

Let us assume that the initial time is the 1st of January 2017 and that the maturity T is equal to the 31st of December 2017. The consumption is chosen as its seasonality during the year 2017: this seasonality is estimated using German hourly load data between the 1st of January 2012 and the 31st December 2016 using the same parametric form as Γ_1 and Γ_2. The spot and the wind are simulated using the two different models. Q is chosen as 1% and the strike to be 30. Results for the various risk measures are given in Table 10.6. These values are computed with the Monte Carlo method using $N = 100,000$ simulated paths. The confidence intervals for the various risk measures are given at level 95%. The confidence intervals for the value at risk and the expected shortfall are computed using bootstrap, with $M = 100,000$

Table 10.6 Different risk measures for the portfolio with strike equal to 30

Model	One-state	Two-states
Expectation	[742451.32, 754842.31]	[642085.23, 654479.20]
VaR 95%	[−906630.39, −880902.99]	[−1009660.00, −978858.45]
VaR 99%	[−1596797.78, −1553512.84]	[−1697907.39, −1647565.64]
ES 95%	[−1323982.81, −1293678.85]	[−1426258.31, −1395788.60]
ES 99%	[−1932779.14, −1876820.83]	[−2036072.86, −1979485.25]

new samples. The bootstrap method consists of randomly drawing M new samples of size N from the simulated one, using the empirical distribution function as the probability measure, and to compute the different quantities we are interested in, here the risk measures, for each sample. We then obtain M values for these quantities and the confidence intervals for these quantities at level 95% is defined by the values between the quantile of order 2.5% and the quantile of order 97.5% of these M values. The reader can refer to [18], see Chap. 23, p. 326, for more details about the bootstrap method.

We observe that modelling the intensity of the negative spikes as a function of wind penetration has an impact on the expectation of P, the values at risk at levels 95 and 99% and the expected shortfalls at levels 95 and 99% of the portfolio: they are lower than in the case when the intensity is constant. Indeed, the portfolio is a function of WP_tS_t, and negative spikes for S_t appear more often when WP_t is high.

Our intuition gained from the above application is that if one were to use a more complex parametric model than the two-state model for the intensity function considered above, then the impact of the dependence modelling could potentially increase which could lead to more extreme values in the corresponding electricity spot prices. It will be worth exploring this aspect in more detail in future research.

Acknowledgements We would like to thank Olivier Féron for his constructive comments on an earlier draft of this article. This research is supported by the department OSIRIS (Optimization, SImulation, RIsk and Statistics for Energy Markets) of EDF and by the FiME (Finance for Energy Markets) Research Initiative which is greatfully acknowledged.

References

1. Y. Aït-Sahalia, J. Jacod, Testing for jumps in a discretely observed process. Ann. Stat. 184–222 (2009)
2. Y. Aït-Sahalia, J. Jacod, *High-Frequency Financial Econometrics* (Princeton University Press, Princeton, 2014)
3. O.E. Barndorff-Nielsen, N. Shephard, M. Winkel, Limit theorems for multipower variation in the presence of jumps. Stoch. Process. Appl. **116**(5), 796–806 (2006)
4. F.E. Benth, J.Š. Benth, *Modeling and Pricing in Financial Markets for Weather Derivatives*, vol. 17 (World Scientific, Singapore, 2012)
5. T. Deschatre, Dependence modeling between continuous time stochastic processes: an application to electricity markets modeling and risk management, Ph.D. Thesis, Université Paris-Dauphine, 2017
6. T. Deschatre, O. Féron, M. Hoffmann, Estimating fast mean-reverting jumps in electricity market models, arXiv:1803.03803 (2018)
7. S. Forrest, I. MacGill, Assessing the impact of wind generation on wholesale prices and generator dispatch in the Australian national electricity market. Energy Policy **59**, 120–132 (2013)
8. A. Goldenshluger, O. Lepski, Bandwidth selection in kernel density estimation: oracle inequalities and adaptive minimax optimality. Ann. Stat. 1608–1632 (2011)
9. N. Haldrup, M.Ø. Nielsen, A regime switching long memory model for electricity prices. J. Econom. **135**(1), 349–376 (2006)
10. R. Huisman, C. Huurman, R. Mahieu, Hourly electricity prices in day-ahead markets. Energy Econ. **29**(2), 240–248 (2007)

11. T. Jónsson, P. Pinson, H. Madsen, On the market impact of wind energy forecasts. Energy Econ. **32**(2), 313–320 (2010)
12. T. Jónsson, P. Pinson, H.A. Nielsen, H. Madsen, T.S. Nielsen, Forecasting electricity spot prices accounting for wind power predictions. IEEE Trans. Sustain. Energy **4**(1), 210–218 (2013)
13. J.C. Ketterer, The impact of wind power generation on the electricity price in Germany. Energy Econ. **44**, 270–280 (2014)
14. C. Klüppelberg, T. Meyer-Brandis, A. Schmidt, Electricity spot price modelling with a view towards extreme spike risk. Quant. Financ. **10**, 963–974 (2010)
15. C. Lacour, P. Massart, V. Rivoirard, Estimator selection: a new method with applications to kernel density estimation. Sankhya A 1–38 (2016)
16. J. Li, V. Todorov, G. Tauchen, Volatility occupation times. Ann. Stat. **41**(4), 1865–1891 (2013)
17. J. Li, V. Todorov, G. Tauchen, Jump regressions. Econometrica **85**(1), 173–195 (2017)
18. A.W. Van der Vaart, *Asymptotic Statistics*, vol. 3 (Cambridge University Press, Cambridge, 1998)
19. A.E.D. Veraart, Inference for the jump part of quadratic variation of Itô semimartingales. Econom. Theory **26**(2), 331–368 (2010)
20. A.E.D. Veraart, Modelling the impact of wind power production on electricity prices by regime-switching Lévy semistationary processes, in *Stochastics of Environmental and Financial Economics* ed. by F.E. Benth, G. Di Nunno (Springer, Berlin, 2016), pp. 321–340
21. A.E.D. Veraart, L.A.M. Veraart, Modelling electricity day-ahead prices by multivariate Lévy semi-stationary processes, in *Wolfgang Pauli Proceedings* ed. by F.E. Benth, V. Kholodnyi, P. Laurence (Springer, Berlin, 2014), pp. 157–188

Chapter 11
The Optimal Control of Storage for Arbitrage and Buffering, with Energy Applications

James Cruise and Stan Zachary

Abstract We study the optimal control of storage which is used for both arbitrage and buffering against unexpected events (shocks), with particular applications to the control of energy systems in a stochastic and typically time-heterogeneous environment. Our philosophy is that of viewing the problem as being formally one of stochastic dynamic programming (SDP), but of recasting the SDP recursion in terms of functions which, if known, would reduce the associated optimisation problem to one which is deterministic, except that it must be re-solved at times when shocks occur. In the case of a perfectly efficient store facing linear buying and selling costs the functions required for this approach may be determined exactly; otherwise they may typically be estimated to good approximation. We provide characterisations of optimal control policies. We consider also the associated deterministic optimisation problem, outlining an approach to its solution which is both computationally tractable and—through the identification of a running forecast horizon—suitable for the management of systems over indefinitely extended periods of time. We give examples based on Great Britain electricity price data.

Keywords Storage · Buffering · Optimisation · Control

11.1 Introduction

How should one optimally control storage which is used simultaneously for a number of different purposes? We study this problem in the case of a single store which is used for both price arbitrage, i.e. for buying and selling over time, and for buffering against unexpected events, or *shocks*. Here an optimal control must balance the some-

J. Cruise · S. Zachary (✉)
School of Mathematical and Computer Sciences, Heriot-Watt University, Edinburgh EH14 4AS, UK
e-mail: s.zachary@gmail.com

J. Cruise
e-mail: r.cruise@hw.ac.uk

P. Drobinski et al. (eds.), *Renewable Energy: Forecasting and Risk Management*, Springer Proceedings in Mathematics & Statistics 254,
https://doi.org/10.1007/978-3-319-99052-1_11

times conflicting controls which would apply to these two uses of storage considered individually. Of particular interest is the control of an energy store in a stochastic and typically time-heterogeneous environment, where at any time a full stochastic description of that environment may not be available over more than a relatively short future time horizon. The shocks correspond, for example, to the loss of a generator or transmission line, or a sudden surge in demand. Our philosophy is that of viewing the problem as being formally one of stochastic dynamic programming (SDP), but of recasting the SDP recursion in terms of functions which may be determined in advance, either exactly or approximately, and which reduce the associated optimisation problem to one which is deterministic, except that it must be re-solved at those times at which shocks occur.

There is considerable literature on the control of storage for each of the above two purposes considered on its own. There have been numerous studies of the use of storage for buffering against both the increased variability and the increased uncertainty in electrical power systems due to the higher penetration of renewable generation—the former due to the natural variability of such resources as wind power, and the latter due to the inherent uncertainty of forecasting. These studies have considered many different more detailed objectives; these range from the sizing and control of storage facilities co-located with the renewable generation so as to provide a smoother supply and so offset the need for network reinforcement [9, 12, 20], to studies on storage embedded within transmission networks so as to increase wind power utilisation and so reduce overall generation costs [17, 26, 32]. In addition there have been a number of studies into the more general use of storage for buffering, for example, so as to provide fast frequency response to power networks [23, 24, 30], or to provide quality of service as part of a microgrid [6, 19].

In the case of the use of storage for arbitrage, and with linear cost functions for buying and selling at each instant in time, the problem of optimal control is the classical *warehouse problem* (see [5, 8, 14] and also [27] for a more recent example). Cruise et al. [10] consider the optimal control of storage in the case where the store is a price-maker (i.e. the size of the store is sufficiently large that its activities influence prices in the market in which it operates) and is subject to both capacity and rate constraints; they develop the associated Lagrangian theory, and further show that the optimal control at any point in time usually depends only on the cost functions associated with a short future time horizon. Recent alternative approaches for studying the value and use of storage for arbitrage can be found in the papers [21, 25, 29, 33, 34]–see also the text [35], and the further references given in [10]. For an assessment of the potential value of energy storage in the UK electricity system see [30].

In general the problem of using a store for buffering is necessarily stochastic. The natural mathematical approach is via stochastic dynamic programming. This, however, is liable to be computationally intractable, especially in the case of long time horizons and the likely time heterogeneity of the stochastic processes involved. Therefore much of the literature considers necessarily somewhat heuristic but nevertheless plausible control policies—again often adapted to meeting a wide variety of objectives. For example, for large stores operating within transmission networks, the buffering policies studied have included that of a fixed target level policy [4], a dynamic target level policy [16], and a two-stage process with day ahead generation scheduling and an online procedure to adapt load levels [1].

Control policies have been studied via a range of analytic and simulation-based methods. Examples of an analytic approach can be found in [18], where partial differential equations are utilised to model the behaviour and control of a store, and in [2, 3], where spectral analysis of wind and load data is used with models which also incorporate turbine behaviour. Simulation-based studies include [4, 16], which use a bootstrap approach based on real wind forecast error data, and [1], which uses Monte Carlo simulation of the network state.

In the present paper we use an economic framework to study the optimal control of a store which, as previously stated, is used both for price arbitrage and for buffering against occasional and unpredictable *shocks* whose occurrence is described by some stochastic process. The store seeks to operate in such a way as to minimise over time the expected total cost of its operation. We believe such an economic framework to be natural when the store operates as part of some larger and perhaps very complex system, provided the price signals under which the store operates are correctly chosen—see, for example, [11]. The store may be sufficiently large as to have market impact, leading to nonlinear cost functions for buying and selling, may be subject to rate (as well as capacity) constraints, and, as will typically be the case, may suffer from round-trip inefficiencies. We formulate a stochastic model which is realistic in many circumstances and characterise some of the properties of an optimal control, relating the results to the existing experimental literature.

Our approach is that of re-expressing the traditional SDP recursion so as to reduce the associated optimisation problem to one which is deterministic, except only that it must be dynamically re-solved whenever shocks occur. The specification of the associated optimisation problem requires that the cost functions C_t, which give the costs of buying or selling at each successive time t, are supplemented by further functions A_t associated with the expected costs of possible shocks. The cost functions C_t are formally assumed to be deterministic. However, when prices are stochastic, deterministic approximations may be used and updated at successive time steps; this deterministic re-optimisation approach is common when storage is used for price arbitrage alone—see, for example, [22, 27, 28, 36] and, for the case where storage is sufficiently large as to have market impact, see [10]. The functions A_t (which, although defined in terms of the stochastic process of shocks, are also deterministic) are formally introduced in Sect. 11.2. We show that in the case of a perfectly efficient store facing linear buying and selling costs the functions A_t may be determined exactly, and that otherwise they may typically be estimated to a good approximation.

The optimal control up to the time of the first shock is given by the solution, at the start of the control period, of an optimisation problem which can be regarded as that of minimising the costs associated with the store buying and selling added to those of notionally "insuring" for each future instant in time against the effects of the random fluctuations, i.e. the shocks, resulting from the provision of buffering services. The cost of such "insurance" depends on the absolute level of the store at the relevant time. Thus the deterministic problem is that of choosing the vector of successive levels of the store so as to minimise a total cost function $\sum_t [C_t(x_t) + A_t(s_t)]$, subject to rate and capacity constraints. Here $C_t(x_t)$ is the cost of incrementing the level of the store (positively or negatively) at time t by x_t, and the function A_t is such that $A_t(s_t)$ is the expected additional cost of dealing with any shock which may occur at the time t when the level of the store is then s_t. We define this optimisation problem more carefully in Sect. 11.2 and discuss various possible approaches to its solution. In the stochastic environment in which the store operates, the solution of this problem determines the future control of the store until such time as its buffering services are actually required, following which the level of the store is perturbed and the optimisation problem must be re-solved starting at the new level. The continuation of this process provides what is in principle the exactly optimal stochastic control of the store on a potentially indefinite time scale.

In Sect. 11.2 we formulate the relevant stochastic model, discuss its applicability, and give various approaches to the determination of the optimal control. These approaches require the availability of good estimates of the above functions A_t, and in Sect. 11.4 we show how these may be obtained. In Sect. 11.3 we provide some characteristic properties of optimal solutions, which we relate to empirical work in the existing literature. In Sect. 11.5 we give examples.

11.2 Model and Determination of Optimal Control

Consider the management of a store over a finite time interval which is divided into a succession of periods indexed by $1, \dots, T$. At the start of each time period t the store makes a decision as to how much to buy or sell during that time period; however, the level of the store at the end of that time period may be different from that planned if, during the course of the period, the store is called upon to provide buffering services to deal with some unexpected event or random *shock*. Such a shock might be the need to supply additional energy during the time period t due to an unexpected failure—for example that of a generator—or might simply be the difference between forecast and actual renewable generation or demand.

We suppose the store has a capacity of E units of energy. Similarly we suppose that the total energy which may be input or output during any time period is subject to rate (i.e. power) constraints P_I and P_O respectively. This slotted-time model corresponds, for example, to real world energy markets where energy is typically traded at half-hourly or hourly intervals, with the actual delivery of that energy occurring in the intervening continuous time period. Detailed descriptions of the operation of the UK market can be found in [15, 31]. The theory developed here easily extends to the case where the above storage parameters are time dependent.

Define also the set $X = [-P_O, P_I]$. Both buying and selling prices associated with any time period t may be represented by a convex function C_t defined on X such that $C_t(x)$ is the cost of a planned change of x to the level of the store during the time period t. Typically each function C_t is increasing and $C_t(0) = 0$; then, for positive x, $C_t(x)$ is the cost of buying x units and, for negative x, $C_t(x)$ is the negative of the reward for selling $-x$ units. Then the convexity assumption corresponds, for each time t, to an increasing cost of buying each additional unit, a decreasing reward obtained for selling each additional unit, and every unit buying price being at least as great as every unit selling price. When, as is usually the case, the store is not perfectly *efficient* in the sense that only a fraction $\eta \leq 1$ of the energy input is available for output, then this may be captured in the cost function by reducing selling prices by the factor η; under the assumption that the cost functions C_t are increasing it is easily verified that this adjustment preserves the above convexity of the functions C_t. We thus assume that the cost functions are so adjusted so as to capture any such round-trip inefficiency. The functions C_t are taken to be deterministic but, as discussed in the Introduction, in a stochastic environment a deterministic re-optimisation approach is possible.

A further form of possible inefficiency of a store is *leakage*, whereby a fraction of the contents of the store is lost in each unit of time. We do not explicitly model this here. However, only routine modifications are required to do so, and are entirely analogous to those described in [10].

Suppose that at the end of each time period $t - 1$ the level of the store is given by the random variable S_{t-1}, where we take S_0 to be given by the initial level s_0 of the store. We assume that one may then choose a *planned* adjustment (contract to buy or sell) $x_t \in X$ and such that $S_{t-1} + x_t \in [0, E]$ to the level of the store during the time period t. The planned adjustment x_t is a (deterministic) function of the level S_{t-1} and the cost of this adjustment is $C_t(x_t)$. Subsequent to this, during the course of the time period t, the store may be subject to a *shock* or random disturbance, corresponding to the need to provide unexpected buffering services. This shock has an associated cost, typically due to the store not being able to provide the required services, and may further disturb the final level of the store at the end of the time period t. We assume that the cost of any shock occurring during the time period t and the resulting *actual* level of the store at the end of the time period t are given by random variables whose joint distribution is a function of the *planned* final level $S_{t-1} + x_t$ of the store for the end of that time period, but that, given this planned final level, these random variables are otherwise independent of all else. Thus we may assume that there are given T independent stochastic processes $(D_t(s), S_t(s))_{s \in [0,E]}$ each taking values in $\mathbb{R} \times \mathbb{R}$ (where each such "process" is indexed by the possible levels s of the store rather than by time), and that the shock cost and actual store level at the end of each time period t are then given respectively by $D_t(S_{t-1} + x_t)$ and $S_t(S_{t-1} + x_t)$; in the absence of any shock during the time period t, we have $D_t(s) = 0$ and $S_t(s) = s$ for all s. The assumption that the joint distribution of the shock cost and store level disturbance associated with any time period t depend only on the planned level $S_{t-1} + x_t$ of the store at the end of the time period t is likely to be most accurate in applications where the store is able to adjust to its target level quickly within each time period, or where the level of the store does not change too much within a single time period; its relaxation–for example, by allowing a dependence of these random variables on a more general function of S_{t-1} and x_t—simply complicates without essentially changing the analysis below. Note that the model further assumes that disturbances do not persist beyond the end of the time periods in which they occur. Under any given control policy for the management of the store satisfying the above conditions (i.e. under any specification, for each time t, of the planned increment x_t as a function of the realised value of S_{t-1}), the levels S_t of the store at the end of the successive time periods t form a Markov process.

For each t, and conditional on each possible value s_{t-1} of the level S_{t-1} of the store at the end of the time period $t-1$, define $V_{t-1}(s_{t-1})$ to be the expected future cost of subsequently managing the store under an optimal control—where, here and elsewhere, by an optimal control we mean a control defined as above under which the expected cost of managing the store is minimised. We then have the SDP recursion

$$V_{t-1}(s_{t-1}) = \min_{\substack{x_t \in X \\ s_{t-1}+x_t \in [0,E]}} \left[C_t(x_t) + \mathbf{E}[D_t(s_{t-1}+x_t) + V_t(S_t(s_{t-1}+x_t))]\right], \quad (11.1)$$

where $\mathbf{E}$ denotes expectation and where, as above, $s_{t-1}+x_t$ and $S_t(s_{t-1}+x_t)$ $(= S_t)$ are respectively the planned and actual levels of the store at the end of the time period t. (The assumed independence of the "processes" of paired random variables $\{(D_t(s), s_t(s))\}_{s\in[0,E]}$ defining shock costs and disturbances in successive time periods ensures that it is sufficient to consider unconditional expectations in (11.1).) We further have the terminal condition

$$V_T(s_T) = 0 \quad (11.2)$$

for all possible levels s_T of the store at the end of the time period T. The recursion (11.1) and the terminal condition (11.2) may in principle be used to determine an optimal control. In particular, given the level s_{t-1} of the store at the end of any time period $t-1$, the optimal planned increment to the level of the store for the time period t is given by $\hat{x}_t(s_{t-1})$ where this is defined to be the value of x_t which achieves the minimisation in the recursion (11.1).

However, as discussed in the Introduction, an SDP approach may frequently be computationally intractable and is further not suitable for the management of a store over indefinite time horizons. Thus, for each t, let the (deterministic) function A_t on $[0, E]$ be such that, for any *planned* level $s_t = s_{t-1} + x_t$ of the store for the end of the time period t,

$$A_t(s_t) = \mathbf{E}[D_t(s_t) + V_t(S_t(s_t))] - V_t(s_t), \quad (11.3)$$

where again the random variable $S_t(s_t)$ is the *actual* level of the store at the end of the time period t. Given the planned level s_t of the store for the end of the time period t, the quantity $A_t(s_t)$ is the difference between the expected cost $\mathbf{E}(D_t(s_t) + V_t(S_t(s_t)))$ of optimally managing the store during and subsequent to the time period t and the corresponding expected cost $V_t(s_t)$ which would be incurred in the guaranteed absence of any shock during that time period. We shall show in Sect. 11.4 that in many cases the functions A_t may be efficiently determined either exactly or to a very good approximation even in the absence of any knowledge of the functions V_t.

It now follows from (11.3) that the recursion (11.1) may be rewritten as

$$V_{t-1}(s_{t-1}) = \min_{\substack{x_t \in X \\ s_{t-1}+x_t \in [0,E]}} \left[C_t(x_t) + A_t(s_{t-1}+x_t) + V_t(s_{t-1}+x_t), \right], \tag{11.4}$$

where we again require the terminal condition (11.2). Further, given the level s_{t-1} of the store at the end of any time period $t-1$, the optimal planned increment $\hat{x}_t(s_{t-1})$ to the level of the store for the time period t is given by the value of x_t which achieves the minimisation in the recursion (11.4).

The (backwards) recursion (11.4) is entirely deterministic. *Given a knowledge of the functions* A_t (see Sect. 11.4), a complete solution of the recursion (11.4) would determine, for all $t = 1, \dots, T$ and for all possible levels s_{t-1} of the store at the end of the time period $t-1$, both the minimised expected future cost $V_{t-1}(s_{t-1})$ and the optimal planned increment $\hat{x}_t(s_{t-1})$ to the level of the store for the time period t. Now let the (Markov) process $(\hat{S}_0, \dots, \hat{S}_T)$, with $\hat{S}_0 = s_0$, correspond to the sequence of levels of the optimally controlled store. Then, since this process is random, the optimal planned increment $\hat{x}_t(\hat{S}_{t-1})$ for each time period t is not known until the end of the time period $t-1$.

However, the solution of the recursion (11.4) as above, typically require the determination of each of the functions V_t for all possible values of its argument. We therefore define a deterministic optimisation problem whose solution $s^* = (s_0^*, \dots, s_T^*)$, with $s_0^* = s_0$, coincides with the optimal control of the store up to the time of the first shock. As we discuss below, the solution of this optimisation problem is typically computationally much simpler than the complete solution of recursion (11.4). However, it is necessary to re-solve this optimisation problem at the end of each time period in which a shock occurs.

For any vector $s = (s_0, \dots, s_T)$ of possible store levels, where s_0 is constrained to be the initial level of the store, and for each $t = 1, \dots, T$, define

$$x_t(s) = s_t - s_{t-1}. \tag{11.5}$$

Define also the optimisation problem:

$\mathbb{P}$: choose $s = (s_0, \dots, s_T)$, where again s_0 is the initial level of the store, so as to minimise

$$\sum_{t=1}^{T} [C_t(x_t(s)) + A_t(s_t)] \tag{11.6}$$

subject to the capacity constraints

$$0 \le s_t \le E, \quad 1 \le t \le T, \tag{11.7}$$

and the rate constraints

$$x_t(s) \in X, \qquad 1 \le t \le T. \tag{11.8}$$

Let $s^* = (s_0^*, \ldots, s_T^*)$, with $s_0^* = s_0$, denote the solution to the above problem $\mathbb{P}$. The recursion (11.4) is the dynamic programming recursion for the solution of the problem $\mathbb{P}$ and it follows straightforwardly from iteration of (11.4), using also the terminal condition (11.2), that $x_1(s^*)$ achieves the minimisation in (11.4) for $t = 1$, i.e. that $x_1(s^*) = \hat{x}_1(s_0)$ is planned first increment in the optimal control of the store. Thus, from (11.5), provided no shock occurs during the time period 1 so that $\hat{S}_1 = s_0 + \hat{x}_1(s_0)$, we have also that $\hat{S}_1 = s_1^*$. More generally, let the random variable T' index the first time period during which a shock does occur. Then repeated application of the above argument gives immediately the following result.

Theorem 11.1 *For all $t < T'$, we have $\hat{S}_t = s_t^*$.*

The solution to the problem $\mathbb{P}$ therefore defines the optimal control of the store up to the end of the time period T' defined above. At that time it is necessary to reformulate the problem $\mathbb{P}$, starting at the end of the time period T', instead of at time 0, and replacing the initial level of the store s_0 by the perturbed level $\hat{S}_{T'}$ at that time. Iterative application of this process at the times of successive shocks leads to the dynamically determined stochastic optimal control—which is exact to the extent that the functions A_t are known exactly.

Given that the functions A_t are known, either exactly or to a sufficiently good approximation (again see Sect. 11.4), the deterministic optimisation problem $\mathbb{P}$ may be solved by using strong Lagrangian techniques to derive a *forward* algorithm which is computationally much simpler than the use of a dynamic programming approach, and which further identifies a running *planning* or *forecast horizon*. The latter is such that, for each time t there exists a time $t' > t$ such that the optimal decision at time t does not depend on the functions C_u and A_u for $u > t'$. This is proved in [10] for the case in which the functions A_t are zero, but the more general result and algorithm may be derived along the same lines. The existence of such a running forecast horizon further reduces the computation required in the solution of the problem $\mathbb{P}$ and makes the present approach particularly suitable for the management of storage over a very long or indefinite time period. It further means that, in an environment in which prices—and so the cost functions C_t—are uncertain, in order to make the optimal decision at any time t as above it is only necessary to estimate the cost functions C_u for values of u up to the associated forecast horizon t'. In the case where, as in the fairly realistic examples of Sect. 11.5, the store fills and partially empties on an approximate daily cycle, the length of this forecast horizon is typically of the order of a day or two. In practice electricity prices in particular may often be estimated accurately on such time scales, and a deterministic re-optimisation approach, as discussed in the Introduction, is likely to suffice for the optimal control of the store.

11.3 Characterisation of Optimal Solutions

In this section we establish some properties of the functions $\hat{x}_t(\cdot)$ defined in the previous section (as achieving the minimisation in the recursion (11.4)) and determining the optimal control of the store.

One case of particular interest is that where the store is a price-taker (i.e. the store is not so large as to impact itself on market prices), so that, for each t, the cost function C_t is given by

$$C_t(x) = \begin{cases} c_t^{(b)}x, & \text{if } x \geq 0 \\ c_t^{(s)}x, & \text{if } x < 0, \end{cases} \tag{11.9}$$

where the unit "buying" price $c_t^{(b)}$ and the unit "selling" price $c_t^{(s)}$ are such that $c_t^{(s)} \leq c_t^{(b)}$ (possible inequality resulting, for example, from the round-trip inefficiency of the store—see the discussion of Sect. 11.2.)

Theorem 11.2 below is a simple result which shows that in the case where buying and selling prices are equal, and provided rate constraints are nonbinding, the optimal policy is a "target" policy. That is, for each time period t there exists a target level $\hat{s}_t$ such that, given that the level of the store at the end of the immediately preceding time period is s_{t-1}, the optimal planned level $s_{t-1} + x_t$ of the store to be achieved during the time period t is set equal to $\hat{s}_t$, independently of s_{t-1}.

Theorem 11.2 *Suppose that, for each t, we have* $c_t^{(b)} = c_t^{(s)} = c_t$; *define*

$$\hat{s}_t = \arg\min_{s \in [0,E]} [c_t s + A_t(s) + V_t(s)], \tag{11.10}$$

where the functions A_t *and* V_t *are as introduced in Sect. 11.2. Then, for each t and for each* s_{t-1}, *we have* $\hat{x}_t(s_{t-1}) = \hat{s}_t - s_{t-1}$ *provided only that this quantity belongs to the set X.*

Proof The recursion (11.4) here becomes, for each t,

$$V_{t-1}(s_{t-1}) = \min_{\substack{x_t \in X \\ s_{t-1}+x_t \in [0,E]}} \left[c_t x_t + A_t(s_{t-1} + x_t) + V_t(s_{t-1} + x_t)\right], \tag{11.11}$$

and the above minimisation is achieved by x_t such that $s_{t-1} + x_t = \hat{s}_t$, provided only that $x_t \in X$.

In order to deal with the possibility of rate constraint violation, or the more general price-taker case where $c_t^{(s)} < c_t^{(b)}$, or the general case where the cost functions C_t are merely required to be convex, we require the additional assumption of convexity of the functions A_t. This condition, while not automatic, is reasonably natural in many applications—see the examples of Sect. 11.5.

Theorem 11.3 *Suppose that, in addition to convexity of the functions C_t, each of the functions A_t is convex. Then, for each t:*

(i) the function V_{t-1} is convex;
(ii) $\hat{x}_t(s_{t-1})$ is a decreasing function of s_{t-1};
(iii) $s_{t-1} + \hat{x}_t(s_{t-1})$ is an increasing function of s_{t-1}.

Proof To show (i) we use backwards induction in time. The function V_T is convex. Suppose that, for any given $t \le T$, the function V_t is convex; we show that the function V_{t-1} is convex. For any given values $s_{t-1}^{(i)}$, $i = 1, \dots, n$, of s_{t-1} and for any convex combination $\bar{s}_{t-1} = \sum_{i=1}^n \kappa_i s_{t-1}^{(i)}$, where each $\kappa_i \ge 0$ and where $\sum_{i=1}^n \kappa_i = 1$, define also $\bar{x}_t = \sum_{i=1}^n \kappa_i \hat{x}_t(s_{t-1}^{(i)})$. Note that $\bar{x}_t \in X$ and that $\bar{s}_{t-1} + \bar{x}_t \in [0, E]$. Then, from (11.4),

$$
\begin{aligned}
V_{t-1}(\bar{s}_{t-1}) &\le C_t(\bar{x}_t) + A_t(\bar{s}_{t-1} + \bar{x}_t) + V_t(\bar{s}_{t-1} + \bar{x}_t) \\
&\le \sum_{i=1}^n \kappa_i \left(C_t(\hat{x}_t(s_{t-1}^{(i)})) + A_t(s_{t-1}^{(i)} + \hat{x}_t(s_{t-1}^{(i)})) + V_t(s_{t-1}^{(i)} + \hat{x}_t(s_{t-1}^{(i)})) \right) \\
&= \sum_{i=1}^n \kappa_i V_{t-1}(s_{t-1}^{(i)}),
\end{aligned}
$$

where the second inequality above follows from the convexity of the functions C_t, A_t and V_t (the latter by the inductive hypothesis). Thus V_{t-1} is convex as required.

To show (ii) and (iii) we make use of the following result: let f and g be functions defined on the real line $\mathbb{R}$ such that g is convex, and suppose that, for each fixed s, the function of $x \in \mathbb{R}$ given by $f(x) + g(s + x)$ is minimised by $\hat{x}(s)$; then $\hat{x}(s)$ is a decreasing function of s. To see this, suppose that $s_1 < s_2$ and note that, under the given assumptions,

$$
f(\hat{x}(s_1)) + g(s_1 + \hat{x}(s_1)) \le f(x) + g(s_1 + x), \qquad x \in \mathbb{R}. \tag{11.12}
$$

The convexity of g implies straightforwardly that

$$
g(s_2 + \hat{x}(s_1)) - g(s_2 + x) \le g(s_1 + \hat{x}(s_1)) - g(s_1 + x), \qquad \text{for all } x > \hat{x}(s_1). \tag{11.13}
$$

It follows from (11.12) and (11.13) that

$$
f(\hat{x}(s_1)) + g(s_2 + \hat{x}(s_1)) \le f(x) + g(s_2 + x), \qquad \text{for all } x > \hat{x}(s_1).
$$

It now follows from the above that $\hat{x}(s_2)$ is (or, in the absence of uniqueness, may be taken to be) less than or equal to $\hat{x}(s_1)$.

The result (ii) of the theorem now follows by applying the above result with the function f given by C_t and the function g given by $A_t + V_t$, since A_t is assumed convex and, from (i), V_t is also convex. (That the minimisation in (11.4) is taken over those x within a closed interval of the real line causes no problems: for example, this restriction may be formally dropped by extending the domains of definition of C_t, A_t and V_t to the entire real line, taking them to be infinite outside the intervals on which they are naturally defined.)

The result (iii) of the theorem similarly follows by applying the above general result with the function f given by $A_t + V_t$ and the function g given by the convex function C_t (in the recursion (11.4) writing $C_t(x_t) = C_t(-s_{t-1} + (s_{t-1} + x_t))$ and, for each fixed value of s_{t-1}, regarding the minimisation in (11.4) as being over the variable $s_{t-1} + x_t$).

Remark 11.1 Given initial levels $s_0^{(1)}$ and $s_0^{(2)}$ of the store, let $\{S_t^{(1)}\}$ and $\{S_t^{(2)}\}$ (with $S_0^{(1)} = s_0^{(1)}$ and $S_0^{(2)} = s_0^{(2)}$) be the respective optimally controlled stochastic processes of levels of the store—coupled with respect to the underlying stochastic process of shocks. Suppose we additionally assume that the level of the store immediately following any shock is an increasing function of the level immediately prior to that shock. It then follows from (iii) of Theorem 11.3, that under the conditions of the theorem, if $s_0^{(1)} \le s_0^{(2)}$ then $S_t^{(1)} \le S_t^{(2)}$ for all subsequent t. This monotonicity property proves useful in Sect. 11.4.

We now return to the price-taker case, in which the cost functions are as defined by (11.9), and which corresponds to a store which is not sufficiently large as to have market impact. Here we prove a strengthened version of Theorem 11.3. For each t, given that the function A_t is convex, define

$$s_t^{(b)} = \arg\min_{s \in [0,E]} [c_t^{(b)} s + A_t(s) + V_t(s)] \tag{11.14}$$

and similarly define

$$s_t^{(s)} = \arg\min_{s \in [0,E]} [c_t^{(s)} s + A_t(s) + V_t(s)]. \tag{11.15}$$

Note that the above convexity assumption and the condition that, for each t, we have $c_t^{(s)} \le c_t^{(b)}$ imply that $s_t^{(b)} \le s_t^{(s)}$. We now have the following result.

Theorem 11.4 *Suppose that the cost functions C_t are as given by* (11.9) *and that the functions A_t are convex. Then the optimal policy is given by: for each t and given s_{t-1},*

$$\hat{x}_t(s_{t-1}) = \begin{cases} \min(s_t^{(b)} - s_{t-1},\ P_I) & \text{if } s_{t-1} < s_t^{(b)}, \\ 0 & \text{if } s_t^{(b)} \le s_{t-1} \le s_t^{(s)}, \\ \max(s_t^{(s)} - s_{t-1},\ -P_O) & \text{if } s_{t-1} > s_t^{(s)}. \end{cases} \tag{11.16}$$

Proof For each t, it follows from the convexity of the functions C_t, A_t and V_t (the latter by the first part of Theorem 11.3) that, for $s_{t-1} < s_t^{(b)}$ the function $C_t(x_t) + A_t(s_{t-1} + x_t) + V_t(s_{t-1} + x_t)$ is minimised by $x_t = s_t^{(b)} - s_{t-1}$, for $s_t^{(b)} \le s_{t-1} \le s_t^{(s)}$ it is minimised by $x_t = 0$, while for $s_{t-1} > s_t^{(s)}$, it is minimised by $x_t = s_t^{(s)} -$

s_{t-1}. The required result now follows from the recursion (11.4) (on again using the convexity of the functions C_t, A_t and V_t to account for the rate constraint in that recursion).

Thus in general in the price-taker case there exists, for each time period t, a "target interval" $[s_t^{(b)}, s_t^{(s)}]$ such that, if the level of the store at the end of the previous time period is s_{t-1}, the optimal policy is to chose $\hat{x}_t(s_{t-1})$ so that $s_{t-1} + \hat{x}_t(s_{t-1})$ is the nearest point (in absolute distance) to s_{t-1} lying within, or as close as possible to, the above interval. In the case where $c_t^{(b)} = c_t^{(s)} = c_t$, the above interval shrinks to the single point $\hat{s}_t$ defined by (11.10).

These results shed some light on earlier, more applied, papers of Bejan et al. [4] and Gast et al. [16], in which the uncertainties in the operation of a energy store result from errors in wind power forecasts. The model considered in those papers is close to that of the present paper, as we now describe. The costs of operating the store result (a) from round-trip inefficiency, which in the formulation of the present paper would be captured by the cost functions C_t as defined by (11.9) with $c_t^{(s)} < c_t^{(b)}$ and with C_t the same for all t, and (b) from buffering events, i.e. from failures to meet demand through insufficient energy available to be supplied from the store when it is needed, and from energy losses through store overflows. In the formulation of the present paper these costs would be captured by the functions A_t. In contrast to the present paper decisions affecting the level of the store (the amount of conventional generation to schedule for a particular time) are made n time steps—rather than a single time step—in advance, when wind power is forecast and conventional generation scheduled. The underlying arguments leading to Theorems 11.2–11.4 continue to apply, at least to a good approximation. In particular sample path arguments suggest that the reduction of round-trip efficiency slows the rate at which the store-level trajectories—started from different initial levels but with the same stochastic description of future shock processes—converge over subsequent time. In particular Gast et al. [16] confirm these results empirically, considering round-trip efficiencies less than 1 and noting that in this case simple "target" policies such as that described by Theorem 11.2 (which is applicable in the case of round-trip efficiencies equal to 1) are here suboptimal.

11.4 Determination of the Functions A_t

We described in Sect. 11.2 how, given a knowledge of the functions A_t defined by (11.3), the optimal control of the store may be reduced to the solution of an optimisation problem which must be re-solved at those randomly occurring times at which shocks occur. In this section we consider conditions under which the functions A_t may be thus known, either exactly or to good approximations—in all cases without the need for the prior determination of the functions V_t.

It is convenient to rewrite slightly the definition (11.3) of each of the functions A_t as

$$A_t(s_t) = \mathbf{E}D_t(s_t) + \mathbf{E}[V_t(S_t(s_t)) - V_t(s_t)], \tag{11.17}$$

and to regard $A_t(s_t)$ as the sum of the two given expectations on the right side of (11.17). We shall argue below that in many applications it the first of these two expectations, i.e. $\mathbf{E}D_t(s_t)$, that is likely to be much the dominant term on the right side of (11.17). Since, for each t and for each s_t, the distribution of $D_t(s_t)$ is part of the model specification, the computation of $\mathbf{E}D_t(s_t)$ is straightforward. We do, however, consider below how one might reasonably *obtain* this major part of the model specification, i.e. the cost of dealing with a random shock as a function of the level of the store at the time at which the shock occurs.

The second expectation on the right side of (11.17) is the difference between the expected cost $\mathbf{E}V_t(S_t(s_t))$ of optimally managing the store subsequent to the time period t (when the actual level of the store at the end of that time period is then given by the random variable $S_t(s_t)$) and the corresponding expected cost $V_t(s_t)$ which would be incurred in the absence of any shock during the time period t (so that the level of the store at the end of that time period was then its planned value s_t). This difference $\mathbf{E}V_t(S_t(s_t)) - V_t(s_t)$ may also be understood in terms of a coupling of optimally controlled processes, started at the end of the time period t at the levels $S_t(s_t)$ and s_t, and is the expectation of the difference of the costs of their optimal control up to the time at which the coupled processes first agree.

Now consider the somewhat idealised conditions of Theorem 11.2, where the store is a perfectly efficient price-taker, so that each cost function C_t is given by $C_t(x) = c_t x_t$ for some market price c_t, and where each target level $\hat{s}_t$ given by (11.10) is assumed to be always achievable. It follows from Theorem 11.2 that, regardless of any shock which may occur during any given time period t, the planned level of the store for the end of the time period $t+1$ is $\hat{s}_{t+1}$. Hence, from (11.4),

$$V_t(S_t(s_t)) - V_t(s_t) = C_{t+1}(s_t - S_t(s_t)), \qquad (11.18)$$

so that, from (11.17),

$$A_t(s_t) = \begin{cases} \mathbf{E}D_t(s_t) + C_{t+1}(s_t - \mathbf{E}S_t(s_t)), & t < T, \\ \mathbf{E}D_T(s_T), & t = T. \end{cases} \qquad (11.19)$$

Thus the functions A_t may here be determined—in terms of the given distributions of the random variables $D_t(s)$ and $S_t(s)$—without the need to estimate the functions V_t.

More generally, the relations (11.19) correspond to the modified control in which, following any shock and hence store level disturbance during any time period t, the disturbed level $S_t(s_t)$ of the store is immediately returned to the planned level s_t for the end of that period at a cost $C_{t+1}(s_t - S_t(s_t))$; subject to this the store is otherwise optimally managed.

In the absence of this modification, the relations (11.19) may be viewed as providing a reasonable first approximation to the functions A_t—given the difficulties, in applications, of estimating the both the likelihood and the precise consequences of shocks, it is not clear that one could do significantly better. Better approximations, if required, might be made by allowing more time for the disturbed and undisturbed processes to couple as described above, and by reasoning as before so as to obtain a more refined version of (11.19). For example, one might extend the coupling time until a known future time at which it is planned that the store will be full. (This is often realistic for electricity storage which may aim to be full at the end of each night so as to take advantage of much higher daytime prices—we give examples based on real price data and realistic store characteristics in Sect. 11.5.) Then, for any planned level s_t of the store at the end of any time period t, the quantity $V_t(S_t(s_t)) - V_t(s_t)$ may be estimated analogously to (11.18) by considering optimal controls from the end of the time period t up to the first subsequent time at which the store is planned to be full.

Finally in the important special case in which shocks are rare but potentially expensive (as might be the case when the store is required to pay the costs of failing to have sufficient energy to deal with an emergency), then, for each t and s_t, the probability that $S_t(s_t)$ is not equal to s_t is small, and the major contribution to $A_t(s_t)$ as defined by (11.17) is likely to be $\mathbf{E}D_t(s_t)$. In this case either the simple approximation $A_t(s_t) = \mathbf{E}D_t(s_t)$, or the more refined approximation given by (11.19), may well suffice in applications.

In applications there is also a need, as part of the model specification, to realistically estimate—for each possible planned level s_t of the store at the end of each time period t—the joint distribution of the random vector $(D_t(s_t), S_t(s_t))$ modelling the cost of any shock and the corresponding store level disturbance. This joint distribution is in general a function of the amount of energy Y_t required to deal with any shock during the time period t, where in practice the distribution of the random variable Y_t may need to be determined by observation. We consider two particular possibilities, both of which are natural in the context of modelling risk in power systems, where the focus may either be on *loss of load* or on *energy unserved* (see, for example, [7]):

(i) the cost of a shock occurring during the time period t is simply a constant $a_t > 0$ if there is insufficient energy within the store to meet it, and is 0 otherwise; we then have $D_t(s_t) = a_t \mathbf{I}(Y_t > s_t)$, where $\mathbf{I}(\cdot)$ is the indicator function, and $S_t(s_t) = \max(0, s_t - Y_t)$;
(ii) the cost of a shock occurring during the time period t is proportional to the shortfall in the energy necessary to meet that shock; we then have $D_t(s_t) = a'_t \max(0, Y_t - s_t)$, where a'_t is the constant of proportionality, and again $S_t(s_t) = \max(0, s_t - Y_t)$.

Given the model of Sect. 11.2, the functions A_t may be determined (as described in this section) from the specification of the joint distributions of the random vectors $(D_t(s), S_t(s))$, together with the specification of the cost functions C_t and the store characteristics. In Sect. 11.5 we consider some plausible functional forms of the functions A_t.

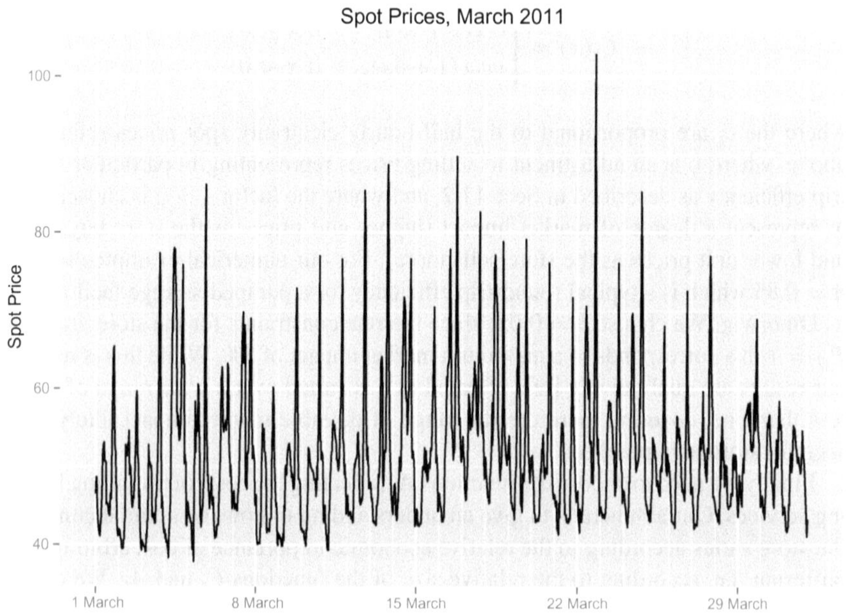

Fig. 11.1 GB half-hourly spots prices (£/MWh) for March 2011

11.5 Examples

We give some examples, in which we solve (exactly) the optimal control problem $\mathbb{P}$ formally defined in Sect. 11.2. We investigate how the optimal solution depends on the cost functions C_t and on the functions A_t which reflect the costs of providing buffering services.

The cost functions C_t are derived from half-hourly electricity prices in the Great Britain spot market over the entire year 2011, adjusted for a modest degree of market impact, as described in detail below. Thus we work in half-hour time units, with the time horizon T corresponding to the number of half-hour periods in the entire year. These spot market prices show a strong daily cyclical behaviour (corresponding to daily demand variation), being low at night and high during the day. This price variation can be seen in Fig. 11.1 which shows half-hourly GB spot prices (in pounds per megawatt-hour) throughout the month of March 2011. There is a similar pattern of variation throughout the rest of the year.

Without loss of generality, we choose energy units such that the rate (power) constraints are given by $P_I = P_O = 1$ unit of energy per half-hour period. For illustration, we take the capacity of the store to be given by $E = 10$ units of energy; thus the store can completely fill or empty over a 5-h period, which is the case, for example, for the large Dinorwig pumped storage facility in Snowdonia [13].

We choose cost functions C_t of the form

$$C_t(x) = \begin{cases} c_t x(1+\delta x), & \text{if } x \geq 0 \\ \eta c_t x(1+\delta x), & \text{if } x < 0, \end{cases} \tag{11.20}$$

where the c_t are proportional to the half-hourly electricity spot prices referred to above, where η is an adjustment to selling prices representing in particular round-trip efficiency as described in Sect. 11.2, and where the factor $\delta > 0$ is chosen so as to represent a degree of market impact (higher unit prices as the store buys more and lower unit prices as the store sells more). For our numerical examples we take $\eta = 0.85$ which is a typical round-trip efficiency for a pumped-storage facility such as Dinorwig. We choose $\delta = 0.05$; since the rate constraints for the store are $P_I = P_O = 1$ this corresponds to a maximum market impact of 5%. While this is modest, our results are qualitatively little affected as δ is varied over a wide range of values less than one, covering therefore the range of possible market impact likely to be seen for storage in practice.

Finally we need to choose the functions A_t reflecting the costs of providing buffering services. Our aim here is to give an understanding of how the optimal control of the store varies according to the relative economic importance of cost arbitrage and buffering, i.e. according to the relative size of the functions C_t and A_t. We choose functions A_t which are constant over time t and of the form $A_t(s) = ae^{-\kappa s}$ and $A_t(s) = b/s$ for a small selection of the parameters a, κ and b. The extent to which a store might provide buffering services in applications is extremely varied, and so the likely balance between arbitrage and buffering cannot be specified in advance. Rather we choose just sufficient values of the above parameters to show the effect of varying this balance. For a possible justification of the chosen forms of the functions A_t, see Sect. 11.4; in particular the form $A_t(s) = ae^{-\kappa s}$ is plausible in the case of light-tailed shocks, while the form $A_t(s) = b/s$ shows the effect of a slow rate of decay in s. (Note that in these examples we allow that the functions A_t should not necessarily be constant for values of their arguments greater than the rate constraint of 1: it is plausible that in practice greater quantities in store than can immediately be discharged to deal with a shock may nevertheless assist in dealing with its ongoing effects at subsequent times and, in the event of such a shock, may be considered as being notionally set aside for this purpose.)

In each of our examples, we determine the optimal control of the store over the entire year, with both the initial level S_0^* and the final level S_T^* given by $S_0^* = S_T^* = 0$. Figure 11.2 shows this optimal control (the sequence of successive levels of the store) for the time window corresponding to the month of March for each of the four cases $A_t(s) = 0$, $A_t(s) = e^{-s}$, $A_t(s) = 10e^{-s}$, and $A_t(s) = 1/s$. In each case the corresponding running forward horizon, as defined in Sect. 11.2, is generally of the order of a day or two. (Recall that the cost functions for March are determined by the prices illustrated in Fig. 11.1. Although the optimal control is determined over the entire year, it may be verified empirically that in every case the restriction of this optimal control to any given time window is independent of the functions C_t and A_t for times t which are outside of a period which includes this time window and a few days on either side of it.)

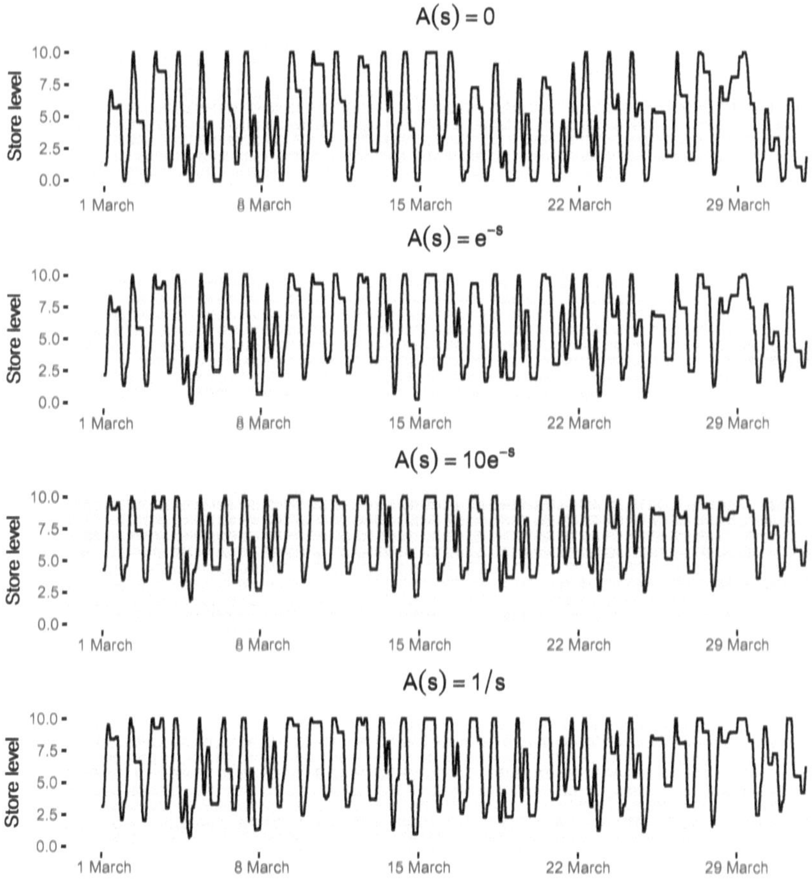

Fig. 11.2 Optimally controlled store level throughout March 2011 for each of the four cases $A_t(s) = 0$, $A_t(s) = e^{-s}$, $A_t(s) = 10e^{-s}$, and $A_t(s) = 1/s$

The case $A_t(s) = 0$ corresponds to the store incurring no penalty for failing to provide buffering services and optimising its control solely on the basis of arbitrage between energy prices at different times. The daily cycle of prices (again see Fig. 11.1) is sufficiently pronounced that here the store fills and empties—or nearly so—on a daily basis, notwithstanding the facts that the round-trip efficiency of 0.85 is considerably less than 1 and that the minimum time for the store to fill or empty is 5 h.

In the case $A_t(s) = e^{-s}$ the store is just sufficiently incentivised by the need to reduce buffering costs that it rarely empties completely (though it does so very occasionally). Otherwise the behaviour of the store is very similar to that in the case $A_t(s) = 0$. In both the cases $A_t(s) = 10e^{-s}$ and $A_t(s) = 1/s$ the costs of failing to provide buffering services are much higher, and so the optimised level of the store rarely falls below 25% of its capacity. Note the very similar behaviour in these two cases despite the very different forms of the "penalty" functions A_t.

Acknowledgements The authors acknowledge the support of the Engineering and Physical Sciences Research Council for the programme (EPSRC grant EP/I017054/1) under which the present research is carried out. They also acknowledge the benefit of very helpful discussions with numerous colleagues, notably: Janusz Bialek, Chris Dent, Lisa Flatley, Richard Gibbens, Frank Kelly and Phil Taylor. They are further most grateful to the referees for many insightful comments and suggestions which have considerably improved the exposition of the paper.

References

1. M. Arnold, G. Andersson, Model predictive control of energy storage including uncertain forecasts, in *Proceedings of the 17th Power Systems Computation Conference* (2011)
2. J.P. Barton, D.G. Infield, Energy storage and its use with intermittent renewable energy. IEEE Trans. Energy Convers. **19**(2), 441–448 (2004)
3. J.P. Barton, D.G. Infield, A probabilistic method for calculating the usefulness of a store with finite energy capacity for smoothing electricity generation from wind and solar power. J. Power Sources **162**, 943–948 (2006)
4. A.I. Bejan, R.J. Gibbens, F.P. Kelly, Statistical aspects of storage systems modelling in energy networks, in *46th Annual Conference on Information Sciences and Systems (invited session on Optimization of Communication Networks)*, Princeton University, USA, 2012
5. R. Bellman, On the theory of dynamic programming–a warehousing problem. Manag. Sci. **2**(3), 272–275 (1956)
6. A. Bernstein, L. Reyes-Chamorro, J.-Y. Le Boudec, M. Paolone, A composable method for real-time control of active distribution networks with explicit power setpoints. Part I: Framework. Electr. Power Syst. Res. **125**, 254–264 (2015)
7. R. Billinton, R.N. Allan, *Reliability Evaluation of Power Systems*, 2nd edn. (Springer, Berlin, 1996)
8. A.S. Cahn, The warehouse problem. Bull. Am. Math. Soc. **54**(11), 1073–1073 (1948)
9. E.D. Castronuovo, J.A. Peças Lopes, Optimal operation and hydro storage sizing of a wind power plant. Int. J. Electr. Power Energy Syst. **26**(10), 771–778 (2004)
10. J.R. Cruise, L.C. Flatley, R.J. Gibbens, S. Zachary, Control of energy storage with market impact: Lagrangian approach and horizons. Oper. Res. (2018). To appear
11. J.R. Cruise, L.C. Flatley, S. Zachary, Impact of storage competition on energy markets. Eur. J. Oper. Res. (2018). To appear
12. P. Denholm, R. Sioshansi, The value of compressed air energy storage with wind in transmission-constrained electric power systems. Energy Policy **37**(8), 3149–3158 (2009)
13. Dinorwig power station, http://www.electricmountain.co.uk/Dinorwig-Power-Station
14. S.E. Dreyfus, An analytic solution of the warehouse problem. Manag. Sci. **4**(1), 99–104 (1957)
15. Electricity balancing services, National Audit Office Briefing, May 2014, http://www.nao.org.uk/wp-content/uploads/2014/05/Electricity-Balancing-Services.pdf
16. N.G. Gast, D.-C. Tomozei, J.-Y. Le Boudec, Optimal storage policies with wind forecast uncertainties, in *Greenmetrics 2012*, Imperial College, London, UK, 2012
17. P. Harsha, M. Dahleh, Optimal management and sizing of energy storage under dynamic pricing for the efficient integration of renewable Energy. IEEE Trans. Power Syst. **30**(3), 1164–1181 (2015)
18. S.D. Howell, P.W. Duck, A. Hazel, P.V. Johnson, H. Pinto, G. Strbac, N. Proudlove, M. Black, A partial differential equation system for modelling stochastic storage in physical systems with applications to wind power generation. IMA J. Manag. Math. **22**, 231–252 (2011)

19. Y. Huang, S. Mao, R.M. Nelms, Adaptive electricity scheduling in microgrids, in *Proceedings of IEEE INFOCOM*, Turin, Italy, 2013
20. M. Korpaas, A.T. Holen, R. Hildrum, Operation and sizing of energy storage for wind power plants in a market system. Int. J. Electr. Power Energy Syst. **25**(8), 599–606 (2003)
21. I. Koutsopoulos, V. Hatzi, L. Tassiulas, Optimal energy storage control policies for the smart power grid, in *Proceedings of IEEE SmartGridComm* (2011), pp. 475–480
22. G. Lai, F. Margot, N. Secomandi, An approximate dynamic programming approach to benchmark practice-based heuristics for natural gas storage valuation. Oper. Res. **58**(3), 564–582 (2010)
23. O. Megel, J.L. Mathieu, G. Andersson, Maximizing the potential of energy storage to provide fast frequency control, in *IEEE PES ISGT Europe 2013* (IEEE, 2013), pp. 1–5
24. A. Oudalov, D. Chartouni, C. Ohler, Optimizing a battery energy storage system for primary frequency control. IEEE Trans. Power Syst. **22**(3), 1259–1266 (2007)
25. D. Pudjianto, M. Aunedi, P. Djapic, G. Strbac, Whole-systems assessment of the value of energy storage in low-carbon electricity systems. IEEE Trans. Smart Grid **5**, 1098–1109 (2014)
26. N. Richmond, P. Jacko, A.M. Makowski, Optimal planning of slow-ramping power production in energy systems with renewables forecasts and limited storage, in *2014 International Conference on Probabilistic Methods Applied to Power Systems (PMAPS)* (IEEE, 2014), pp. 1–6
27. N. Secomandi, Optimal commodity trading with a capacitated storage asset. Manag. Sci. **56**(3), 449–467 (2010)
28. N. Secomandi, Merchant commodity storage practice revisited. Oper. Res. **63**(5), 1131–1143 (2015)
29. R. Sioshansi, P. Denholm, T. Jenkin, J. Weiss, Estimating the value of electricity storage in PJM: arbitrage and some welfare effects. Energy Econ. **31**(2), 269–277 (2009)
30. F. Teng, J. Miles, A. Thomson, G. Strbac, N. Brandon, D. Pudjianto, Potential value of energy storage in the UK electricity system. Proc. ICE Energy **168**(2), 107–117 (2015)
31. The electricity trading arrangements, ELEXON, https://www.elexon.co.uk/wp-content/uploads/2013/11/beginners_guide_to_trading_arrangements_v4.0_cgi.pdf
32. A. Tuohy, M. O'Malley, Impact of pumped storage on power systems with increasing wind penetration, in *2009 IEEE Power & Energy Society General Meeting* (IEEE, 2009), pp. 1–8
33. P.M. van de Ven, N. Hegde, L. Massoulié, T. Salonidis, Optimal control of end-user energy storage. IEEE Trans. Smart Grid **4**, 789–797 (2013)
34. R. Walawalkar, J. Apt, R. Mancini, Economics of electric energy storage for energy arbitrage and regulation in New York. Energy Policy **35**(4), 2558–2568 (2007)
35. J.C. Williams, B.D. Wright, *Storage and Commodity Markets* (Cambridge University Press, Cambridge, 2005)
36. O.Q. Wu, D.D. Wang, Z. Qin, Seasonal energy storage operations with limited flexibility: the price-adjusted rolling intrinsic policy. Manuf. Serv. Oper. Manag. **14**(3), 455–471 (2012)

Chapter 12
Optimal Management of a Wind Power Plant with Storage Capacity

Jérôme Collet, Olivier Féron and Peter Tankov

Abstract We consider the problem of a wind producer who has access to the spot and intraday electricity markets and has the possibility of partially storing the produced energy using a battery storage facility. The aim of the producer is to maximize the expected gain of selling in the market the energy produced during a 24-h period. We propose and calibrate statistical models for the power production and the intraday electricity price, and compute the optimal strategy of the producer via dynamic programming.

Keywords Wind power generation · Battery storage · Intraday electricity market · Stochastic control

12.1 Introduction

Wind power is now widely recognized as an important part of the global energy mix, and the actors of the energy industry must cope with the intermittent and to a large extent unpredictable nature of the wind power production. To deal with this intermittency, various economic and physical tools are available to the agents. On the one hand, intraday markets, where wind power may be traded up to 30 min prior to delivery, allow the wind power producers to adjust their delivery volume estimates as the forecast becomes more precise. On the other hand, physical storage facilities whose cost is constantly declining, may be used to smooth out the production peaks and store the extra power until it can be sold at a profit.

J. Collet · O. Féron
EDF Lab, Palaiseau, France
e-mail: jerome.collet@edf.fr

O. Féron
e-mail: olivier-2.feron@edf.fr

P. Tankov (✉)
CREST-ENSAE ParisTech, Palaiseau, France
e-mail: peter.tankov@ensae.fr

P. Drobinski et al. (eds.), *Renewable Energy: Forecasting and Risk Management*,
Springer Proceedings in Mathematics & Statistics 254,
https://doi.org/10.1007/978-3-319-99052-1_12

In this paper, we therefore consider the problem of a (small) wind producer who has access to the spot (day-ahead) and intraday electricity markets and has the possibility of partially storing the produced energy using a battery storage facility. The role of battery storage is two-fold: on the one hand, it smoothes the variations of wind power production, and on the other hand, it allows to exploit intertemporal price discrepancies in the day-ahead market. The aim of the producer is to maximize the expected gain of selling in the market the energy produced during a 24-h period. The producer first makes a bid in the day-ahead electricity market for the following day, and then, when the intraday markets for that day opens, may adjust her position by trading in the intraday market. The strategy of the producer therefore consists of a static part (position in the day-ahead market) and a dynamic part (trading strategy in the intraday market). The deliveries in the spot and intraday market must be at all times balanced by the wind production and battery injections/withdrawals. The dynamic trading strategy is thus constrained by the finite capacity of the battery. To determine the dynamic part of the strategy, we set up a stochastic model for the intraday market price and the realized power production.

We mention that the majority of wind power producers in Europe still operate within the framework of *guaranteed purchase schemes* whereby all the power they produce is bought by the state-owned operator at a fixed price. However, as the guaranteed purchase schemes are either phased out or replaced with more market-oriented subsidies, the wind power producers face the need to sell the future power production in the open markets in the absence of precise knowledge of the volume to be produced.

In the literature, optimal operation of battery storage facilities has primarily been considered in the context of microgrid control (see [11] for an up-to-date review and, e.g., [8] for an example of using dynamic programming techniques similar in spirit to the ones employed in the present paper). Another interesting reference in this respect is [7], where the impact of forecast errors on optimal sizing of battery storage in an isolated microgrid is evaluated. On the other hand, optimal bidding strategies in intraday electricity markets for wind power producers who do not have access to battery storage, have been studied in a number of papers, see e.g., [1, 2, 6, 10]. However, among wind power producers there is an interest towards investing into battery storage to smooth out the intermittency of the renewable resource. More recently, therefore, several authors have considered optimal bidding strategies for systems consisting of a wind power plant and a storage capacity.

In [4, 5], optimal bidding strategies in the day-ahead market for wind-storage systems are determined. The optimization is in this case static and the presence of intraday market or the dynamic properties of market prices are not taken into account. However, a recent study [9] shows that intraday markets are increasingly used by renewable energy producers to balance the forecasting errors in their production. It is therefore important to evaluate the economic benefits of battery storage facilities for renewable power producers and to determine optimal strategies for their operation in the presence of intraday market. One study of bidding strategies in both day-ahead and intraday markets is [3], where it is considered that market participants can readjust their bids 6 times per day in the intraday market. The price are, however, assumed to be the same in the day-ahead in intraday market. Compared to these references, our contribution is to introduce stochastic dynamic models for both the wind power production and the intraday market price, calibrate them to market price and wind production data, and find globally optimal dynamic bidding and operation strategies for the wind producer using the stochastic control approach.

The paper is structured as follows. After describing the optimization problem faced by the producer in Sect. 12.2, we introduce stochastic models for the intraday price process and the realized production process. The methodology for calibrating these models to real data is described in Sect. 12.3. Finally, in Sect. 12.4, we solve the optimization problem of the agent within the framework of stochastic control and optimal quantization and present numerical applications. The data used for model calibration and numerical examples comes from a power plant in France consisting of three 2MW wind turbines.

12.2 Description of the Model and the Optimization Problem

In this section, we detail our assumptions concerning the structure of electricity markets and formulate the optimization problem faced by the power producer and the models we use for power production and the market prices.

Structure of the Intraday Market

Intraday electricity market is an electricity exchange where blocks of power for delivery on a given day may be traded starting typically from 15 h on the day before, up to a very short time (e.g., 30 or 60 min) before delivery. A block corresponds to the delivery of a certain power throughout a fixed time period, such as an hour, a half-hour or a quarter. The trading day in the intraday market is divided into N such delivery periods of equal length.

Although the trading in the intraday market starts at 15 h of the previous day, at which point purchases/sales can be made for any delivery period of the following day, in practice liquidity becomes sufficient only 2–3 h prior to delivery (see Fig. 12.1). For this reason, and to simplify the analysis, we assume that the power producer may trade in the intraday market only once for each delivery period, at a fixed time interval δ (e.g., one hour) before delivery. There are thus N possible trading times, and we shall denote these moments by $T_1, \ldots, T_N$.

Strategy of the Producer

The producer makes a bid in the spot (day-ahead) market at time $t = 0$, by making an engagement to deliver the amount $\overline{P}_k$ of electricity during the delivery period $[T_k + \delta, T_{k+1} + \delta]$ for each $k = 1, \ldots, N$. These deliveries will be paid at the spot market price denoted by $F(0, T_k + \delta)$, $k = 1, \ldots, N$.

At each time T_k, the producer knows the amount of power, which will be generated during the delivery period $[T_k + \delta, T_{k+1} + \delta]$ (we neglect the forecast uncertainty

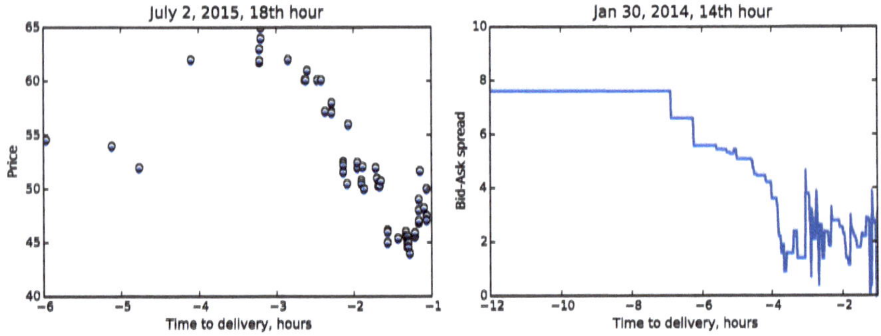

Fig. 12.1 Left: French intraday electricity transaction prices for a fixed delivery hour (each point corresponds to a single transaction). Right: bid-ask spread evolution in the German intraday market for a fixed delivery hour. In both cases, we see a sharp increase in liquidity 2–3 h prior to delivery

at such short time scales), and must decide how much power to buy/sell in the intraday market, and how much power to withdraw from/inject into the battery during this period, under the condition that injections/withdrawals must be balanced by production and market purchases. This decision may be based on the known power production for the upcoming delivery period, but also on the forecasts of power production for future delivery times, as well as the current intraday prices for all future delivery times.

The notation for various quantities is described (recalled) in the following table:

Q_k	Energy stored in the battery at the beginning of delivery period $k+1$ (at time $T_{k+1}+\delta$).
Q_{min}	Minimal energy stored in the battery at all times.
Q_{max}	Maximal battery capacity.
p_k	Energy purchased in the intraday market during kth delivery period $[T_k+\delta, T_{k+1}+\delta]$.
P_k	Energy produced during kth delivery period.
$P(t, T_k), 0 \leq t < T_k$	Forecast at time t of energy production during kth delivery period.
$\overline{P}_k$	Energy delivered during kth delivery period according to the engagements taken in the spot market.
F_k	Intraday market price at time T_k for kth delivery period.
$F(0, T_k)$	Spot market price for kth delivery period.
$F(t, T_k), 0 < t < T_k$	Intraday market price at time t for kth delivery period.

Formulation of the Optimization Problem

The total gain from trading of the wind power producer is given by

$$G = \sum_{k=1}^{N} (\overline{P}_k F(0, T_k) - F_k(p_k + \alpha |p_k|)),$$

where the term $\alpha|p_k|$ models the bid-ask spread in the intraday market. The aim of the producer is to maximize the expected value of this gain under the storage constraint

$$Q_k \in [Q_{min}, Q_{max}], \quad k = 1, \ldots, N.$$

The dynamics of the battery storage is described by

$$Q_k = Q_{k-1} + P_k - \overline{P}_k + p_k.$$

The optimization problem of the producer thus writes:

$$\max_{\overline{P}_1,\ldots,\overline{P}_N,p_1,\ldots,p_N} \left\{ \sum_{k=1}^{N} \overline{P}_k F(0, T_k) - \mathbb{E}\left[\sum_{k=1}^{N} F_k(p_k + \alpha|p_k|) \right] \right\},$$

where $\overline{P}_1, \ldots, \overline{P}_N$ are constants (determined at time 0), and $(p_k)_{1\le k\le N}$ is a dynamic strategy of trading in the intraday market, that is, a discrete-time stochastic process adapted to the filtration generated by the production values (P_k), the price processes in the intraday market (F_k) and $(F(t, T_k))_{t<T_k}$ and the process of forecast updates $(P(t, T_k))_{t<T_k}$.

Modeling the Dynamics of the Intraday Price Process

To understand how the intraday prices for the future delivery times and the power production forecasts affect the strategy of the producer and formulate the optimization problem for the power producer as a stochastic control problem, we need, in principle, to model the dynamics of the intraday price process $(F(t, T_k)_{t\ge 0, T_k > t})$ and of the forecast update process $(P(t, T_k)_{t\ge 0, T_k > t})$ as function of t. Since the dimension of these processes is very large (24 or 48 depending on the number of delivery periods), some form of dimension reduction is necessary. Figure 12.2 shows that, for example, the shape of the forecast curve does not change much in time, and therefore 2–3 stochastic factors should be sufficient to model the dynamics of the entire forecast curve.

Inspired by the modeling approaches for the interest rate curve, and to allow negative prices which are common in electricity markets with strong penetration of renewables, we use a Gaussian factor-based model:

$$F_t = F(0, t) + \bar{\alpha}(t) \sum_{j=1}^{M'} Y_t^j.$$

Here we recall that F_t is the "last" intraday price, $F(0, t)$ is the day-ahead price (where the time 0 correspond to the gate closure time of the day-ahead market) and $(Y^j)_{j=1}^{M'}$ are independent Ornstein–Uhlenbeck processes:

$$dY_t^j = -\bar{\lambda}^j Y_t^j dt + \bar{\sigma}^j d\widehat{B}_t^j,$$

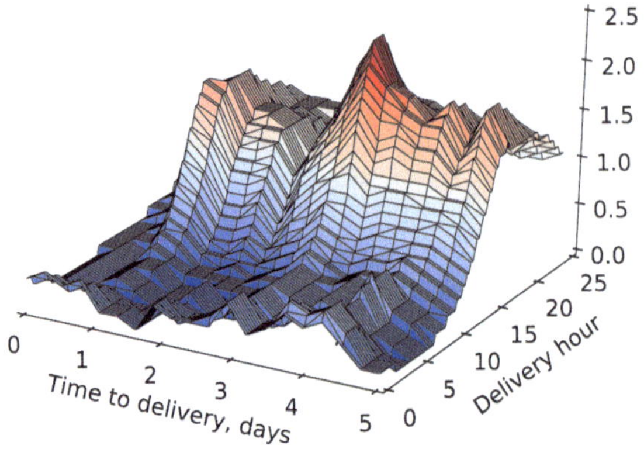

Fig. 12.2 Evolution of the forecasts for all delivery horizons on a single day, as function of time remaining to delivery day

where, $(\widehat{B}^j)_{j=1}^{M'}$ are Brownian motions under the risk-neutral measure $\mathbb{Q}$. Note that we have not included discounting because the effect of interest rates is negligible at intraday time scales. Forward prices are computed by taking risk-neutral expectation:

$$F(t,T) = \mathbb{E}[F_T|\mathscr{F}_t] = F(0,T) + \sum_{j=1}^{M'} \bar{\alpha}(T) e^{-\bar{\lambda}_j(T-t)} Y_t^j. \tag{12.1}$$

To obtain the dynamics of forward prices under the real-world measure, we make a change of probability

$$\frac{d\mathbb{P}}{d\mathbb{Q}}\bigg|_{\mathscr{F}_T} = \exp\left(-\int_0^T \phi_t dB_t - \frac{1}{2}\int_0^T \phi_t^2 dt \right),$$

where the process ϕ is assumed to be deterministic. The process

$$B_t = \widehat{B}_t + \int_0^t \phi_s ds$$

is then a Brownian motion under the historical measure and we can write

$$Y_t^j = \bar{\sigma}^j \int_0^t e^{-\lambda_j(t-s)} dB_s^j + \bar{\sigma}^j \int_0^t e^{-\lambda_j(t-s)} \phi_s^j ds := \bar{\sigma}^j \int_0^t e^{-\lambda_j(t-s)} dB_s^j + \bar{\mu}^j(t).$$

We can then write

$$F_t = F(0,t) + \bar{\alpha}(t) \sum_{j=1}^{M'} (\overline{Y}_t^j + \bar{\mu}^j(t)), \tag{12.2}$$

where $\overline{Y}_t^j = \bar{\sigma}^j \int_0^t e^{-\lambda_j(t-s)} dB_s^j$ is a centered Gaussian factor process.

This model describes the dynamics of the intraday price over a single day. The factors $\bar{\alpha}$ and $\bar{\mu}$ describe the daily seasonality of the price. For estimating this model, we shall assume that the random factors $\overline{Y}^j$ for different days are independent and compute averages over all trading days present in the data. Of course, intraday prices for different days have different distributions due to the presence of the annual seasonality pattern, but we assume that this pattern is fully taken into account by the day-ahead market price.

The estimation of this model, as well as the one for the forecast dynamics, presented in the next paragraph, may be carried out in two different settings:

- The number of factors is smaller than the number of different delivery periods for which the price is available in the market at the same time. In this case, one can use the prices for different delivery periods to reconstruct the factors Y_t^j from the formula (12.1). In other words, the individual factors become observable. It makes sense, then, to model separately the seasonality of each factor with the corresponding mean function $\bar{\mu}^j(t)$ as in formula (12.2).
- The number of factors is larger than the number of different delivery periods for which the price is available at the same time. For instance, one may assume that only the last intraday price is known and the number of factors is greater than one. In this case, the individual factors are inobservable, and one cannot reconstruct the mean function separately for each factor. In practice, we recommend to use the number of factors which is less or equal to the number of delivery periods for which prices are available.

More details on the estimation procedure are given in the following section, where the estimation is based on the last intraday price only and the number of factors is taken equal to one.

Modeling the Forecast Dynamics

Since the electricity price process in the intraday market and the wind power production (forecast) may be negatively correlated, we propose a model for the wind production process, which is of a similar structure to the model for the price processes and includes a possible correlation.

$$P_t = P(0,t) + \alpha(t)(1 + \gamma P(0,t)^{\delta}) \sum_{j=1}^{M} (X_t^j + \mu^j(t)),$$

where $(X^j)_{j=1}^M$ are independent centered Gaussian factors modeled as Ornstein–Uhlenbeck processes:

$$dX_t^j = -\lambda_j X_t^j dt + \sigma_j dW_t^j,$$

α and μ^j are deterministic functions modeling the daily seasonality of the production process, $P(0,t)$ is the forecast at the gate closure time of the intraday market, and W^j are Brownian motions possibly correlated with the Brownian motions B^j driving the price process. Note that this model for the production process may allow for negative production values, but such values may also be possible in practice when the wind speed is very low due to nonzero power consumption of the wind turbine. The factor $1+\gamma P(0,t)^\delta$ reflects the fact that forecast errors are larger in amplitude when the forecast itself is large.

The forecast processes at other times are given by

$$P(t,T)=\mathbb{E}[P_T|\mathscr{F}_t]=P(0,T)+\alpha(T)(1+\gamma P(0,T)^\delta)\sum_{j=1}^M e^{-\lambda_j(T-t)}(X_t^j+\mu^j(t)).$$

In other words, the forecast process follows a Gaussian dynamics and is completely determined by the knowledge of the M factors $X^1,\dots,X^M$.

12.3 Model Calibration

In this section we explain how our models for the production (P_t) and the intraday market price (F_t) are estimated from data. The estimation procedure will be different in the one-factor case (one factor for the price and one for the forecast) and the multifactor case, because in the one-factor case it is enough to observe only the price/production to recover the risk factor. We describe the estimation procedure of the model for the production, the one for the price being very similar.

Estimation in the One-Factor Case

We first focus on the estimation of the model for production. Since there is only one factor, we may omit the index j and take $\sigma=1$ without loss of generality. We assume that the agent observes L realizations of the forecast $(P^l(0,T_k))_{k=1,\dots,N}^{l=1,\dots,L}$ and the production process $(P_k^l)_{1\le k\le N}^{l=1,\dots,L}$ (each realization corresponds to a single production day in the past), and we introduce the forecast error process $(Z_k^l)_{1\le k\le N}^{1\le l\le L}$, where $Z_k^l=P_k^l-P^l(0,T_k)$. We assume that Z_k^l is a Gaussian vector with parameters

$$\mathbb{E}[Z_k^l]:=\tilde{\mu}_k:=\mu_k\alpha_k(1+\gamma P^l(0,T_k)^\delta),\qquad \mathrm{Cov}[Z_i^l,Z_j^m]=0\quad\text{for } l\neq m$$
$$\text{and}\quad \mathrm{Cov}[Z_i^l,Z_j^l]=\alpha_i\alpha_j(1+\gamma P^l(0,T_i)^\delta)(1+\gamma P^l(0,T_j)^\delta)\Omega_{ij}(\lambda),$$
$$\Omega_{ij}(\lambda):=\frac{e^{-\lambda(T_i-T_j)^+}-e^{-\lambda(T_i+T_j)}}{2\lambda},$$

where we denote $\alpha_k:=\alpha(T_k)$ and $\mu_k=\mu(T_k)$. The log-likelihood of $(Z_k^l)_{k=1,\dots,N}^{l=1,\dots,L}$, omitting constant terms, is given by

$$
\begin{aligned}
l(\alpha,\mu,\lambda,\gamma,\delta) &= -\frac{L}{2}\log(\det \Omega(\lambda)) - L\sum_{j=1}^{N}\log\alpha_j - \sum_l\sum_j \log(1+\gamma P^l(0,T_j)^\delta)\\
&\quad -\frac{1}{2}\sum_{l=1}^{L}\left(\frac{Z^l}{\alpha(1+\gamma P^l(0,T_\cdot)^\delta)}-\mu\right)^\top \Omega^{-1}(\lambda)\left(\frac{Z^l}{\alpha(1+\gamma P^l(0,T_\cdot)^\delta)}-\mu\right)\\
&= -\frac{L}{2}\log(\det \Omega(\lambda)) - L\sum_{j=1}^{N}\log\alpha_j - \sum_l\sum_j \log(1+\gamma P^l(0,T_j)^\delta)\\
&\quad -\frac{1}{2}\sum_{l=1}^{L}\left(\frac{Z^l_{\gamma,\delta}}{\alpha}-\mu\right)^\top \Omega^{-1}(\lambda)\left(\frac{Z^l_{\gamma,\delta}}{\alpha}-\mu\right),
\end{aligned}
$$

where we use the shorthand notation $Z^l_{\gamma,\delta} = Z^l/(1+\gamma P^l(0,T)^\delta)$.

In an attempt to partially maximize the likelihood in explicit form, we first compute the derivatives with respect to μ:

$$
\frac{\partial l}{\partial \mu_i} = e_i^\top \Omega^{-1}(\lambda)\sum_{l=1}^{L}\left(\frac{Z^l_{\gamma,\delta}}{\alpha}-\mu\right),
$$

where e_i is a vector with 1 at the ith position and 0 elsewhere. As a result,

$$
\mu = \frac{1}{L\alpha}\sum_{l=1}^{L} Z^l_{\gamma,\delta} := \frac{1}{\alpha}\overline{Z}_{\gamma,\delta},
$$

and we can write the simplified form of the log-likelihood

$$
\begin{aligned}
l(\alpha,\lambda,\gamma,\delta) &= -\frac{L}{2}\log(\det \Omega(\lambda)) - L\sum_{j=1}^{N}\log\alpha_j - \sum_l\sum_j \log(1+\gamma P^l(0,T_j)^\delta)\\
&\quad -\frac{1}{2}\sum_{l=1}^{L}\left(\frac{Z^l_{\gamma,\delta}}{\alpha}-\frac{\overline{Z}_{\gamma,\delta}}{\alpha}\right)^\top \Omega^{-1}(\lambda)\left(\frac{Z^l_{\gamma,\delta}}{\alpha}-\frac{\overline{Z}_{\gamma,\delta}}{\alpha}\right).
\end{aligned}
$$

Differentiating now with respect to α, and making some straightforward simplifications, we find

$$
\frac{\partial l}{\partial \alpha_i} = -\frac{L}{\alpha_i} + \frac{1}{\alpha_i}\sum_{l=1}^{L}\frac{Z^l_{\gamma,\delta,i}-\overline{Z}_{\gamma,\delta,i}}{\alpha_i} e_i^\top \Omega^{-1}(\lambda)\frac{Z^l_{\gamma,\delta}-\overline{Z}_{\gamma,\delta}}{\alpha}
$$

with the corresponding first-order condition

$$1 = \frac{1}{L}\sum_{l=1}^{L}\frac{Z^l_{\gamma,\delta,i}-\overline{Z}_{\gamma,\delta,i}}{\alpha_i}e_i^T\Omega^{-1}(\lambda)\frac{Z^l_{\gamma,\delta}-\overline{Z}_{\gamma,\delta}}{\alpha}. \tag{12.3}$$

Summing up and substituting into the expression for the log-likelihood, we find that the latter is given by (once again without constant terms):

$$l(\lambda,\gamma,\delta) = -\frac{L}{2}\log(\det\Omega(\lambda)) - \sum_l\sum_j \log(1+\gamma P^l(0,T_j)^\delta) - L\sum_{j=1}^N \log\alpha_j^*(\lambda), \tag{12.4}$$

where $\alpha^*(\lambda)$ is the solution of (12.3). This equation can also be written as

$$1 = \frac{1}{\alpha_i}\sum_{k=1}^N \Omega^{-1}(\lambda)_{ik}\frac{\widehat{R}_{ki}}{\alpha_k},\quad \widehat{R}_{ki} = \overline{Z_{\gamma,\delta,k}Z_{\gamma,\delta,i}} - \overline{Z}_{\gamma,\delta,k}\overline{Z}_{\gamma,\delta,i},$$

or, in vector notation, as

$$\alpha = \widehat{M}\alpha^{-1}, \tag{12.5}$$

where $M_{ij} = \Omega^{-1}(\lambda)_{ij}\widehat{R}_{ij}$.

We compute the maximum likelihood estimator by solving numerically the equation (12.5) and then minimizing the function $l(\lambda,\gamma,\delta)$ given by (12.4).

Estimation in the Multifactor Case

In the presence of M random factors we assume that the agent observes not only the production process and the forecast at date zero, but also, at each trading date $(T_k)_{1\le k\le N-1}$, the forecast of production of the next M delivery periods, that is, $P(T_k, T_{k+i})$ for $1\le i\le M\wedge(N-k)$. Let $Z^l_{i,k} = P^l(T_k,T_{k+i}) - P^l(0,T_{k+i})$. Then, the random vector $\{Z^l_{i,k}, 1\le l\le L, 1\le k\le N, 0\le i\le M\wedge(N-k)\}$ is a Gaussian random vector with parameters

$$\mathbb{E}[Z^l_{i,k}] := \tilde{\mu}_{i,k} = \sum_{j=1}^M \alpha(T_{k+i})(1+\gamma P(0,T_{k+i})^\delta)e^{-\lambda_j(T_{k+i}-T_k)}\mu_k^j$$

$$\mathrm{Cov}[Z^l_{i,k}, Z^m_{j,n}] = 0 \quad \text{for } l\ne m$$

$$\mathrm{Cov}[Z^l_{i,k}, Z^l_{j,n}] = \sum_{p=1}^M \alpha(T_{i+k})(1+\gamma P(0,T_{k+i})^\delta)(1+\gamma P(0,T_{n+j})^\delta)$$

$$\times\,\alpha(T_{j+n})e^{-\lambda_p(T_{i+k}-T_k+T_{j+n}-T_n)}\sigma_p^2\frac{e^{-\lambda(T_k-T_n)^+}-e^{-\lambda(T_k+T_n)}}{2\lambda}.$$

Similarly to the one-factor case, one can then write down the explicit likelihood of the model and estimate parameters by numerical maximization of the likelihood function.

Numerical Illustration

In this paragraph we illustrate our estimation procedure on a real data set. For this illustration we use a one-factor specification of the model. For estimating the production model we use a time series of power production from Jan 1st, 2012 to Dec 31st, 2014 from a wind park in France provided by Engie Green/Maïa Eolis, together with a time series of historical forecasts provided by the same producer. The production data had a 10-min frequency and was averaged down to 1 h frequency. The forecast data had 15-min resolution, which was averaged down to 1 h resolution. Every day, 4 forecasts are available, at 0, 6, 12 and 18 h. In this study, we used only the forecast at 12 h, corresponding to the gate closure time of the day-ahead market.

Figure 12.3, left graph illustrates the evolution of the day-ahead forecast and the realized production on September 8, 2014. The right graph of this figure shows the results of estimation. The estimated value of the mean reversion parameter is $\lambda^* \approx 3.85\text{days}^{-1}$; in other words the characteristic length of mean reversion is about 6.2 h. The function $\alpha(t)$ appears to have a slightly increasing profile reflecting the fact that forecast uncertainty grows with time. The function $\mu(t)$ is small and negative, which means that the forecasts in our data set may have a small but statistically significant positive bias.

For estimating the intraday price model we use the day-ahead price for the Germany-Austria region downloaded from the web site of EPEX Spot, and the average intraday price time series at 1 h frequency for the same region, computed from a high-frequency time series provided to us by EPEX Spot, from Jan 1st, 2014 to Dec 31st, 2014. In this study, we construct a proxy for the 'last' intraday price for each delivery hour, by taking, for the delivery hour H, the average intraday price for the hour H-1. The Germany Austria region was chosen for reasons of market liquidity and data availability. Figure 12.4, left graph, illustrates the evolution of the day-ahead price and the corresponding intraday price on September 8, 2014.

Since our price and production data sets are from different regions and do not cover the same period completely, we carry out the estimation procedure independently for the two sets and assume that there is no correlation between the innovations of price and production. In the same region, the two processes are likely to be correlated, and our estimation procedure can be easily adapted to that case.

Figure 12.4, right graph shows the results of the estimation of the intraday price model. The estimated value of the mean reversion parameter is $\bar{\lambda}^* \approx 9.6$, which corresponds to the length of mean reversion of about 2.5 h. The function $\bar{\alpha}$ (price volatility) appears to have peaks at 10th and 20th hours, which correspond, approximately, to morning and evening peaks of electricity demand. The estimator of the function $\bar{\mu}$ (bias) falls within the 5% confidence interval around zero for almost all hours, which means that day-ahead prices are almost equal to expectations of the last intraday prices under the historical measure.

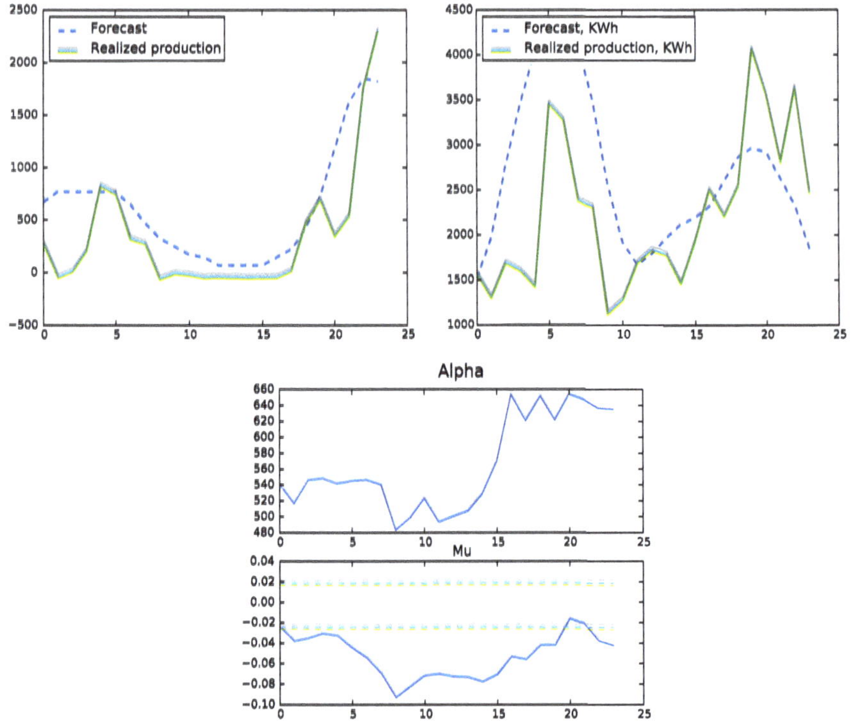

Fig. 12.3 Top: Evolution of the day-ahead forecast and the realized production on September 8, 2014 (left graph) and October 19, 2014 (right graph). Bottom: Results of estimation of the model for realized production. Dashed lines show the 5% confidence interval around zero for the estimator of μ

12.4 Solving the Optimization Problem by Dynamic Programming and Optimal Quantization

The state variables of the problem are the battery charge state $(Q_k)_{1\leq k\leq N}$ and the factor processes for the wind production $(X^j_{T_k})^{1\leq j\leq M}_{1\leq k\leq N}$ and the intraday market price $(Y^j_{T_k})^{1\leq j\leq M'}_{1\leq k\leq N}$. For our optimization problem we consider all processes in discrete time. Define the value function

$$v_k(q, x^1, \ldots, x^M, y^1, \ldots, y^{M'}) = \min_{p_k,\ldots,p_N, Q_{k-1}=q} \mathbb{E}^{T_k, x^1,\ldots,x^M, y^1,\ldots,y^{M'}} \left[\sum_{n=k}^{N} F_n(p_n + \alpha |p_n|) \right].$$

In the following, to save space, we write x for $x^1, \ldots, x^M$ and similarly for other variables. The original optimization problem then writes

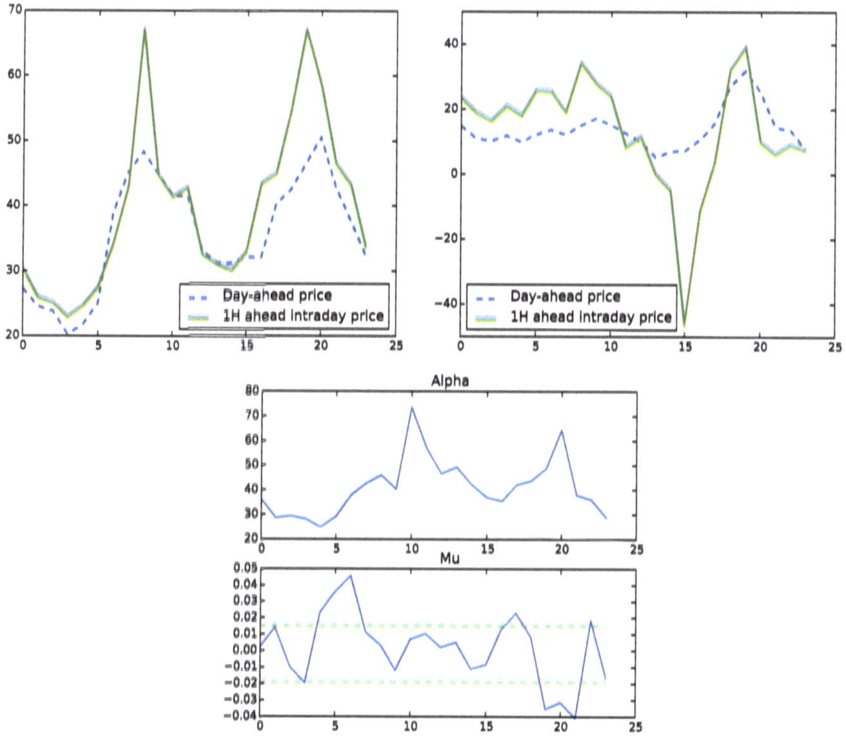

Fig. 12.4 Top: evolution of the day-ahead price and the corresponding last intraday price on September 8, 2014 (left graph) and October 19, 2014 (right graph). Bottom: Results of estimation of the intraday price model. Dashed green lines show the 5% confidence interval around zero for the estimator of $\bar{\mu}$

$$\max_{\overline{P}_1,\dots,\overline{P}_N} \left\{ \sum_{k=1}^{N} \overline{P}_k F(0, T_k) - \mathbb{E}^{t_0, X_{t_0}, Y_{t_0}}[v_1(Q_0, X_{T_1}, Y_{T_1})] \right\}.$$

The dynamic programming principle for the value function writes

$$v_k(q, x, y) = \min_{p_k : q + P_k - \overline{P}_k + p_k \in [Q_{min}, Q_{max}]} \{\phi_k(y)(p_k + \alpha |p_k|) + \mathbb{E}^{T_k, x, y}[v_{k+1}(q + \pi_k(x) - \overline{P}_k + p_k, X_{T_{k+1}}, Y_{T_{k+1}})\}$$

with the terminal condition

$$v_N = v_N(q, x, y) = \min_{p_N : q + P_N - \overline{P}_N + p_N \in [Q_{min}, Q_{max}]} \phi_N(p_N + \alpha |p_N|),$$

where

$$\phi_k(y) = F(0, T_k) + \bar{\alpha}(T_k)\sum_{j=1}^{M'}(y^j + \bar{\mu}^j(T_k)),$$

$$\pi_k(x) = P(0, T_k) + \alpha(T_k)(1 + \gamma P(0, T_k)^\delta)\sum_{j=1}^{M}(x^j + \mu^j(T_k)).$$

We may also impose a constraint on the state of charge of the battery at the terminal date: $Q_{T_N} = Q_{T_0}$. In this case, $v_N(q) = F_N(p_N + \alpha|p_N|)$ with $p_N = Q_{T_0} - q - P_N + \overline{P}_N$.

To compute the value function and the optimal strategy numerically, we start by discretizing the state of charge of the battery, introducing a uniform grid $Q_{min} = q_1 < \cdots < q_J = Q_{max}$. This means that the control p_k also takes a finite number of values. We denote $v_k(q_j, \dots)$ by v_k^j. Then,

$$v_k^j(x, y) = \min_{i=1,\dots,J}\{\phi_k(y)\eta(q_i - q_j + \overline{P}_k - \pi_k(x)) + \mathbb{E}^{T_k,x,y}[v_{k+1}^i(X_{T_{k+1}}, Y_{T_{k+1}})\},$$

where we have used the notation $\eta(p) = p + \alpha|p|$ to simplify the formula.

The second step is to replace the discrete-time Ornstein–Uhlenbeck processes (X, Y) with a finite-state Markov chain. This will be achieved using the method of *optimal quantization*. Let $\mathbb{P}_k$ be the unconditional distribution of $Z := (X_{T_k}, Y_{T_k})$. Note that it is a multivariate Gaussian distribution with zero mean. For every $k = 1, \dots, N$, we define the optimal grid of size N_q by solving

$$\min_{\hat{Z}} \mathbb{E}^{\mathbb{P}_k}[(Z - \hat{Z})^2],$$

where the minimum is taken over all random vectors supported by N_q points (and the variable Z is $M + M'$-dimensional in our setting). It is known (see e.g., [12] for a review) that the solution is the so called optimal Voronoi quantization which is obtained by nearest-neighbor projection of the vector Z on a set of N_q points. We shall denote these points by $\hat{z}_1^k, \dots, \hat{z}_{N_q}^k$ with $\hat{z}_j^k := (\hat{x}_j^k, \hat{y}_j^k)$, the associated Voronoi cells by $C_1^k, \dots, C_{N_q}^k$ and the associated probabilities by $\hat{p}_1^k, \dots, \hat{p}_{N_q}^k$. To find the points, one can simulate a large number of samples from $\mathbb{P}_k$ and use the randomized Lloyd's algorithm (also known as the K-means clustering algorithm). In the numerical illustration below, since the processes X and Y are one-dimensional and uncorrelated, we use the precomputed grids for the multivariate Gaussian distribution, downloaded from the web site `quantize.maths-fi.com`.

Next, we replace the continuous process with a Markov chain $(\hat{Z}_k)_{0\le k\le N}$ with N_q states. The transition probabilities of the chain are defined by

$$\hat{\pi}_i^0 = \mathbb{P}[\hat{Z}_1 = \hat{z}_i^1] = \hat{\pi}_i^1$$
$$\text{and}\quad \hat{\pi}_{ij}^k = \mathbb{P}[\hat{Z}_{k+1} = \hat{z}_j^{k+1}|\hat{Z}_k = \hat{z}_i^k] = \mathbb{P}[Z_{T_{k+1}} \in C_j^{k+1}|Z_{T_k} \in C_i^k].$$

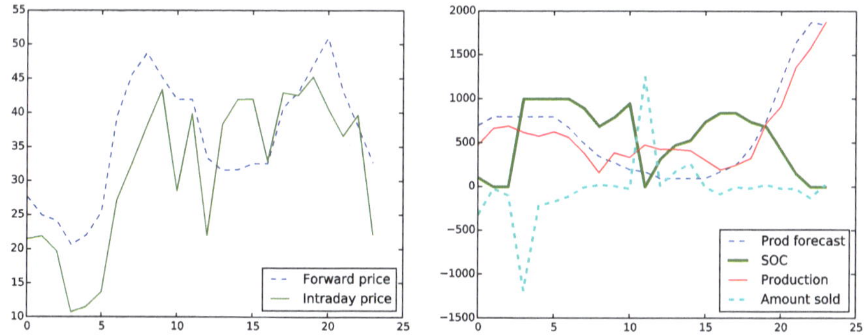

Fig. 12.5 Sample evolution of the modelled quantities. In the left graph, prices are in Euros per MWh. In the right graph, all amounts are shown in KWh, and in the case of production and amount sold correspond to the production and the amount sold during the specified hour

These transition probabilities are evaluated by Monte Carlo.

The value function can then be computed on the quantization grid using the following formula:

$$v_k^j(\hat{z}_m^k) = \min_{i=1,\dots,J}\{\phi_k(\hat{y}_m^k)\eta(q_i - q_j + \overline{P}_k - \pi_k(\hat{x}_m^k)) + \sum_{l=1}^{N_q} \hat{\pi}_{ml}^k v_{k+1}^i(z_l^{k+1})\}.$$

Numerical Illustration

We first illustrate the computation of the value function v_1 and the corresponding optimal strategy. For the numerical illustration we have taken $N_q = 500$ quantization nodes, and the state of charge was discretized over $J = 20$ regularly spaced values. The computation of the value function v_1 takes about 10 s on a MacBookPro with i5-2.90 GHz processor and 8 Gb physical memory (C++ implementation using only a single processor core). The value function depends on the spot market engagements $\overline{P}_k$, and they have been taken equal to production forecasts for the corresponding hour: $\overline{P}_k = P(0, T_k)$. Figure 12.5 illustrates the evolution of various quantities in our model. The forward price and forecast curves are taken from the market data on a specific day (September 8, 2014); the intraday price and production were simulated using our model estimated from market data, and the state of charge of the battery (SOC) and the amount to be sold in the intraday market were computed from the solution of the HJB equation. In the left graph, prices are in Euros per MWh and in the right graph, all values are shown in KWh, and in the case of production (forecast and realized) refer to the power generated during the specified hour. The model parameters are $Q_{min} = 0$, $Q_{max} = 1000$ KWh, $\alpha = 0.2$ (the intraday market spread) and $Q_0 = 0$ (initial state of charge).

We next illustrate the effect of the battery capacity Q_{max} and the intraday market spread α. Still under the assumption that $\overline{P}_k = P(0, T_k)$ for every k, we show in Fig. 12.6 the maximum expected gain of the power producer for different values of the battery capacity Q_{max} and different values of α, that is, the value

Table 12.1 Expected gain from adding a 1000 KWh battery capacity, in euros per day

	$\alpha = 10\%$	$\alpha = 20\%$	$\alpha = 30\%$
September 8, 2014	46.64	41.58	39.28
October 19, 2014	41.35	38.05	36.15

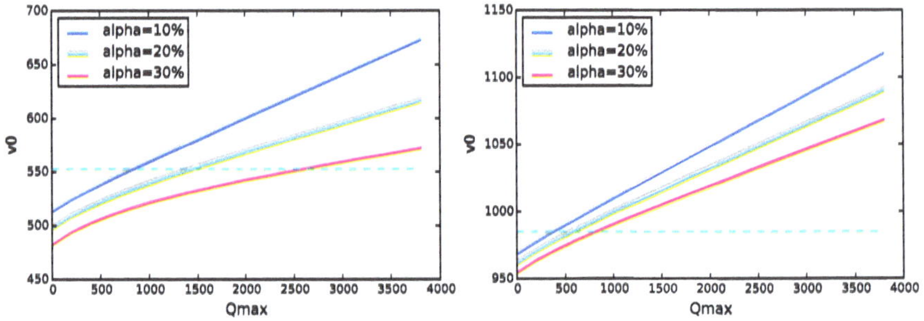

Fig. 12.6 Expected gain (en euros) of the wind power producer as function of the battery capacity Q_{max}, for different values of the spread parameter α. The dotted line shows the theoretical profit of the power producer if the power production were exactly equal to the day-ahead forecast. Left: 8 September 2014. Right: 19 October 2014

$$\sum_{k=1}^{N} \overline{P}_k F(0, T_k) - \min_{p_1,\dots,p_N} \mathbb{E}\left[\sum_{k=1}^{N} F_k(p_k + \alpha|p_k|)\right]. \qquad (12.6)$$

The dotted line shows the theoretical profit of the power producer if the power production were exactly equal to the day-ahead forecast and no trading in the intraday market were allowed. We see that in the absence of battery storage the expected gain is considerably reduced compared to perfect forecast owing to the intermittency of wind power, but that sufficient storage capacity allows to attain the theoretical value and even exceed since it allows both to smooth the variations of power output and trade in the intraday market. Note that the overall expected gain of the power producer depends on the production forecast and the day-ahead price, therefore it will not be the same for different days (19 October 2014 was a day with relatively strong wind, so the gain of the producer was higher than on 8 September 2014, even though the prices were lower). The extra gain from adding battery capacity (defined as the difference between the value function in the presence of a battery and the value function with zero capacity) is more stable, as shown in Table 12.1.

Finally we study the optimal bidding strategies for the producer in the day-ahead market. These are obtained by maximizing the value function of the producer with fixed bids $\overline{P}_k$, with respect to $\overline{P}_k$ with a numerical optimization algorithm (BFGS). Figure 12.7 shows the optimal bids (quantity to be delivered in the intraday market), together with the day-ahead production forecast, and the day-ahead price (thin line with right scale). We see that the producer aims to exploit the intertemporal price discrepancies in the day-ahead market using her battery storage capacity by selling

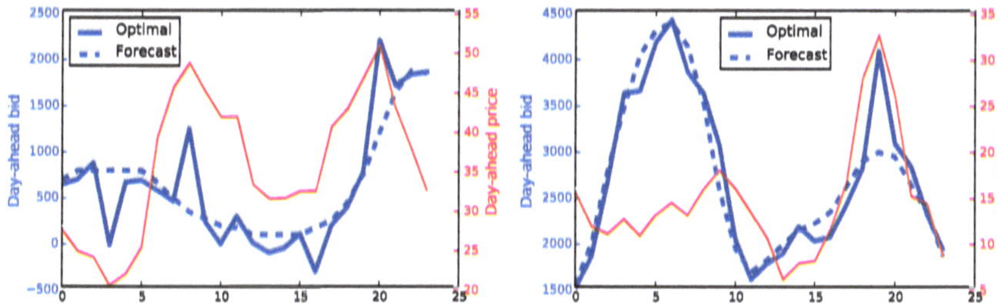

Fig. 12.7 Optimal day-ahead bids for the power producer compared to production forecast (left scale) and the day-ahead market price (right scale). Left graph: 8 September 2014. Right graph: 19 October 2014

more at times when the day-ahead price is high and buying more when the price is low. However, the gain of the producer from this additional trading is limited: on 8 September 2014 the value function (expected gain) increases from 542.12 euros to 554.26 euros, and on 19 October it increases only from 1001.90 to 1009.07 euros.

Acknowledgements This research was supported by the ANR project FOREWER (ANR-14-CE05-0028). The research of Peter Tankov has also benefited the support of the FIME Research Initiative. We are grateful to N. Girard (Maïa Eolis/Engie Green) and Philippe Vassilopoulos (EPEX Spot) for providing the data used in this study.

References

1. R. Aïd, P. Gruet, H. Pham, An optimal trading problem in intraday electricity markets. Math. Financ. Econ. **10**(1), 49–85 (2016)
2. A. Botterud, J. Wang, R.J. Bessa, H. Keko, V. Miranda, Risk management and optimal bidding for a wind power producer, in *2010 IEEE Power and Energy Society General Meeting* (IEEE, 2010), pp. 1–8
3. H. Ding, H. Zechun, Y. Song, Rolling optimization of wind farm and energy storage system in electricity markets. IEEE Trans. Power Syst. **30**(5), 2676–2684 (2015)
4. H. Ding, P. Pinson, H. Zechun, Y. Song, Integrated bidding and operating strategies for wind-storage systems. IEEE Trans. Sustain. Energy **7**(1), 163–172 (2016)
5. H. Ding, P. Pinson, H. Zechun, Y. Song, Optimal offering and operating strategies for wind-storage systems with linear decision rules. IEEE Trans. Power Syst. **31**(6), 4755–4764 (2016)
6. E. Garnier, R. Madlener, Balancing forecast errors in continuous-trade intraday markets. Energy Syst. **6**(3), 361–388 (2015)
7. P. Haessig, B. Multon, H.B. Ahmed, S. Lascaud, P. Bondon, Energy storage sizing for wind power: impact of the autocorrelation of day-ahead forecast errors. Wind Energy **18**(1), 43–57 (2015)
8. B. Heymann, J.F. Bonnans, P. Martinon, F.J. Silva, F. Lanas, G. Jiménez-Estévez, Continuous optimal control approaches to microgrid energy management. Energy Syst. 1–19 (2015)
9. R. Kiesel, F. Paraschiv, Econometric analysis of 15-minute intraday electricity prices. Energy Econ. **64**, 77–90 (2017)
10. J.M. Morales, A.J. Conejo, J. Pérez-Ruiz, Short-term trading for a wind power producer. IEEE Trans. Power Syst. **25**(1), 554–564 (2010)

11. D.E. Olivares, A. Mehrizi-Sani, A.H. Etemadi, C.A. Cañizares, R. Iravani, M. Kazerani, A.H. Hajimiragha, O. Gomis-Bellmunt, M. Saeedifard, R. Palma-Behnke, Trends in microgrid control. IEEE Trans. Smart Grid **5**(4), 1905–1919 (2014)
12. G. Pages, B. Wilbertz, Optimal Delaunay and Voronoi quantization schemes for pricing American style options, in *Numerical methods in Finance* (Springer, Berlin, 2012), pp. 171–213

FSC
www.fsc.org
MIX
Papier aus verantwortungsvollen Quellen
Paper from responsible sources
FSC® C105338

If you have any concerns about our products,
you can contact us on
ProductSafety@springernature.com

In case Publisher is established outside the EU,
the EU authorized representative is:
Springer Nature Customer Service Center GmbH
Europaplatz 3, 69115 Heidelberg, Germany

Printed by Libri Plureos GmbH
in Hamburg, Germany